Handbook of Photochemistry

Second Edition, Revised and Expanded

Steven L. Murov

Modesto Junior College
Modesto, California

Ian Carmichael

University of Notre Dame
Notre Dame, Indiana

Gordon L. Hug

University of Notre Dame
Notre Dame, Indiana

MARCEL DEKKER, INC.   NEW YORK · BASEL

Library of Congress Cataloging-in-Publication Data

Murov, Steven L.
    Handbook of photochemistry / Steven L. Murov, Ian Carmichael, Gordon
  L. Hug. -- 2nd ed., rev. and expanded.
        p.   cm.
    Includes bibliographical references and index.
    ISBN 0-8247-7911-8 (acid-free paper)
    1. Photochemistry--Tables.   I. Carmichael, Ian.   II. Hug, Gordon L.
  III. Title.
  QD719.M87   1993
  541.3'5'0212--dc20   93-4764
                                                                           CIP

The publisher offers discounts on this book when ordered in bulk quantities. For more information, write to Special Sales/Professional Marketing at the address below.

This book is printed on acid-free paper.

MARCEL DEKKER, INC.
270 Madison Avenue, New York, New York 10016

Current printing (last digit):
10 9 8 7 6 5

PRINTED IN THE UNITED STATES OF AMERICA

Preface

More than twenty years have passed since the publication of the first edition of this handbook. In 1972 the laser was still an infant on the photochemical scene whereas today its routine use has diminished the time scale of commonly studied transients to the picosecond range. The total number of abstracts in *Chemical Abstracts* has approximately doubled during this period and probably along with it the number of photochemical papers. This productivity coupled with the fact that the number of citations to the *Handbook of Photochemistry* in *Science Citation Index* continues to increase annually suggested that the time was ripe for a second edition. Preparation of a second edition became possible with the generous offer of support and assistance from the Radiation Chemistry Data Center at the University of Notre Dame.

The Radiation Chemistry Data Center collects and indexes current literature within its scope which now includes photochemical and photophysical processes in solution. Most of the citations used to update this book were located using the RCDC Bibliographic Database (RCDCbib). Without the availability of this database, preparation of this update would have presented an insurmountable task.

As with the first edition, the goal of this photochemical data compilation is to put at your fingertips the information needed to set up an optimal photochemical system on the first try. The types of information needed are many and varied depending on the nature of your experiment. We have attempted to devise tables organized according to the type of experiment you are likely to perform. Because of the advances in the types of equipment available such as tunable dye lasers, the already mentioned doubling of accumulated data, and the facility of referencing using RCDCbib's reference codes, a considerable reorganization of the tables was deemed appropriate.

In the first edition, the first section was divided into three parts. The first and second parts were alphabetical listings of compounds that included important photochemical parameters such as singlet and triplet energies and lifetimes. Because of frequency of use as photosensitizers and quenchers, 42 of these compounds were distinguished by inclusion in the first table along with additional parameters such as quantum yields of fluorescence, intersystem crossing, and extinction coefficients at the commonly utilized mercury emission lines. Because of the number of parameters and the basic goal of the book to include a citation for every entry, it was necessary to reference the first table in a rather unusual roadmap fashion. Because such a system would be inconvenient with the RCDCbib reference codes, a different type of table organization seemed more appropriate. The compounds of the first two tables were combined, along with a new selection of other compounds, into one table in the second edition.

Left out of this new table are the extinction coefficients of the commonly used sensitizers and quenchers. However, absorption spectra of selected sensitizers have been vastly expanded from five compounds to 35 compounds in the new Section 5. This makes approximate extinction coefficients of these compounds available over a broad wavelength range.

The first edition focused on a critical compilation of parameters for photochemical sensitizers and quenchers. The philosophy for the second edition has been substantially different. Selection of entries has not been as narrow, and the search for data has been more extensive, but still not exhaustive. Whereas the first edition avoided compounds such as nitrogen heterocycles because of strong solvent and sometimes wavelength dependencies, these compounds have often been included in this edition. This means that much more of the burden of choice has been left to the user. This has been necessitated by the considerable expansion in the nature of the systems being studied. In this edition, as with the first, it has been our perhaps nearsighted choice to de-emphasize aqueous systems. When a photochemical parameter depends on solvent, wavelength, and pH, it almost becomes necessary to include a table for each compound and that was not within the scope of our goals.

For the first edition, most of the compounds included could be purchased. Now photochemical parameters have been determined for many compounds that are not commercially available and in some cases are rather exotic and/or difficult to synthesize. Selection of entries has in those cases often been dictated by need. If a gap existed in the table of compounds arranged according to triplet energies, an attempt was made to find compounds to fill the gap. Often this meant inclusion of the exotic compounds. Commonly for these compounds, parameters such as quantum yields of intersystem crossing have not been determined and their actual utility will be subject to experimentation. It is also highly probable that we have excluded some exotic compounds that may be useful.

While pleased to see an ever-increasing number of citations in the *Science Citation Index*, we are also somewhat concerned. This book is intended as a guide, and whenever possible the original literature should eventually be read and referenced.

Science Citation Index was used by us as a guide for the updating of this book. Page citations to the *Handbook* indicated that the first section on sensitizers and quenchers received frequent usage, and this section has been vastly expanded. Surprisingly however, the actinometry sections and the oxygen solubility section were also often cited. For this reason considerable ferrioxalate quantum yield data have been added, and the oxygen solubility table has been expanded.

Other tables have been included primarily to decrease the number of trips to the library. An effort has been made to update all tables and in many cases little resemblance to the corresponding table in the first edition will be found. As with the first edition, a photochemical reaction section has not been included.

A Compound Name Index and a Molecular Formula Index are included for Tables 1 through 12, excepting the solvents in Table 7 and the glasses in Table 12-4. They were computer-generated with programs operating simultaneously on the data files and a local file containing registry numbers, formulas, and chemical names (RCDCreg). The Table of Contents is intended to allow access to the material in the other sections of the book. Similarly, the bibliographic references were obtained semiautomatically for the above-mentioned data tables and manually added for the remainder.

The publication of an excellent methods and techniques book by Rabek [1] (and another handbook edited by Scaiano [2] that appeared during the preparation of this second edition) has enabled us to focus on data rather than techniques, and a few techniques items from the first edition have been dropped. Others have been retained for convenience.

Many people deserve credit for contributions and useful discussions during the preparation of the second edition. Among them Drs. Alberta Ross, W.P. Helman, and P.K. Das deserve special mention. Several other members, Christa Wardlow, Joyce Vogler, and Alice Hohulin, of the Radiation Laboratory Data Center scurried after references and entered data. Ms. Wardlow also served as a copy editor for portions of the manuscript. The Notre Dame Chemistry/Physics librarians, especially Dori Ringhofer, went beyond the call of duty to help.

The contributions from the Radiation Chemistry Data Center were supported by the Office of Basic Energy Sciences of the United States Department of Energy, and by the National Institute of Standards and Technology, Standard Reference Data. This is Document No. NDRL-3569 from the Notre Dame Radiation Laboratory.

Photochemistry has come a long way since the pioneering work of Ciamician and Silber. Laser photolysis and computer data acquisition have opened up the field and changed the limits of our time scales. But the dream of Ciamician [3] quoted in the first edition is still a goal and worth repeating:

On the arid lands there will spring up industrial colonies without smoke and without smokestacks; forests of glass tubes will extend over the plains and glass buildings will rise everywhere; inside of these will take place the photochemical processes that hitherto have been the guarded secret of the plants, but that will have been mastered by human industry which will know how to make them bear even more abundant fruit than nature, for nature is not in a hurry and mankind is.

References

[1] Rabek, J.F., *Experimental Methods in Photochemistry and Photophysics.* Wiley, Chichester, UK, 1982.

[2] Scaiano, J.C., *CRC Handbook of Organic Photochemistry.* CRC Press, Inc., Boca Raton, FL, 1989.

[3] Ciamician, G. *Science* **36**: 385-94 (1912).

Steven L. Murov
Ian Carmichael
Gordon L. Hug

Contents

Section 1

Photophysics of Organic Molecules in Solution

Three states play a dominant role in the photophysics of organic molecules in solution. They are the ground state, the lowest triplet state, and the lowest excited singlet state. In condensed media, vibrational relaxation is usually so fast that excited molecules quickly relax to one of the two excited states. For such systems, a three-state model, with the transitions between the states governed by first-order competing kinetics, forms an adequate framework to understand much of the data on unimolecular decays of the excited states.

In particular, radiative and radiationless transitions between these two excited states and the ground state (see Fig. 1-1) delimit the extent to which photochemical processes can occur. For singlet photochemistry, the quantum yield can be written in terms of the fundamental kinetic parameters as

$$\phi_{pc}^S = \frac{k_{pc}^S}{k_S} \tag{1-1}$$

where k_{pc}^S is the first-order rate constant for singlet photochemistry and k_S is the overall first-order rate constant for the singlet decay. Thus, in order to see singlet photochemistry, k_{pc}^S must be competitive with other decay channels of the lowest excited singlet state. Non-sensitized triplet photochemistry, on the other hand, is limited both by analogous processes to those in Eq. (1-1) and also by the triplet quantum yield.

In Table 1, the energies of the two excited states are given. Both singlet and triplet energies are estimates of the lowest excited level of each multiplicity, namely the 0–0 vibronic band. For the singlet level, the fluorescence 0–0 band was chosen when it could be distinguished; otherwise, the 0–0 band in absorption was chosen. If neither was distinguishable, the midpoint in energy between the fluorescence and absorption maxima was taken as an estimate of the lowest excited singlet vibronic level. Triplet energy measurements are discussed in the introduction to Section 2.

Table 1 displays two important quantum yields: the yield of fluorescence per photon absorbed (the fluorescence quantum yield, ϕ_{fl}) and the triplet quantum yield, ϕ_T (also called the intersystem crossing yield). They are related by

$$\phi_{fl} + \phi_T + \phi_{ic}^S + \phi_{pc}^S = 1 \tag{1-2}$$

where the last two quantum yields in Eq. (1-2) are the internal conversion and the singlet photochemistry quantum yields, respectively. Since ϕ_{ic}^S is often small due to the large singlet energy gaps, the sum of ϕ_{fl} and ϕ_T is often close to one when there is no singlet photochemistry. This rule-of-thumb breaks down often, even for relatively simple sets of compounds such as polyenes.

Lifetimes of the two excited states are also given in Table 1. The overall fluorescence lifetime is the reciprocal of the decay rate constant of the excited singlet state which in the three-state model is given by

$$k_S = \frac{1}{\tau_S} = k_{isc}^S + k_{ic}^S + k_r^S + k_{pc}^S \tag{1-3}$$

where k_{isc}^S, k_{ic}^S, and k_r^S are the intersystem crossing, the internal conversion, and the natural radiative rate constants, respectively, from the excited singlet. The natural radiative rate constant can be calculated from

$$k_r^S = k_S \, \phi_{fl} \tag{1-4}$$

The other lifetime given in Table 1 is the triplet lifetime, τ_T, which is the reciprocal of the total decay rate constant for all processes out of the triplet state. The triplet lifetimes are usually several orders of magnitude longer than those of the singlet which makes triplet states susceptible to bimolecular quenching with impurities, cell walls, self-quenching with ground state species, or even weak reactions with solvents. For example, triplet lifetimes of aromatic ketones in hydroxylic solvents can reflect the progress of hydrogen-abstraction reactions between the triplet state and solvent. The triplet lifetimes in Table 1 should be taken as lower limits of a unimolecular lifetime. At room temperature, the unimolecular process contributing to the finite lifetime of triplet states is usually just the intersystem crossing to the ground state. However, in thioketones, the decay has a radiative component.

In Table 1 the molecules are listed primarily by inverted name. Many synonyms are listed in the Compound Name Index, and there is also a Formula Index at the end of this *Handbook*. The photophysical measurements were initially classified by solvent media: non-polar, polar, aromatic, gas, polar crystalline, nonpolar crystalline, etc. In Table 1, these categories have been condensed into a simple bimodal (*nonpolar* and *polar*) classification, with only one value chosen for a particular excited state/solvent type pair. Some remnants of the more detailed media classification are retained as footnotes attached to the values themselves. The nonpolar/polar classification allows for a limited range of measurements to be displayed, yet it keeps the format succinct. Because of the wide-spread use of cyclohexane and ethanol, they were chosen whenever possible as the representative nonpolar and polar solvents, respectively.

The bimodal classification of solvents is not meant to imply that the unimolecular decays of molecules are solely, or even strongly, dependent on the solvent polarity. Polar solvents can blue-shift n, π^* transitions leading to energy shifts in the spectra. If these blue-shifts of the n, π^* transitions are large enough, they can lead to inversion of n, π^* and π, π^* transitions which can dramatically affect the kinetic parameters characterizing radiative and radiationless transitions. However, barring inversion of excited states, solvent couplings are generally weak enough that they are routinely ignored as a first approximation in theories of radiationless transitions. These considerations are the rationalization for choosing to display unimolecular photophysical properties within a nonpolar/polar scheme. Caution should be exercised when using the scheme in systems where different ionization states are possible, when heterocyclic compounds (which have low-lying n, π^* transitions) are considered, or when there are strong specific solvent interactions (such as H-abstraction) with the excited molecule.

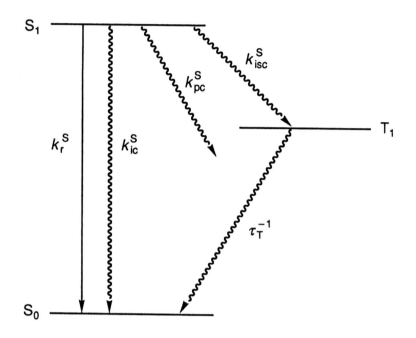

Fig. 1-1

Table 1

Photophysical Parameters of Organic Species

No.	Solv	E_S (kJ/mol)	Ref.	τ_S (ns)	Ref.	ϕ_{fl}	Ref.	ϕ_T	Ref.	E_T (kJ/mol)	Ref.	τ_T (μs)	Ref.
1	**Acenaphthene**												
	n	374	71Z001	46	71Z001	0.50	706229	0.46	69E202	250	64E037		
	p	376	68Z003			0.39	66E098	0.58	68E098	248	67E112	3300	68Z003
2	**Acetone**												
	n			1.7	716017	0.0009	716017	0.90	707357			6.3	84B051
	p			2.1	707538	0.01	66E095	1	66F206	332	66E095	47	717489
3	**Acetophenone**												
	n	330	64E028			<10^{-6}	736002	1^b	85A268	310	81E561	0.23	717179
	p	338	81E561					1	85A268	311	81E561	0.14	717179
4	**Acetophenone, 4'-methoxy-**												
	n	340^b	66E094					1^b	85A268	300	67F523		
	p							1	85A268	299	67F521		
5	**Acetylene, diphenyl-**												
	n					0.0028	67F510	0.033	67F510	262	57B004		
	p	396	566004			0.0016	67F510	0.056	67F510	262	566004		
6	**Acridan**												
	p	356	88E763	10	73E359	0.32	73E359			291	73E359		
7	**Acridine**												
	n			0.045	84E322	1×10^{-4}	84E322	0.5	78E222	190	81E147	10000^b	677498
	p	315	779025	0.35	84E322	0.0079	84E322	0.82^s	85E408	188	84E803	14	78E394

No.	Compound												
8	Acridine-d_9												
	n												
	p			11.5	81E552	0.31	81E552	0.75^b	776258	190	81E552		
9	Acridine, 9-amino-	280	71Z014	15.2	64E031	0.99	57E003			193	81E552		
10	Acridine Orange	234	69E235	4.4	71Z001	0.4	85E800	<0.02	82E660	206	69E235	285	79E219
11	Acridine Orange, conjugate monoacid	228	71Z014			0.46	68Z003	0.10	727073	191	68Z003	105	79E219
12	Acridine Yellow	251	71Z001	5.1	71Z001	0.47	79E560			220	83Z077		
13	Acridinium	270	71Z014	33.9	81E552	0.66	81E552					670	80B057
14	Acridinium, 3,6-diamino-	250	71Z014			0.40	68Z003	0.22	727073	205	68Z003	20	80F373
15	Acridinium, 3,6-diamino-10-methyl-	247	69E235	4.3	85E800	0.16	85E800			214	69E235		
16	9-Acridinone												
	n	304	78E542	0.78^b	766377	0.015	78E542	0.99^b	766377	244	78E542	20^b	766377
	p	290	78E542	10.8	80E596	0.97	78E542			252	78E878	9.2	89A120

Table 1—**Photophysical Parameters**—Continued

No.	Solv	E_S (kJ/mol)	Ref.	τ_S (ns)	Ref.	ϕ_{fl}	Ref.	ϕ_T	Ref.	E_T (kJ/mol)	Ref.	τ_T (μs)	Ref.
17	**9-Acridinone, 10-methyl-**												
	n	302	78E542	9.9	80E596	0.017	78E542	0.96[b]	79E677	247	78E542		
	p	283	78E542	11.0	80E596	0.98	78E542	0.008	79E677	252	78E542	36	89A120
18	**9-Acridinone, 10-phenyl-**												
	n	305	78E542			0.018	78E542			244	78E542		
	p	289	78E542	7.7	80E596	0.99	78E542			254	78E542		
19	**Adenosine 5′-monophosphate**												
	p	421	673066			5×10^{-5}	743202			319	673066		
20	**Angelicin**												
	n							0.009[b]	78E157				
	p	330	73E372					0.031	84E794	264	73E372	2.6	84E794
21	**Aniline**												
	n	398	82B140	6.9	87B098	0.17	87B098	0.75	80E253	297	67E116	0.72	87B098
	p	384	80E253	16.6	87B098	0.10	80E253	0.90	80E253	321	44E001		
22	**Aniline, N,N-dimethyl-**												
	n	383	71Z001	2.4	71Z001	0.11	71Z001						
	p	375	71Z001	2.8	71Z001					317	61E013		
23	**Aniline, N,N-diphenyl-**												
	n	362	71Z001			0.045[b]	706243	1.1[b]	706243	291	69Z002	0.06[b]	706243
	p												

No.	Compound		(1)	ref	(2)	ref	(3)	ref	(4)	ref	(5)	ref	(6)	ref
24	**Aniline, N-phenyl-**													
		n					0.05[b]	706243	0.32[b]	706243			0.3[b]	706243
		p	372	73E359	2.4	73E359	0.11	73E359	0.47	706243	301	73E359	0.5	706243
25	**Anthracene**													
		n	318	85E096	5.3	85E096	0.30	78A307	0.71	757282	178	57B004	670	85E555
		p	319	779025	5.8	62E014	0.27	68E114	0.66	85E408	178	779025	3300	747049
26	**Anthracene-d_{10}**													
		n	320	71Z001	4.9	71Z001	0.32	71Z001						
		p					0.24	68E114						
27	**Anthracene, 1-amino-**													
		n	280	71Z001	22.8	71Z001	0.61	71Z001			184	69E240		
		p	259	69E240										
28	**Anthracene, 2-amino-**													
		n	264	71Z001	25.8[b]	71Z001	0.57	87B098			184	69E240		
		p			30.8	71Z001								
29	**Anthracene, 9-amino-**													
		p	257	69E240	10.0	566007	0.29	566007			184	69E240		
30	**Anthracene, 9-bromo-**													
		n	306	79E108	1.1	79E108	0.011	78A307					43	62E009
		p		566007	1.1	566007	0.020	61E010	0.98	89B155	173	84B110	19.5	84B110

7

Table 1—**Photophysical Parameters**—Continued

No.	Solv	E_S (kJ/mol)	Ref.	τ_S (ns)	Ref.	ϕ_{fl}	Ref.	ϕ_T	Ref.	E_T (kJ/mol)	Ref.	τ_T (μs)	Ref.
31	**Anthracene, 9-chloro-**												
	n	306	79E108	2.0	78A307	0.11	78A307						
	p			2.8	566007	0.11	566007						
32	**Anthracene, 9-cyano-**												
	n	297	81E183	15.6	85E096	0.93	85E096	0.040	737140			600	737140
	p	287	737140	11.9	88E219			0.021	737140			1800	737140
33	**Anthracene, 9,10-dibromo-**												
	n	295[b]	84E393	1.3	82E303	0.094	78A307	0.70[b]	84E393	168	566006	36	62E009
	p	295	756125	1.8	84E785	0.15	84E785	0.82	84E785			11	78E394
34	**Anthracene, 9,10-dichloro-**												
	n	298	71Z001	8.5	71Z001	0.46	706229	0.29[b]	84E393	169	566006	100	62E009
	p	297	66E086	8.7	65E046	0.56	706054	0.45	706054	169	66E086		
35	**Anthracene, 9,10-dicyano-**												
	n	284	85A009	11.7	85A009	0.90	78A307	0.23[b]	84E393	175	78E414	100	78E414
	p	280	85A009	15.1	85A009	0.87	766170						
36	**Anthracene, 9,10-dimethoxy-**												
	n	296	79E108	14.7	78A307	0.87	78A307						
	p			9.2	566007	0.41	566007						
37	**Anthracene, 9,10-dimethyl-**												
	n	300	71Z014	14.0	78A307	0.93	78A307	0.02[b]	68F286				
	p	297	68Z003	11.0	566007	0.89	67E129	0.032	68Z003			8000	67E129

8

No.	Compound / phase											
38	**Anthracene, 9,10-diphenyl-**											
	n	304	85E096	7.7	85E096	0.91	83E527	0.02	83E281	171	68E131	2500	746270
	p	305	71Z014	8.2	766100	0.95	766100	0.02	83E281			3000	746270
39	**Anthracene, 9-methoxy-**											
	n		78A307	4.9	78A307	0.34	78A307	0.26	89E424				
	p		566007	3.9	566007	0.17	566007						
40	**Anthracene, 9-methyl-**											
	n	310	85E096	4.6	71Z001	0.29	706229	0.48	756505	173	57B004	10000	67E129
	p	306	68Z003	5.8	66E086	0.33	67E129	0.67	67E129	170	66E086		
41	**Anthracene, 1,2,3,4,5,6,7,8-octahydro-**											
	n	414	82E344	17.1	82E344	0.27	82E344			322	82E344		
42	**Anthracene, 9-phenyl-**											
	n	309	71Z001	6.5	71Z001	0.41	706229	0.37	65F031			15000	67E129
	p	304	68Z003	5.1	566007	0.49	67E129	0.51	67E109				
43	**9-Anthracenecarboxylic acid**											
	p		566007	4.1	566007	0.04	566007			177	82E338		
44	**1,8-Anthracenedisulfonate ion**											
	p		80E233	1.5	80E233		85E452	0.87	85E452			71	85E452
45	**1-Anthracenesulfonate ion**											
	p		80E233	7.3	80E233		85E452	0.98	85E452			103	85E452

Table 1—**Photophysical Parameters**—Continued

No.	Solv	E_S (kJ/mol)	Ref.	τ_S (ns)	Ref.	ϕ_{fl}	Ref.	ϕ_T	Ref.	E_T (kJ/mol)	Ref.	τ_T (μs)	Ref.
46	**9,10-Anthraquinone**												
	n	284[g]	85E712					0.90[b]	65F030	261	64E021	0.11[b]	81F130
	p									263	67E115		
47	**9,10-Anthraquinone, 1-amino-**												
	n			1.8[b]	82E343	0.058[b]	82E343	0.02[b]	86Z077			5[b]	720392
	p	220	71Z014	0.46	82E343	0.0082	82E343						
48	**9,10-Anthraquinone, 2-amino-**												
	n			6.5[b]	82E343	0.21[b]	82E343	0.4[b]	86Z077			5[b]	720392
	p			0.054	82E343	0.0006	82E343						
49	**Azulene**												
	n	170	71Z014	1.4 S_2	71Z001	0.02 S_2	72E287			163[b]	757247	11[b]	84E491
	p											3	81F275
50	**Benz[a]acridine**												
	n			0.7	85E054	0.05	85E054	0.82	85E054				
	p	295	78E761	7.2	85E054	0.32	85E054	0.42	85E054	209	78E761		
51	**Benz[c]acridine**												
	n			3.6	85E054	0.09	85E054	0.83	85E054				
	p	312	779025	5.6	85E054	0.18	85E054	0.65	85E054	213	779025		
52	**Benz[a]acridinium**												
	p	276	71Z014	14.2	85E054	0.50	85E054	0.27	85E054				

#	Compound		P1 value	P1 ref	P2 value	P2 ref	P3 value	P3 ref	P4 value	P4 ref	P5 value	P5 ref	P6 value	P6 ref
53	**Benz[b]acridin-12-one**														
		n	245	78E878	19.1	78E878	0.14	78E878			191	85F172		
		p									193	78E878		
54	**Benzaldehyde**														
		n	323	72E315			$<10^{-6}$	736002			301	72E315		
		p									298	706018		
55	**Benz[a]anthracene**														
		n	311	71Z014	45	64E034	0.19	64E034	0.79	69E208	197	54E002	100	64E015
		p	307	78E761	40.0	706216	0.22	68E098	0.79	727047	198	78E761	9400	64E026
56	**Benzene**														
		n	459	71Z014	34	726120	0.06	726120	0.25	69E202	353	67F523		
		p	459	506002	28	726120	0.04	726120	0.15	736048	353	67F523		
57	**Benzene-d_6**														
		n	449	71Z001	30	737138	0.042	737138	0.25	737138				
58	**Benzene, chloro-**														
		n	440	71Z014	0.74	84E529	0.0070	84E529	0.6	83F178	342[g]	59E009	1.6	84E529
		p			0.79	83E156			0.7	85E488			0.715	85E488
59	**Benzene, 1,4-dichloro-**														
		n	424	82B140					0.8	83F178	335	67E112	430	85E406
		p	427	79B163					0.95	85E406			330	85E406

Table 1—**Photophysical Parameters**—Continued

No.	Solv	E_S (kJ/mol)	Ref.	τ_S (ns)	Ref.	ϕ_{fl}	Ref.	ϕ_T	Ref.	E_T (kJ/mol)	Ref.	τ_T (μs)	Ref.
60	**Benzene, 1,4-dicyano-**												
	n	415	86E917							294	62E018		
	p	412	86E917	9.7	767370					295	70E318		
61	**Benzene, 1,4-dimethoxy-**												
	n	397	71Z001	2.9	71Z001	0.21	71Z001			314[b]	72D316		
	p	396	71Z001	2.7	71Z001								
62	**Benzene, fluoro-**												
	n	449	71Z014	7.6	71Z001	0.11	706229	0.80	69E202	353[b]	57B004	0.67	707561
63	**Benzene, hexachloro-**												
	n	397	82B140					0.5	83F178	307	67E112		
64	**Benzene, methoxy-**												
	n	437	496002	8.3	71Z001	0.24	706229	0.64	82E060	338	67F523		
	p	431	82B140	7.5	82E060	0.24	82E060			338	67F523	3.3	757161
65	**Benzene, 1-methoxy-4-methyl-**												
	n	416	71Z001	8.7	71Z001	0.26	71Z001			326[b]	78D082		
66	**Benzene, nitro-**												
	n	372	71Z014					0.67[b]	687061	243	84E530	0.0008	84E530
	p									252	44E001		
67	**Benzene, 1,2,3,4-tetramethyl-**												
	n	426	82E344	36.2	82E344	0.14	82E344			331	82E344		

No.	Compound													
68	**Benzene, 1,3,5-trichloro-**													
		n												
		p	427	79B163					0.8	83F178	331	67E112		
69	**Benzene, 1,3,5-trimethyl-**													
		n	434	71Z001	36.5	71Z001	0.088	72E313			335	61E014		
		p	439	71Z014							336	64E022		
70	**Benzene, 1,3,5-triphenyl-**													
		n	375	71Z001	42.6	71Z001	0.27	71Z001			269	67E112		
		p												
71	**Benzidine, N,N,N',N'-tetramethyl-**													
		n			10.1	84E405	0.38	84E405	0.52	84E405	261	767177		
		p			9.4	84E405	0.36	84E405	0.41	84E405			5	84F074
72	**Benzil**													
		n	247[x]	69E223	2.0	85F138	0.0013	67E120	0.92[b]	65F030	223	67E119	150	717447
		p									227	67E119	1500	68E128
73	**Benzimidazole**													
		n									321[x]	755396		
		p	423	63E012					0.67	66E102	318	82E624		
74	**Benzo[a]carbazole**													
		p	341	779025							256	779025		

Table 1—**Photophysical Parameters**—Continued

No.	Solv	E_S (kJ/mol)	Ref.	τ_S (ns)	Ref.	ϕ_{fl}	Ref.	ϕ_T	Ref.	E_T (kJ/mol)	Ref.	τ_T (μs)	Ref.
75	**Benzo[b]carbazole**												
	p	296	78E761							218	78E761		
76	**Benzo[b]carbazole, N-methyl-**												
	n	302	71Z001	26.2	71Z001								
	p	300	71Z001	23.5	71Z001	0.60	71Z001						
77	**Benzo[def]carbazole**												
	p	317	78E761							230	78E761		
78	**Benzo[a]coronene**												
	n	277[b]	59E013			0.27	68E104	0.55	68E104	215	78E314		
	p												
79	**Benzo[b]fluoranthene**												
	n			44.3	82E059	0.53	82E059						
	p	328	779025							228	82E059		
80	**Benzo[ghi]fluoranthene**												
	n	285	71Z001	45	71Z001	0.30	71Z001						
	p	282	779025							226	779025		
81	**Benzo[k]fluoranthene**												
	n			11.3	82E059	1.0	82E059						
	p	299	779025							211	779025		
82	**Benzo[a]fluorene**												
	n	346	71Z001	63	71Z001	0.51	71Z001						

No.	Compound	n/p										
82	**Benzo[a]fluorene**—Continued	p	326	779025							241	779025
83	**Benzo[b]fluorene**	n	352	82B140	32.3	737591						
		p	326	779025							240	779025
84	**Benzo[c]fluorene**	p	332	779025							231	779025
85	**Benzoic acid**	n					0.0068	86E481			324	63E014
		p	428	71Z014							326	767546
86	**Benzoic acid, 4-(dimethylamino)-, ethyl ester**	n			1.0	85E006	0.29	85E006	0.14	85E006	284	83E440
87	**Benzonitrile, 4-amino-**	n	396	83E440	3.3	83E440	0.14	83E440			298	83E440
		p	365	79E415							293	756176
88	**Benzonitrile, 4-(diethylamino)-**	n			3.8	83E440	0.12	83E440			288	83E440
		p	365	79E415							288	79E415
89	**Benzonitrile, 4-(dimethylamino)-**	n			2.9	83E890	0.2	87E518	0.18	80E262	290	83E440
		p	368	79E415							288	79E415

Table 1—**Photophysical Parameters**—Continued

No.	Solv	E_S (kJ/mol)	Ref.	τ_S (ns)	Ref.	ϕ_{fl}	Ref.	ϕ_T	Ref.	E_T (kJ/mol)	Ref.	τ_T (μs)	Ref.
90	**Benzonitrile, 4-methoxy-**												
	n	422	83E440	3.4	83E440	0.10	83E440			315	83E440		
	p	421	756304							315	756304		
91	**Benzo[*ghi*]perylene**												
	n	293[b]	696020	203[b]	696020	0.38[b]	696020			194	566005	150[b]	71E361
	p	294	779025	188	696020	0.25	68E114	0.53	696020				
92	**Benzo[*c*]phenanthrene**												
	n	323	71Z014	76	71Z001	0.12	71Z001			237	54E002		
	p	321	779025							239	779025		
93	**Benzo[*f*][4,7]phenanthroline**												
	n	352	81E545							284	81E545		
	p	350	755398			0.07	755398			285	755398		
94	**Benzo[*a*]phenazine**												
	n	295	81E433			0.0008	85E054	1.1	85E054	209	85E370		
	p	289	81E433			0.02	85E054	0.96	85E054	202	81E433		
95	**Benzo[*a*]phenazinium**												
	p	248	81E433	2.7	85E054	0.02	85E054	0.81	85E054	187	81E433		
96	**Benzophenone**												
	n	316	64E028	0.030[b]	747025	4×10^{-6}	736002	1.0	69E202	287	67F523	6.9[b]	776016
	p	311	63Z003	0.016	747025			1	85A268	289	80E223	50	766276

No.	Compound											
97	**Benzophenone, 4,4′-bis(dimethylamino)-**											
	n	295	79E210			0.91	777004	275	68E135	25	777603
	p					0.47	85A268	255	79E210	20	777603
98	**Benzophenone, 4,4′-dibromo-**											
	n	316[b]	66E094	4×10^{-6}	736002			288	66E094		
	p										
99	**Benzophenone, 4,4′-dimethoxy-**											
	n	328[b]	66E094	0.0065[b]	78E489	1[b]	85A268	293[b]	81E099	3.6	776016
	p					1	85A268	292	80E223		
100	**Benzophenone, 4,4′-dimethyl-**											
	n	324	697155			1[b]	85A268	288	67F523	11	776016
	p	328	697155			1	85A268	290	80E223		
101	**Benzophenone, 4-methoxy-**											
	n					1[b]	85A268	287	68F291		
	p					1	85A268	290	80E223		
102	**Benzophenone, 4-methyl-**											
	n	323	697155			1[b]	85A268	287	67F523	7.2	82A082
	p	327	697155			1	85A268	290	80E223		
103	**Benzophenone, 4-phenyl-**											
	n					1.0	69E202	254	60E014		
	p	321	60E014								

Table 1—Photophysical Parameters—Continued

No.	Solv	E_S (kJ/mol)	Ref.	τ_S (ns)	Ref.	ϕ_{fl}	Ref.	ϕ_T	Ref.	E_T (kJ/mol)	Ref.	τ_T (μs)	Ref.
104	**Benzo[g]pteridine-2,4-dione**												
	p			1.0	777617	0.033	777617	0.31	777617			13	737439
105	**Benzo[g]pteridine-2,4-dione, 7,8-dimethyl-**												
	p	282	69E235	0.85	777617	0.036	777617	0.61	777617	232	69E235	12	777617
106	**Benzo[g]pteridine-2,4-dione, 7,8,10-trimethyl-**												
	n											30	68E100
	p	239	69E235	5.5	777617	0.40	777617	0.30	777617	209	69E235	320	70E295
107	**Benzo[g]pteridine-2,4-dione, 7,8,10-trimethyl-, conjugate monoacid**												
	p			2.5	777617	0.16	777617	0.42	777617			29	757078
108	**Benzo[a]pyrene**												
	n	295	71Z014	49	706229					175	57B004	8700	78A345
	p	297	779025	27.3	88E804	0.42	68Z003			177	779025	8800	64E026
109	**Benzo[e]pyrene**												
	n	325	71Z014										
	p	327	779025							221	779025	120[b]	71E361
110	**Benzo[f]quinoline**												
	n	348	479001							262	59E009		
	p									262	59E011		
111	**Benzo[g]quinoline**												
	n	307	81E444	2.1	81E444	0.10	81E444	0.67	81E444	180	59E009		
	p	293	81E444	11.6	81E444	0.62	81E444	0.32	81E444	184	81E444		

#	Compound													
112	Benzo[h]quinoline	n	346	71Z014	4.0	86A205	0.11	86A205	0.88	86A205	260	59E009		
		p									261	59E011		
113	1,4-Benzoquinone	n	262	69E247							224	69E247	0.53	80B112
		p												
114	1,4-Benzoquinone, tetrachloro-	n	266	69E247					0.98	79B061	206	69E247	2.0	697272
		p											1.2	697272
115	1,4-Benzoquinone, tetramethyl-	.n							1.0	767144			21	767144
		p							1.0	767144			15	767144
116	Benzo[f]quinoxaline	n	329	83E176							254	83E176		
		p	323	63E018							251	63E018		
117	Benzo[b]thiophene	n									287	59E009		
		p	394	85E272							288	85E272		
118	Benzotriazole	p	401	63E012	0.02	66E102					295	63E012		

Table 1—Photophysical Parameters—Continued

No.	Solv	E_S (kJ/mol)	Ref.	τ_S (ns)	Ref.	ϕ_{fl}	Ref.	ϕ_T	Ref.	E_T (kJ/mol)	Ref.	τ_T (μs)	Ref.
119	**Benzo[b]triphenylene**												
	n	319	71Z014	53.5	70E288					213	57B004	90[b]	71E361
	p	320	779025	43	737591					213	779025		
120	**Benzoxazole, 2-phenyl-**												
	n					0.66	88E862			260	736125		
	p	371	88E862			0.71	88E862			262	88E862		
121	**Benzyl alcohol**												
	n	448	71Z001	29.0	71Z001	0.070	87E509	0.51	87E509				
	p	427	82B140										
122	**Benzylamine**												
	n			27.8	87E509	0.072	87E509	0.33	87E509	345	69Z002		
123	**Benzyl cyanide**												
	n			26.9	87E509	0.046	87E509	0.41	87E509				
124	**Biacetyl**												
	n			11.5	726211	0.0027	726211	1.0[b]	60E012			638	726211
	p	267[x]	55E008	7.7	696078					236[x]	55E008	145[p]	80F269
125	**9,9'-Bianthryl**												
	n	304[b]	71Z001	6.5	85E246	0.57	85E246						
	p			45	85E246	0.22	85E246						

No.	Compound														
126	**Bibenzyl**														
		n	448	71Z001	35.0	71Z001	0.13	71Z001							
127	**1,1′-Binaphthyl**														
		n	368	71Z001	3.0	71Z001	0.77	71Z001					14[b]	771048
		p	365	776226											
128	**2,2′-Binaphthyl**														
		n	359	71Z001	35.2	71Z001	0.41	71Z001							
		p									234	566005			
129	**9,9′-Biphenanthryl**														
		p	340	776226							257	776226			
130	**Biphenyl**														
		n	418	71Z001	16.0	71Z001	0.15	706229	0.84	757282	274	64E037	130	69E208
		p	391	776226							274	735067			
131	**Biphenyl, 4,4′-diamino-**														
		p	346	71Z001	13.5	84E405	0.25	84E405							
132	**Biphenyl, 4,4′-dibromo-**														
		n			0.030	82E303									
133	**Biphenyl, 4,4′-dimethoxy-**														
		n	386	71Z001	11.0	71Z001	0.21	71Z001					4[b]	80A235
		p	383	71Z001	11.7	71Z001					264	735067			

Table 1—**Photophysical Parameters**—Continued

No.	Solv	E_S (kJ/mol)	Ref.	τ_S (ns)	Ref.	ϕ_{fl}	Ref.	ϕ_T	Ref.	E_T (kJ/mol)	Ref.	τ_T (μs)	Ref.
134	**Biphenyl, 3,3′-dimethyl-**												
	n	410	71Z001	13.2	71Z001	0.18	85F488	0.32	85F488				
	p									271	89D071		
135	**Biphenyl, 4,4′-dimethyl-**												
	n	419	71Z014	16.5	85F488	0.22	85F488	0.47	85F488				
	p									269	735067		
136	**Biphenyl, 4-methoxy-**												
	n	401	71Z001	9.4	71Z001	0.26	71Z001						
	p	399	71Z001	9.6	71Z001								
137	**1,3-Butadiene, 1,4-diphenyl-**												
	n			0.60	82E365	0.42	82E365	0.020	82E365	177	707199	1.6	82E365
	p	334	80B135	0.060	82F056	0.042	82F056	≤0.002	82E365			5.0	84E319
138	**1,3-Butadiene, 1,1,4,4-tetraphenyl-**												
	n	308	71Z001	1.8	71Z001	0.60	71Z001					0.665[b]	84E144
139	**C$_{60}$**												
	n	193	91E003	1.2[b]	91E302			1	92E260	151[b]	91E368	250[b]	92E205
140	**C$_{70}$**												
	n	185	91E594	0.67[b]	91A349			0.97[b]	92E142	148[b]	91D034	250[b]	92E205
141	**Carbazole**												
	n	361	71Z001	16.1	71Z001	0.31	706229	0.36[b]	65F030	294	58E005	170	77A178
	p	347	78E761	15.2	71Z001	0.42	73E359						

#		v1	ref1	v2	ref2	v3	ref3	v4	ref4	v5	ref5	v6	ref6
142	**Carbazole, N-methyl-**												
	n	349	71Z001	18.3	71Z001	0.51	71Z001			292	81E648		
	p	347	71Z001	16.0	71Z001	0.42	766474			292	68Z005		
143	**Carbazole, N-phenyl-**												
	n	352	71Z001	10.3	71Z001	0.37	71Z001			294	68Z005		
	p												
144	**4,4′-Carbocyanine, 1,1′-diethyl-**												
	p					0.007	736051	$<6\times10^{-4}$	736051			1100	736051
145	**β-apo-14′-Carotenal**												
	n	262	78E432	0.8	78E432	0.0098	78E432	0.54	79E546			8	83E026
	p	257	78E432			0.0015	78E432	-0.033	79E546			10.3	79E546
146	**β-Carotene**												
	n	228	71Z014	0.0084[b]	86E320			<0.001	776412	88[b]	757247	70	66E089
	p	170	92N199							85	92N199	9	81B115
147	**Chlorophyll a**												
	n	177[b]	81F121	7.8[b]	57E004	0.32[b]	57E003	0.53	86R013	125	79E838	1500[b]	58R001
	p	178	79E838	5.5	86R013	0.33	86R013					800	68Z003
148	**Chlorophyll b**												
	n	181	79E838	6.3[b]	57E004	0.11[b]	57E003	0.81	86R013	130	79E838	2500[b]	59B002
	p	179	68Z003	3.5	86R013	0.12	86R013			136	68Z003	1500	68Z003

Table 1—Photophysical Parameters—Continued

No.	Solv	E_S (kJ/mol)	Ref.	τ_S (ns)	Ref.	ϕ_{fl}	Ref.	ϕ_T	Ref.	E_T (kJ/mol)	Ref.	τ_T (μs)	Ref.
149	**Chrysene**												
	n	331	71Z014	44.7	71Z001	0.12	706229	0.85	757282			710	69E208
	p	332	779025	42.6	71Z001	0.17	66E098	0.85	68E098	239	779025		
150	**Cinnoline**												
	n	270	736174	0.24	86B042	0.0018	736174						
151	**Coronene**												
	n	279[b]	64Z007	307	70E288								
	p	279	779025	320	69E216	0.23	68E098	0.56	68E098	228	566005		
152	**Coumarin**												
	n					$<10^{-4}$	707531			258[b]	775025	3.8[b]	79E282
	p	350	73E372			$<10^{-4}$	707531	0.054	79E282	261	73E372	1.3	79E282
153	**Coumarin, 7-(diethylamino)-4-methyl-**												
	n			2.8	80E296	0.49	85E025	0.30	85E025				
	p			3.1	80E296	0.73	85E025	0.006	84F375			3300	747049
154	**Coumarin, 7-(diethylamino)-4-(trifluoromethyl)-**												
	n			4.1	80E296	1.0	85E025	0.043	84F375				
	p			0.85	85E025	0.090	85E025						
155	**Coumarin, 5,7-dimethoxy-**												
	n					0.003	87E187					10[b]	79E282
	p	339	73E372			0.65	87E187	0.072	79E282	253	73E372		

No.	Compound														
156	1,3-Cyclohexadiene	n	410	399001								219	65E036	30[b]	80B021
157	Cyclopentadiene	n										243	65E036	1.7[b]	81E270
158	Cytidine 5'-monophosphate	p	403	673066			0.0001	743202				334	673066		
159	Deoxythymidine 5'-monophosphate	p	408	673066			0.0001	743202	0.055	79B087		315	673066	25	79B087
160	1,3-Diazaazulene	n	265[b]	73E353	0.80	85E519	0.0024	85E519	0.63[b]	85E519		228[b]	73E353		
		p			0.87	85E519	0.0025	85E519							
161	Dibenz[a,h]acridine	n	304	706135			0.24	706135				225	706135		
		p	306	779025			0.25	706135				229	779025		
162	Dibenz[a,j]acridine	n	305	706135	8.8	706135	0.42	706135							
		p	303	779025			0.51	706135				223	779025		
163	Dibenz[a,h]anthracene	n	303	71Z014	37[b]	71Z001	0.12	68E114	0.90[b]	727047					
		p	303	779025	27	737591			0.9	83F075		218	779025		

Table 1—Photophysical Parameters—Continued

No.	Solv	E_S (kJ/mol)	Ref.	τ_S (ns)	Ref.	ϕ_{fl}	Ref.	ϕ_T	Ref.	E_T (kJ/mol)	Ref.	τ_T (μs)	Ref.
164	**Dibenz[a,j]anthracene**												
	n	303[b]	64Z006										
	p	303	779025	80	737591					221	779025		
165	**Dibenzo[def,mno]chrysene**												
	n	279[b]	71Z001	5.0[b]	71Z001	0.62[b]	68E114	0.21[b]	69F388	141	72B007		
	p	276	779025							143	779025		
166	**Dibenzofuran**												
	n	398	81E648	7.3	71Z001	0.53	71Z001	0.39	706049	287	81E648		
	p									293	58E005		
167	**Dibenzothiophene**												
	n	367	81E648	0.90	71Z001	0.09	71Z001	0.97	706049	285	81E648		
	p									288	766267		
168	**2,2'-Dicarbocyanine, 1,1'-diethyl-**												
	p					0.0028	736051	<3×10⁻⁴	736051			480	736051
169	**1,4-Dioxin, 2,3,5,6-tetraphenyl-**												
	n											0.630[b]	79A241
	p									232	79A241	0.535	79A241
170	**Ethene, tetraphenyl-**												
	n	333	71Z014							209[b]	82E204	0.18[b]	82E204
171	**Flavanone**												
	n									305	86F287		

No.	Compound														
171	Flavanone—Continued														
		p							<0.001	86F287					
172	Flavone														
		n							0.9[b]	86E567	259	86E567	4.5[b]	86E567
173	Fluoranthene														
		n	295	76E696	53	71Z001	0.35	82E059			221	57B004	8500	64E026
		p	295	779025			0.21	68Z003			221	779025		
174	Fluorene														
		n	397	71Z014	10	71Z001	0.68	85F488	0.22	706049	282	64E037	150	69E208
		p	397	779025			0.68	68E114	0.32	68E098	284	779025		
175	9-Fluorenone														
		n			2.8[b]	78E495	0.0005	69E239	0.94[b]	757282			500	78E495
		p	266	706018	21.5	78E495	0.0027	69E239	0.48	78E495	211	78E060	100	78E495
176	Fluorescein dianion														
		p	230	69E235	3.6	82E660	0.97	776251	0.02	82E660	197	69E235	20000	60A001
177	Fluorescein dianion, 4',5'-dibromo-2',7'-dinitro-														
		p	218	69E235	4.5	71Z001					190	69E235		
178	Fluorescein dianion, 2',4',5',7'-tetrabromo-														
		p	209	68Z004	3.6	776251	0.69	776251	0.33	82E660	177	68Z004		
179	Fluorescein dianion, 2',4',5',7'-tetrabromo-3,6-dichloro-														
		p		82E660	3.0	82E660			0.19	82E660			1700[p]	61E011

Table 1—Photophysical Parameters—Continued

No.	Solv	E_S (kJ/mol)	Ref.	τ_S (ns)	Ref.	ϕ_{fl}	Ref.	ϕ_T	Ref.	E_T (kJ/mol)	Ref.	τ_T (μs)	Ref.
180	**Fluorescein dianion, 2',4',5',7'-tetrabromo-3,4,5,6-tetrachloro-**												
	p	210	68Z004	3.3	82E660			0.22	82E660	167	68Z004		
181	**Fluorescein dianion, 3,4,5,6-tetrachloro-2',4',5',7'-tetraiodo-**												
	p	213	71Z001	0.82	776251	0.11	776251	0.61	82E660	164	68Z004	30	84E216
182	**Fluorescein dianion, 2',4',5',7'-tetraiodo-**												
	p	212	68Z004	0.56	776251	0.08	776251	0.83	82E660	184	69E203	630	64E016
183	**Fluorescein monoanion**												
	p					0.45	85A166	0.10	85A166				
184	**Formaldehyde**												
	n	337^g	80Z097							303^g	80Z097		
185	**Furan, 2,5-diphenyl-**												
	n	349	71Z001	1.2	71Z001	1.0	71Z001						
	p	350	71Z001	1.6	71Z001								
186	**Guanosine 5'-monophosphate**												
	p	407	673066			8×10^{-5}	743202			325	673066		
187	**(E,E,E)-2,4,6-Heptatrienal, 5-methyl-7-(2,6,6-trimethyl-1-cyclohexen-1-yl)-**												
	n	299	78E431					0.66	79E546	150	84E180	6.2	79E546
	p							0.41	79E546			10.9	79E546
188	**Hexahelicene**												
	n	291	62E008	14.5	68E130	0.041	68E130			228	62E008		
	p	290	779025							228	779025		

No.	Compound																	
189	**1,3,5-Hexatriene, 1,6-diphenyl-**																	
		n	300	78B145	12.9	82E365	82E365	0.65	82E365	82E365	0.029	82E365	82E365	149	72B007	20	82E365
		p			5.2	82E365	82E365	0.27	82E365	82E365	0.020	761088				30	82E365
190	**1-Indanone**																	
		n	331	86E058										314	86E058		
		p						$<10^{-6}$	736002					317	66E097		
191	**2-Indanone**																	
		p	367[x]	85E384										345[x]	85E384		
192	**Indazole**																	
		n												284[b]	755396		
		p	396	63E012				0.46	66E102					284	63E012		
193	**Indeno[2,1-a]indene**																	
		n			2.1	73E366		0.92	73E366					199	80E113		
		p												218	89E090		
194	**Indole**																	
		n	415	51Z002	7.9	80E014		0.33	84E843		0.43	81E082		301	84E843	16	777037
		p	401	63E012	4.6	71Z001		0.42	71Z001		0.23	81E082		296	84E843	11.6	757163
195	**Indole, 1,2-diphenyl-**																	
		n	364	71Z001	2.0	71Z001		0.90	71Z001								
		p	363	71Z001	2.4	71Z001											

29

Table 1—**Photophysical Parameters**—Continued

No.	Solv	E_S (kJ/mol)	Ref.	τ_S (ns)	Ref.	ϕ_{fl}	Ref.	ϕ_T	Ref.	E_T (kJ/mol)	Ref.	τ_T (μs)	Ref.
196	**Indole, 1-methyl-**												
	n									289	771021	1.8[b]	771021
	p	413	71Z014	8.5	696102	0.38	69E244			292	771021	29	83A213
197	**Indole, 3-methyl-**												
	n			3.7	80E014	0.31	80E014						
	p	374	59E008	8.5	84E335	0.45	84E843			285	84E843		
198	**Indole, 2-phenyl-**												
	n	361	71Z001	2.0	71Z001	0.86	71Z001						
	p	354	71Z001	2.6	71Z001								
199	**Indolo[2,3-b]quinoxaline**												
	n			0.8	86E207	0.02	86E207	0.88	86E207				
	p			1.4	86E207	0.01	86E207	0.92	86E207				
200	**β-Ionone**												
	n	295[b]	85E293					0.49[b]	85E293	230[b]	85E293	0.16	78E721
201	**Isoquinoline**												
	n							0.21[b]	87E642	254	59E009		
	p			0.25	766314	0.012	766314			254	59E008		
202	**Methylene Blue cation**												
	p	180	63E027			0.04	79E560	0.52	69E203	138	67F524	450	756162
203	**Naphthalene**												
	n	385	71Z014	96	71Z001	0.19	706229	0.75	757282	253	57B004	175	62E009

No.	Compound												
203	Naphthalene—Continued												
	p	384	506002	105	68E129	0.21	66E098	0.80	68E098	255	44E001	1800	747049
204	Naphthalene-d_8												
	n	382	71Z001	96	71Z001	0.27	71Z001	>0.38[b]	65F030				
	p					0.22	68E114						
205	Naphthalene, 2-acetyl-												
	n							0.84[b]	65F030	249	67F523	300[b]	66E092
	p	325	60E014							249	67F523		
206	Naphthalene, 1-amino-												
	n	348	71Z001	6.0	71Z001	0.47	83E543	>0.15[b]	65F030				
	p	324	71Z001	19.6	71Z001	0.57	70E321			229	766421		
207	Naphthalene, 2-amino-												
	n			6.9	70E321	0.33	70E321	0.58	727047				
	p	306	766421	16.6	70E321	0.46	70E321	0.32	727047	239	766421		
208	Naphthalene, 1-bromo-												
	n	373	71Z014	0.075	82E303	0.003[b]	80E429			247	57B004	270	62E009
	p									247	63Z003	830	61E005
209	Naphthalene, 2-bromo-												
	n	372	71Z014	0.15	82E303	0.004[b]	80E429			252	44E001	150	62E009

Table 1—**Photophysical Parameters**—Continued

No.	Solv	E_S (kJ/mol)	Ref.	τ_S (ns)	Ref.	ϕ_{fl}	Ref.	ϕ_T	Ref.	E_T (kJ/mol)	Ref.	τ_T (μs)	Ref.
210	**Naphthalene, 1-chloro-**												
	n	375	79B163	2.4	82E303	0.014[b]	80E429	0.79	757282	245[b]	78E067	280	62E009
	p									248	44E001		
211	**Naphthalene, 2-chloro-**												
	n	372	71Z001	4.2	71Z001								
	p	373	79B163							251	44E001	180	62E009
212	**Naphthalene, 1-cyano-**												
	n	398[b]	84F449					>0.17[b]	65F030				
	p	373	71Z014	8.9	767370					240	767370		
213	**Naphthalene, 2-cyano-**												
	p	363	79B163							248	44E001		
214	**Naphthalene, 1,4-dicyano-**												
	n	356	84F449	3.4	84E236								
	p			10.1	767370			0.19	84B066	232	767370	40	84B066
215	**Naphthalene, 1,6-dimethyl-**												
	n	373	82B140	50	66E106	0.25	65E043						
	p			55	66E099	0.20	66E099						
216	**Naphthalene, 2,6-dimethyl-**												
	n	370	71Z001	38	71Z001	0.37	706229						
	p			44	66E099	0.30	66E099						

No.													
217	**Naphthalene, 1-(dimethylamino)-**												
	n			0.13	83E543	0.011	83E543						
	p			2.1	83E543	0.13	83E543			243	44E001		
218	**Naphthalene, 1,4-diphenyl-**												
	n	357	71Z001	1.3	71Z001	0.40	71Z001						
219	**Naphthalene, 1-hydroxy-**												
	n	372	71Z001	10.6	71Z001	0.17	706229	>0.27[b]	65F030	245	44E001		
	p	371	71Z014	7.5	71Z001								
220	**Naphthalene, 2-hydroxy-**												
	n	362	71Z001	13.3	71Z001	0.27	706229			252	44E001	67	737113
	p	362	71Z014	8.9	71Z001								
221	**Naphthalene, 1-methoxy-**												
	n	374	79B163			0.36[b]	68E133	0.45	757282	250	44E001		
	p					0.53	66E098	0.50	68E098			5500	68Z003
222	**Naphthalene, 1-methyl-**												
	n	377	71Z001	67	71Z001	0.21	726120	0.58	757282	254	44E001		
	p	377	71Z014	97	726120	0.19	726120						
223	**Naphthalene, 2-methyl-**												
	n	376	71Z001	59	71Z001	0.27	706229	0.56	757282	254	67E112		
	p	374	71Z014	47	66E099	0.16	66E099					25[b]	767159

Table 1—Photophysical Parameters—Continued

No.	Solv	E_S (kJ/mol)	Ref.	τ_S (ns)	Ref.	ϕ_{fl}	Ref.	ϕ_T	Ref.	E_T (kJ/mol)	Ref.	τ_T (µs)	Ref.
224	**Naphthalene, 1-nitro-**												
	n	313	71Z014	0.012[b]	747307			0.63[b]	687061			0.93	81B064
	p			0.008	747307					231	71F587	4.9	81B064
225	**Naphthalene, 2-nitro-**												
	n	315	71Z014	0.010[b]	747307			0.83[b]	71F587			0.53	767269
	p			0.022	747307					238	71F587	1.70	767269
226	**Naphthalene, 1-phenyl-**												
	n	379	71Z001	13	71Z001	0.37	71Z001	0.52	757282	246	67E112		
227	**Naphthalene, 2-phenyl-**												
	n	368	71Z001	114	71Z001	0.26	71Z001	0.43	757282	245	67E112		
228	**Naphthalene, 1-styryl-, (E)-**												
	n			1.9	776378	0.64	776378	0.04[b]	84E237			0.39[b]	84E237
	p					0.25	776378						
229	**Naphthalene, 2-styryl-, (E)-**												
	n			15	776378	0.71	776378					0.14[b]	84E237
	p					0.51	776378						
230	**Naphthalene, 1,2,3,4-tetrahydro-**												
	n	429	82E344	29.2	82E344	0.20	82E344			339	82E344		

231	**1,5-Naphthyridine**												
	n	324[x]	81E790	0.27	86B042			0.55	82E203	278[x]	73E356		
	p			0.25	86B042								
232	**1,8-Naphthyridine**												
	n	370	71Z014	0.21	86B042			0.04	82E203				
	p	388	71Z014	0.24	86B042								
233	**1,3,5,7-Octatetraene, 1,8-diphenyl-**												
	n	270	72E330	6.7	82E365	0.085	82E365	0.005	82E365	132	72B007	40	82E365
	p			6.8	82E365	0.091	82E365	0.006	761088			34	82E365
234	**1,3,4-Oxadiazole, 2-(4-biphenylyl)-5-phenyl-**												
	n	362	71Z001	1.0	71Z001	0.83	71Z001					0.460[b]	777265
	p	360	71Z001	1.1	71Z001								
235	**1,3,4-Oxadiazole, 2,5-diphenyl-**												
	n	385	71Z001	1.4	71Z001	0.89	71Z001					0.300[b]	777265
	p	386	71Z001	1.5	71Z001								
236	**2,2′-Oxadicarbocyanine, 3,3′-diethyl-**												
	p			1.1	82E086	0.49	82E086	<0.005	726156			5000	726156
237	**Oxazole, 2,5-bis(4-biphenylyl)-**												
	n	319[b]	71Z001	1.2[b]	71Z001	0.84	88E754					0.285[b]	777265
	p			1.6	88E754								

Table 1—**Photophysical Parameters**—Continued

No.	Solv	E_S (kJ/mol)	Ref.	τ_S (ns)	Ref.	ϕ_{fl}	Ref.	ϕ_T	Ref.	E_T (kJ/mol)	Ref.	τ_T (μs)	Ref.
238	**Oxazole, 2,5-diphenyl-**												
	n	357	71Z001	1.4	71Z001	0.85	80E439	0.12	80E439			1700	80E439
	p	356	71Z001	1.6	71Z001	0.70	88E754					2500	747049
239	**Oxazole, 2-(1-naphthyl)-5-phenyl-**												
	n	329	71Z001	2.1	71Z001	0.94	71Z001					0.215[b]	777265
	p	326	71Z001	2.3	71Z001	0.78	88E754						
240	**Oxazole, 2,2′-(1,4-phenylene)bis[5-phenyl-**												
	n	315	71Z001	1.0[b]	86E128	0.98	776387	0.054[b]	86E128			1750[b]	86E128
	p	310	71Z001	1.3	88R129	0.91	776387			232	86E128		
241	**Pentacene**												
	n	205[b]	71Z014			0.08[b]	727073	0.16[b]	727073	75[b]	716279	110	61E005
242	**(E,E)-2,4-Pentadienal, 3-methyl-5-(2,6,6-trimethyl-1-cyclohexen-1-yl)-**												
	n	334	78E431					0.20	84E036	188	78E721	0.1	78E721
	p							0.45	79E546			0.19	79E546
243	**Pentahelicene**												
	n			25	79A237	0.070	79A237					0.31	79A237
	p	303	64Z006							237	566005		
244	**2,3-Pentanedione**												
	n	264[g]	727134							226[g]	727134		
	p									233	67E114		

36

No.	Compound		C1	C2	C3	C4	C5	C6
245	**2-Pentanone**	n		1.8 70E304		0.59 84E581		0.21 84E581
		p		3.3 84E582				
246	**Perylene**	n	275 71Z014	6.4 71Z001	0.75 79E560	0.014 66E101	148[x] 69E238	5000 68Z003
		p	273 68Z003	6.0 62E014	0.87 61E010	0.0088 66E101	151 68Z003	
247	**Phenanthrene**	n	346 71Z014	57.5 71Z001	0.14 67E031	0.73 757282	260 63E024	145 62E009
		p	345 79E505	60.7 68E124	0.13 66E098	0.85 68E098	257 89E090	910 61E005
248	**Phenanthrene, 9,10-dihydro-**	n	395 71Z001	6.6 71Z001	0.55 71Z001	0.13 69E208		120 69E208
249	**Phenanthridine**	p	342 78E761				264 78E761	
250	**6-Phenanthridone**	p	352 80E792		0.25 80E792		286 80E792	
251	**1,10-Phenanthroline**	n	350 83E180		0.001 777201			26 777201
		p	353 63E018	2.1 777201	0.004 777201		264 63E018	35 777201
252	**Phenazine**	n	273[x] 79E967	0.014 766469	0.0015 85B074	0.21 716169	186 69E229	42 716169
		p	299 779025	0.020 766469		0.45 85B074	187 85B074	770 85B074

Table 1—Photophysical Parameters—Continued

No.	Solv	E_S (kJ/mol)	Ref.	τ_S (ns)	Ref.	ϕ_{fl}	Ref.	ϕ_T	Ref.	E_T (kJ/mol)	Ref.	τ_T (μs)	Ref.
253		**Phenazinium, 3,7-diamino-5-phenyl-**											
	p		89A343	4.1	89A343	0.32	89A343	0.10	89A343			25	89A343
254		**Phenol**											
	n	431	71Z014	2.1	71Z001	0.066	706229						
	p	423	84E090	7.0	82E060	0.19	62E022	0.32	82E060	342	84E090	3.3	757161
255		**Phenol, 4-methyl-**											
	n	417	71Z001	2.3	71Z001	0.09	71Z001						
	p	412	71Z001	4.8	71Z001							3.4	757161
256		**Phenothiazine**											
	n												
	p	323	71Z014	1.5	83E835	0.0034	83E835	0.54	83E835	253	757279		
257		**Phenoxazine**											
	n	330	72E303	1.5	72E303	0.027	72E303			262	72E303	32	72E303
	p	327	72E303	1.1	72E303	0.023	72E303			260	72E303	44	707186
258		**Phenoxazine, 10-phenyl-**											
	n	328	72E303	3.2	72E303	0.047	72E303			265	72E303	49	72E303
	p	327	72E303	2.9	72E303	0.040	72E303			264	72E303		
259		**Phenoxazinium, 3,7-diamino-, conjugate monoacid**											
	p							≤0.003	767661			55	767246
260		**p-Phenylenediamine, N,N,N',N'-tetramethyl-**											
	n	334	71Z001	4.3	71Z001	0.18	71Z001					1.4	82E474

260 *p*-Phenylenediamine, *N*,*N*,*N*′,*N*′-tetramethyl—Continued

No.	Compound						
260	*p*-Phenylenediamine, *N*,*N*,*N*′,*N*′-tetramethyl—Continued						
	p	329 / 71Z001	7.1 / 71Z001				0.5 / 84B061
261	*o*-Phenylenepyrene						
	n	260 / 71Z001	9.1 / 71Z001	0.17 / 71Z001			
262	Phenyl ether						
	n	426 / 71Z001	2.0 / 71Z001	0.03 / 71Z001			
	p	426 / 84E090				339 / 84E090	
263	Pheophytin *a*						
	n	179 / 79E838					
	p	179 / 70E296			0.95 / 70E296	130 / 79E838	750 / 70E296
264	Pheophytin *b*						
	n	182 / 79E838					
	p	183 / 70E296			0.75 / 70E296	134 / 79E838	1050 / 70E296
265	Phthalazine						
	n		0.19 / 81F509		0.29 / 82E203	264[b] / 82E585	2.7[b] / 87E642
	p	309 / 59E008			0.44 / 87E642	275 / 59E008	21.3 / 757309
266	Phthalocyanine						
	n	170[b] / 69E231		0.67[b] / 70E319	0.14[b] / 78A378	120[b] / 78A378	130[b] / 78A378
267	Phthalocyanine, magnesium(II)						
	n	174[b] / 69E231	6.5 / 573002	0.48[b] / 69E231			100 / 65A001
	p		7.6 / 573002	0.76 / 70E319	0.23 / 70E319		430 / 86E784

Table 1—Photophysical Parameters—Continued

No.	Solv	E_S (kJ/mol)	Ref.	τ_S (ns)	Ref.	ϕ_{fl}	Ref.	ϕ_T	Ref.	E_T (kJ/mol)	Ref.	τ_T (µs)	Ref.
268	**Phthalocyanine, zinc(II)**												
	n	175[b]	71E386			0.30[b]	69E231	0.65[b]	81E457	109[b]	71E386		
	p					0.45	70E319	0.04	86E784			270	86E784
269	**Picene**												
	n	318[b]	64Z006									160[b]	71E361
	p	318	779025							240	779025		
270	**Porphine**												
	n	195[b]	753056	12.5[b]	70E319	0.055[b]	70E319	0.90	82F161	151[b]	753056		
	p	196	743135							152	743135		
271	**Porphine, magnesium(II)**												
	n	208	753056			0.066	71E357			164	753056		
	p	207	753056							163	753056		
272	**Porphine, octaethyl-**												
	n	193	753056							155	753056		
	p	194	743135							156	743135		
273	**Porphine, octaethyl-, zinc(II)**												
	n			1.9[b]	86E020	0.05[b]	86E020					5000[b]	86E020
	p	208	71E357	2.1	86E020	0.045	86E020			170	71E357	2100	86E020
274	**Porphine, 2,7,12,17-tetraethyl-3,8,13,18-tetramethyl-**												
	n	192[b]	753056			0.09[b]	69E231			155[b]	753056		
	p	194	753056							156	753056		

No.	Compound		C1		C2		C3		C4		C5		C6	
275	Porphine, 2,7,12,17-tetraethyl-3,8,13,18-tetramethyl-, magnesium(II)	n	207	753056			0.25^b	71E357			162	753056		
		p	203	753056										
276	Porphine, 2,7,12,17-tetraethyl-3,8,13,18-tetramethyl-, zinc(II)	n	209^b	69E231			0.04^b	69E231			170	71E357		
		p	208	71E357			0.04	71E357						
277	Porphine, tetrakis(1-methylpyridinium-4-yl)-	p	177	84E346	6.0	84E346	0.047	84E346	0.92	82E622			170	84E346
278	Porphine, tetrakis(1-methylpyridinium-4-yl)-, zinc(II)	p	191	84E346	1.3	82E622	0.025	84E346	0.90	82E622			2000	84E346
279	Porphine, tetrakis(4-sulfonatophenyl)-	p			10.4	82E622	0.16	84E203	0.78	82E622			420	82E622
280	Porphine, tetrakis(4-sulfonatophenyl)-, zinc(II)	p			1.4	84E203	0.041	84E203	0.84	82E622			80	82N068
281	Porphine, tetrakis(4-trimethylammoniophenyl)-	p			9.3	83E462	0.07	83E462	0.80	83E462			540	83E462
282	Porphine, tetrakis(4-trimethylammoniophenyl)-, zinc(II)	p			1.8	83E462			0.82	83E462			1200	83E462
283	Porphine, tetraphenyl-	n	179	82Z053	13.6	80E540	0.11^b	84E203	0.82	82F161	138	82Z053	1500^b	84E203
		p	185	743135	10.1	70E319	0.15	81E375	0.88	83F182	140	743135		

Table 1—**Photophysical Parameters**—Continued

No.	Solv	E_S (kJ/mol)	Ref.	τ_S (ns)	Ref.	ϕ_{fl}	Ref.	ϕ_T	Ref.	E_T (kJ/mol)	Ref.	τ_T (μs)	Ref.
284	**Porphine, tetraphenyl-, magnesium(II)**												
	n	196	82Z053	9.2	81E271	0.15	81E271			143	82Z053	1400	81E271
	p	193	71E357					0.85	83F182	143	71E357		
285	**Porphine, tetraphenyl-, zinc(II)**												
	n	198	82Z053	2.7	81E271	0.04	81E271	0.88	82F161	153	756229	1200	81E271
	p	199	71E357					0.90	83F182	153	71E357		
286	**Porphine, zinc(II)**												
	n					0.022[b]	78E504						
	p	210	71E357			0.023	71E357			166	71E357		
287	**Porphine-2,12-dipropanoic acid, 7,17-diethyl-3,8,13,18-tetramethyl-, dimethyl ester**												
	n	191[b]	69E231			0.08[b]	69E231	0.81[b]	80E200			220[b]	80E200
288	**Porphine-2,18-dipropanoic acid, 7,12-diethenyl-3,8,13,17-tetramethyl-, dimethyl ester**												
	n			23[b]	771078			0.80	82F161			550[b]	80B017
	p	191	743135			0.06	743135			150	743135		
289	**Porphine-2,18-dipropanoic acid, 2,3-dihydro-3,3,7,12,17-pentamethyl-**												
	p	187	80E593	6.3	80E593	0.07	80E593	0.85	80E593	180	80E593	430	80E593
290	**Porphycene**												
	n			10.2[b]	86E633	0.36[b]	86E633	0.42[b]	86E633			200[b]	86E633
291	**Propiophenone**												
	n	336	86E058							312	86E058		
	p									313	706018		

No.	Compound		val	ref	val	ref	val	ref	val	ref	val	ref	val	ref
292	Psoralen	n							0.034[b]	78E157				
		p	327	73E372	0.92[m]	84Z353	0.01	84Z353	0.06	79E678	262	73E372	5	79E678
293	Psoralen, 3-carbethoxy-	n							0.30[b]	82E133			20[b]	82E133
		p	319	73E372	0.93	84Z353	0.025	84Z353	0.44	82E133	254	83E324	5.5	82E133
294	Psoralen, 5-methoxy-	n			0.15	88E643	0.0012	88E643	0.067[b]	78E157				
		p	309	73E372	0.93	88E643	0.010	88E643	0.1	83E324	254	83E324	4.2	83E324
295	Psoralen, 8-methoxy-	n							0.011[b]	78E157			1.1[b]	78E157
		p					0.0020	87E187	0.04	83E324	262	73E372	10	79E678
296	Psoralen, 4,5',8-trimethyl-	p	321	73E372					0.093	79B042	268	73E372	7.1	79B042
297	Purine, 2-amino-	n	362	74E524	1.6	756270	0.11	74E524					83	756270
		p	355	74E524	6.0	756270	0.64	74E524						
298	Purine, 2-(dimethylamino)-	n	345	74E524	7.1	756270	0.13	74E524					59	756270
		p	333	74E524	9.4	756270	0.75	74E524						

Table 1—**Photophysical Parameters**—Continued

No.	Solv	E_S (kJ/mol)	Ref.	τ_S (ns)	Ref.	ϕ_{fl}	Ref.	ϕ_T	Ref.	E_T (kJ/mol)	Ref.	τ_T (μs)	Ref.
299	**Pyrazine**												
	n	365	83E031			0.0004	67E117	0.33	67E117	315	78E312		
	p							0.87	757309	311	58E006	4.5	757309
300	**Pyrene**												
	n	322	71Z014	650	79E109	0.65	79E109	0.37	82E042	203	57B004	180	70E295
	p	321	78E761	190	88E804	0.72	66E098	0.38	68E098	202	78E761	11000	68Z003
301	**1-Pyrenecarboxaldehyde**												
	n					0.0002	83E387	0.78	83E387	180[b]	83E387	50	83E387
	p			1.7	83E387	0.084	83E387	0.65	83E387			38	83E387
302	**Pyridazine**												
	n	318	67E117	2.6	65E051	0.0002	67E117	0.07	92E049	297	67E117	0.065	92E049
	p							0.08	92E049	290[x]	67B017	0.10	92E049
303	**Pyridine, 2-amino-**												
	n			24.2	84E583	0.04	82E341						
	p			5.1	84E583	0.19	82E341						
304	**Pyrimidine**												
	n	360	67E117	1.5	65E051	0.0029	67E117	0.12	67E117	338	70E310		
	p							1.0	757309			1.4	757309
305	**Pyronine cation**												
	p	215	77E801	2.3	82E660			<0.02	82E660	178	77E801		

No.	Compound														
306	**Pyruvic acid**														
		n							0.65[b]	81F070				
		p							0.88	81F070				
307	**p-Quaterphenyl**														
		n	362	71Z001	0.8	71Z001	0.92	83E027					0.5	81F070
308	**Quinazoline**														
		n	326	78E894	0.079	80B040			0.70	82E203	262	78E894		
		p	327	59E008							262	59E008		
309	**Quinine bisulfate**														
		p	305	71Z001	19.2	71Z001	0.55	61E010						
310	**Quinoline**														
		n	383	80E627					0.31[b]	65F030	258	80E627		
		p	381	80E627							261	83E417		
311	**Quinoxaline**														
		n	314	82E355	0.023	80B040			0.99	82E203	255	82E355		
		p	319	59E008					0.90	85E408	254	59E008	29.4	757309
312	**(all-E)-Retinal**														
		n	281	717003					0.43[w]	75E529	123	84E180	9.3	82A288
		p							0.12[w]	78E467			18	62E007
313	**(all-E)-Retinol**														
		n	321	69E251	5.0[b]	85E190	0.02	69E251	0.017	776412	140	84E180	25	716113

Table 1—Photophysical Parameters—Continued

No.	Solv	E_S (kJ/mol)	Ref.	τ_S (ns)	Ref.	ϕ_{fl}	Ref.	ϕ_T	Ref.	E_T (kJ/mol)	Ref.	τ_T (µs)	Ref.
313	(all-E)-Retinol—Continued												
	p	327	69E251	2.3	85E190	0.006	69E251	−0.003	85E190				
314	Rhodamine, inner salt												
	p	230	77E801	4.4	65E042	0.88	65E042	0.0024	777316	191	77E801		
315	Rhodamine, inner salt, N,N'-diethyl												
	p	224	77E801			0.94	65E042	0.005	777041	184	77E801		
316	Rhodamine B, inner salt												
	p	213	77E801	2.7	86E458	0.65	86E458	0.0024	777316	178	77E801	1.6	777041
317	Rhodamine 6G cation												
	p	219	77E801	3.8	82E660	0.86[w]	79E560	0.0021	747050	181	77E801	3500	747050
318	Riboflavine, conjugate monoacid												
	p	263	71Z014	2.3	777617	0.12	777617	0.40	777617			19	777617
319	Rubrene												
	n	221[b]	81E716	16.5[b]	71Z001	0.98[b]	68E113	0.0092	86E782	110[b]	81E346	120	86E782
	p			10.8	88E219			0.023	89B155			80	68E103
320	Spirilloxanthin												
	n	219	71Z014					0.028	761035			6.2	761035
321	(E)-Stilbene												
	n	358[x]	62E012	0.075	79E086	0.036	80E169			206	80E113	14	680379
	p					0.016	82Z102			206	80E113	62	81E214

No.	Compound							
322	**(E)-Stilbene, 4-bromo-**							
	347[x] 62E012							
		n		0.038 79E640		201 89E090	0.03 79E378
		p					
323	**(E)-Stilbene, 4,4′-dinitro-**							
		n		<10^{-4b} 84F488	0.81[b] 84F488	189 89E090	0.080 747022
		p		<10^{-4} 84F488	0.69 84F488		0.13 79Z027
324	**(E)-Stilbene, 4,4′-diphenyl-**							
	322[b] 71Z001		1.1[b] 71Z001				
		n		0.81 88E481			
		p		0.82 88E481			
325	**(E)-Stilbene, 4-nitro-**							
		n		<10^{-4b} 84F488	0.86[b] 84F488	195 89E090	0.063 747022
		p		<10^{-4} 84F488	0.71 84F488	208 44E001	0.083 79Z027
326	**(Z)-Stilbene**							
	360[x] 62E012							
		n				227[b] 69F410	17 680379
327	**Styrene**							
	415 776060		13.9 79E265	0.25 79E265	0.40 82E181	258 776060	0.025 82E181
		n					
328	**Styrene, β-(2-pyridyl)-, (E)-**							
	349 71Z014						
		n		0.0008 80E169		205 89E090	
		p		0.001 82F252			
329	**Styrene, β-(4-pyridyl)-, (E)-**							
		n		0.0016 80E169		208 89E090	

47

Table 1—**Photophysical Parameters**—Continued

No.	Solv	E_S (kJ/mol)	Ref.	τ_S (ns)	Ref.	ϕ_{fl}	Ref.	ϕ_T	Ref.	E_T (kJ/mol)	Ref.	τ_T (μs)	Ref.
329	— **Styrene, β-(4-pyridyl)-, (E)**—Continued												
	p					0.0015	82F252			206	89E090		
330	**m-Terphenyl**												
	n	393	71Z001	28.5	71Z001	0.29	71Z001	0.41	85F488				
	p									269	67E112		
331	**p-Terphenyl**												
	n	385	71Z001	0.95	71Z001	0.77	706229	0.11	69E208			450	69E208
	p									244	67E112		
332	**Tetrabenzoporphine, zinc(II)**												
	n					0.23[b]	70E319			149	71E357	525[b]	73E345
	p	191	71E357			0.11	71E357						
333	**Tetracene**												
	n	254	71Z014	6.4	65Z001	0.17	706229	0.62[b]	757282	123	64E038	400	85E555
	p	254	779025			0.16	717459	0.66	717459				
334	**Thianthrene**												
	n	407	71Z014			0.036	77E638	0.94	77E638				
335	**Thiobenzophenone**												
	n	191	756061					1.0[wb]	84A221	165	756061	1.7[b]	84A221
	p	191	726174							170	726174		
336	**Thiobenzophenone, 4,4'-bis(dimethylamino)-**												
	n	205	726174					1.0[b]	84A221	176	726174	1.3[b]	84A221

No.	Compound	Solvent										
337	Thiobenzophenone, 4,4'-dimethoxy-	n	197	726174			1.0[wb]	84A221	172	726174	1.4[b]	84A221
		p							176	726174		
338	Thiocoumarin	n					1[b]	86A240	212[b]	86A240	>2.5[b]	86A240
339	Thiofluorenone	n	173	756061					159	756061		
340	Thioindigo	n	11.0	84E107	0.30	84E107					0.158	78F030
		p	7.5	84E107	0.18	84E107					0.20	79E543
341	Thionine cation	p	196	697141			0.62	69E203	163	697141	72	697141
342	Thionine cation, conjugate monoacid	p	0.37	776118	0.047	776118					16	777315
343	Thiophene, 2,2'-(1,3-phenylene)bis-	n	366	86A357	6	86A357			238	86A357	31	86A357
344	Thioxanthene	n			0.008	77E638	0.78	77E638				
345	Thioxanthen-9-one	n							265	86E676	95[b]	81A294

Table 1—**Photophysical Parameters**—Continued

No.	Solv	E_S (kJ/mol)	Ref.	τ_S (ns)	Ref.	ϕ_{fl}	Ref.	ϕ_T	Ref.	E_T (kJ/mol)	Ref.	τ_T (μs)	Ref.
345	**Thioxanthen-9-one**—Continued												
	p			2.0	89A120	0.12	746190					73	737190
346	**Thioxanthione**												
	n							1.0[wb]	84A221	166	82E214	0.83[b]	84A221
347	**Thymidine**												
	n	421	80E025										
	p	417	80E025					0.069	79B087			25	79B087
348	**Thymine**												
	n	422	80E025										
	p	423	80E025					0.06	757510			10	757510
349	**α-Tocopherol**												
	n			0.8	89N023	0.16	89N023						
	p			1.8	89N023	0.34	89N023						
350	**Toluene**												
	n	445	71Z014	34	71Z001	0.14	82E344	0.53	69E202	346	61E014		
	p			35	766441	0.13	746251			347	67F523		
351	**Toluene-d_8**												
	n	447	71Z001	35	71Z001	0.21	71Z001						
	p			36	766441	0.14	766441						
352	**Triphenylene**												
	n	349	71Z001	36.6	71Z001	0.066	706229	0.86	69E202			55	61E005

#	Compound		idx											
352	Triphenylene—Continued	p	352	779025	37.0	766276	0.09	66E098	0.89	66E098	280	84E612	1000	61E005
353	Triphenylene-d_{12}	n	351	71Z001	38	71Z001	0.11	71Z001						
		p					0.068	68E114						
354	Triphenylene, dodecahydro-	n	415	82E344	51.5	82E344	0.21	82E344			321	82E344		
355	Triphenylmethane	n	444	71Z014			0.028	677472	0.16	677472				
356	Tryptophan	n					0.07	84E843						
		p	399	62Z002	2.5	733187	0.13	84E843	0.18	777432			14.3	757163
357	Uracil	n	430	80E025										
		p	430	80E025					0.1	79B087			2	757510
358	Uracil, 1,3-dimethyl-	n	423	80E025					0.02	81E042				
		p	423	80E025										
359	Uridine	p							0.078	79B087			20	79B087

Table 1—**Photophysical Parameters**—Continued

No.	Solv	E_S (kJ/mol)	Ref.	τ_S (ns)	Ref.	ϕ_{fl}	Ref.	ϕ_T	Ref.	E_T (kJ/mol)	Ref.	τ_T (μs)	Ref.
360	**Uridine 5′-monophosphate**												
	p	417	673066			3×10^{-5}	743202	0.044	79B087	328	737541	33	79B087
361	**Xanthene**												
	p	408	88E763							331	88E763		
362	**9-Xanthione**												
	n	193	81F218					0.8[wb]	84A221	181	81F218	1.8[b]	84A221
	p											0.70[p]	74E514
363	**Xanthone**												
	n	324	717449	0.013	79B007					310	717449	0.02	767171
	p			0.008	78E489					310	717449	17.9	767171
364	**m-Xylene**												
	n	439	71Z014	30.8	71Z001	0.14	706229			339	61E014		
	p									336	60E016		

No.	Compound											
365	o-Xylene	n	438	82E344	32.2	71Z001	0.18	82E344	0.28	69E243	343	61E014
		p			38	726120	0.14	726120			344	60E016
366	p-Xylene	n	435	71Z014	30	71Z001	0.22	72E313	0.63	69E202	337	61E014
		p			33.7	766441	0.24	766441			337	60E016
367	p-Xylene-d_{10}	n	435	71Z001	30.3	71Z001	0.26	71Z001				
		p			32.9	766441	0.22	766441				
368	Zinc(II) chlorophyll a	p	180	79E838							131	79E838
369	Zinc(II) chlorophyll b	p	186	79E838							140	79E838

[b] Aromatic solvent, benzene-like; [g] Gas-phase measurement; [m] Major component of multiexponential; [p] Phosphorescence decay; [s] Very solvent-dependent; [w] Wavelength dependent; [x] Crystalline medium

Section 2

Triplet-State Energies: Ordered

In Table 2, the compounds are listed in order of increasing triplet energy. This table has been designed to facilitate selection of photosensitizers and quenchers. The energies are given in kJ/mol.

The triplet energies have been taken mainly from three types of experiments. First, triplet energies have been estimated from *singlet-triplet absorption* spectra. These radiative transitions are highly spin forbidden, but they can borrow intensity from singlet-singlet transitions through the use of oxygen as a external, paramagnetic perturber which enhances spin-orbit coupling in the molecules. A large fraction of the singlet-triplet absorption spectra show no clear 0–0 bands. However, empirical methods have been used to successfully estimate the location of the 0–0 bands. When triplet energies taken from singlet-triplet absorption have been chosen in Table 2, either spectra showing distinct 0–0 bands or estimations of 0–0 bands have been quoted whenever possible.

Second, the *short-wavelength band of the phosphorescence spectra* usually represents the 0–0 band of a transition between the lowest triplet state and the ground state. Even though this is a spin-forbidden transition between the same vibronic levels as that seen in absorption, luminescence detection is much more sensitive than absorption methods. Phosphorescence is the most widely used method for determining triplet energies, and of the three common methods, it gives the most definitive value. However, phosphorescence spectra are almost always taken in rigid glasses or polymers; whereas one of the most common uses of the data is to design photosensitization experiments in fluid solution.

Third, *photosensitized formation of triplet acceptors* can be used to estimate the triplet energy of a molecule by measuring the energy-transfer rate constant of a series of triplet donors being quenched by (or acceptors quenching) the molecule in question.

$$^3D* + {}^1A \rightarrow {}^1D + {}^3A* \qquad (2\text{-I})$$

When the triplet energy of the potential triplet donor is below the triplet energy of the potential triplet acceptor, the rate constant for energy transfer falls off very rapidly as the energy difference increases. Since triplet-triplet energy transfer is a spin-allowed process, as long as the triplet donor's energy is above (or within easy thermal assess to) that of the triplet acceptor's, the quenching normally proceeds at a rate constant approaching that of diffusion controlled. Of the three methods, this one best simulates applications to photosensitization, but it usually gives only a range of values.

When available, 0–0 bands of the phosphorescence are quoted in Table 2 in preference to the other two methods. Only as a last resort are the *maxima* of either singlet-triplet absorption or phosphorescence quoted. A spectral maximum from phosphorescence represents a lower limit to the lowest triplet energy, and a spectral maximum from singlet-triplet absorption represents an upper limit to the lowest triplet energy, see Fig. 2-1.

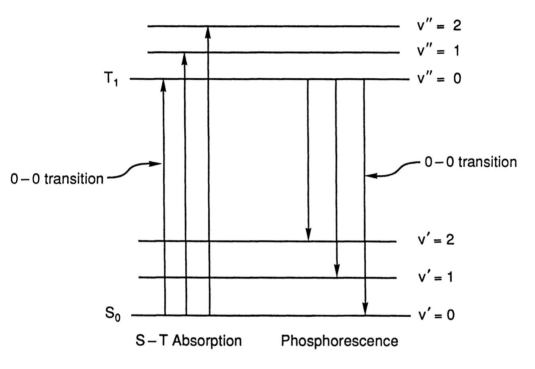

Fig. 2-1

Table 2

Triplet State Energies of Organic Compounds

No.	Compound	$E_T(n)$ kJ/mol	cm^{-1}	Ref.	$E_T(p)$ kJ/mol	cm^{-1}	Ref.
1	Pentacene	75[b]	6300[b]	716279			
2	β-Carotene	88[b]	7300[b]	757247	85	7100	92N199
3	Dibenzo[a,o]perylene, 7,16-diphenyl-	92	7700	85E555			
4	Phthalocyanine, zinc(II)	109[b]	9150[b]	71E386			
5	Rubrene	110[b]	9200[b]	81E346			
6	Phthalocyanine	120[b]	10000[b]	78A378			
7	(Z)-Azobenzene	121[b]	10100[b]	81E012			
8	Tetracene, 5-methyl-	122	10170	72B007			
9	Tetracene	123	10250	64E038			
10	(all-E)-Retinal	123	10300	84E180			
11	Chlorophyll a				125	10400	79E838
12	Phthalocyanine, platinum(II)	127[b]	10590[b]	71E386			
13	Pheophytin a	130	10800	79E838			
14	Chlorophyll b	130	10900	79E838	136	11400	68Z003
15	Zinc(II) chlorophyll a	132	11050	60E013	131	11000	79E838
16	2,4,6,8,10-Dodecapentaenal	132	11050	72B007			
17	1,3,5,7-Octatetraene, 1,8-diphenyl-	134	11200	79E838			
18	Pheophytin b						
19	Methylene Blue cation				138	11500	67F524
20	Porphine, tetraphenyl-	138	11500	82Z053	140	11700	743135
21	(all-E)-Retinol	140	11700	84E180			
22	Zinc(II) chlorophyll b				140	11700	79E838
23	Dibenzo[def,mno]chrysene	141	11800	72B007	143	12000	779025

No.	Name						
24	Isobenzofuran, 1,3-diphenyl-	142[b]	11900[b]	81E346			
25	Porphine, tetraphenyl-, magnesium(II)	143	11900	82Z053	143	12000	71E357
26	Benzo[c]cinnoline				144	12000	617012
27	Dibenzo[b,def]chrysene	144[b]	12040[b]	62E015			
28	(E)-Azobenzene	146[b]	12200[b]	81E012			
29	C$_{70}$	148[b]	12300[b]	91D034			
30	Perylene	148[x]	12373[x]	69E238	151	12600	68Z003
31	1,3,5-Hexatriene, 1,6-diphenyl-	149	12450	72B007			
32	Tetrabenzoporphine, zinc(II)				149	12500	71E357
33	(E,E,E)-2,4,6-Heptatrienal, 5-methyl-7-(2,6,6-trimethyl-1-cyclohexen-1-yl)-	150	12500	84E180			
34	Porphine-2,18-dipropanoic acid, 7,12-diethenyl-3,8,13,17-tetramethyl-, dimethyl ester				150	12600	743135
35	C$_{60}$	151[b]	12600[b]	91E368			
36	Porphine	151[b]	12600[b]	753056	152	12700	743135
37	Porphine-2,18-dipropanoic acid, 7,12-diethenyl-3,8,13,17-tetramethyl-	151	12600	753056			
38	2,4,6,8-Decatetraenal	152	12700	60E013			
39	Porphine, tetraphenyl-, zinc(II)	153	12800	756229	153	12800	71E357
40	Porphine-2,18-dipropanoic acid, 7,12-diethyl-3,8,13,17-tetramethyl-	154[b]	12900[b]	753056			
41	Porphine-2,18-dipropanoic acid, 7,12-diethyl-3,8,13,17-tetramethyl-, dimethyl ester				155	12900	743135
42	Porphine, octaethyl-	155	13000	753056	156	13000	743135

Table 2—**Triplet State Energies**—Continued

No.	Compound	$E_T(n)$			$E_T(p)$		
		kJ/mol	cm^{-1}	Ref.	kJ/mol	cm^{-1}	Ref.
43	Porphine, 2,7,12,17-tetraethyl-3,8,13,18-tetramethyl-	155[b]	13000[b]	753056	156	13000	753056
44	Porphine, tetrakis(4-chlorophenyl)-, zinc(II)	156	13000	756229			
45	Porphine, tetrakis(2-methylphenyl)-, zinc(II)	157	13100	756229			
46	Porphine, tetrakis(2-chlorophenyl)-, zinc(II)	158	13200	756229			
47	Cycloheptatriene	159[b]	13300[b]	85E281			
48	Ferrocene	159[b]	13300[b]	757247			
49	Thiofluorenone	159	13300	756061			
50	Benzo[*rst*]pentaphene, 5-methyl-	161[b]	13480[b]	62E015			
51	Porphine, 2,7,12,17-tetraethyl-3,8,13,18-tetramethyl-, magnesium(II)				162	13600	753056
52	Azulene	163[b]	13600[b]	757247			
53	Thionine cation				163	13600	697141
54	Anthrathione, 9,9-dimethyl-	163	13640	83E230			
55	Fluorescein dianion, 3,4,5,6-tetrachloro-2',4',5',7'-tetraiodo-				164	13700	68Z004
56	Porphine, magnesium(II)	164	13700	753056	163	13700	753056
57	2,2'-Thiacarbocyanine, 3,3'-diethyl-				164	13750	78E021
58	Thiobenzophenone	165	13760	756061	170	14200	726174
59	Anthracene, 1,5,10-trichloro-				165	13800	566006
60	Tetrabenzoporphine, magnesium(II)				165	13800	753056
61	Thioxanthione	166	13800	82E214			
62	1-Aceanthrylenone, 2-methyl-				165	13830	82E338
63	Porphine, zinc(II)				166	13900	71E357
64	Fluorescein dianion, 2',4',5',7'-tetrabromo-3,4,5,6-tetrachloro-				167	13950	68Z004

No.	Compound						
65	Porphine, 2,8,12,18-tetraethyl-3,7,13,17-tetramethyl-, magnesium(II)	167	13990	63E011	169	14100	779025
66	Benzo[rst]pentaphene	168[b]	14050[b]	62E015			
67	Anthracene, 9,10-dibromo-	168	14060	566006			
68	Dibenzo[a,d]cycloheptene-5-thione	169	14100	83E230			
69	2,4,6-Octatriene, 2,6-dimethyl-	169[b]	14100[b]	83E258			
70	Anthracene, 1,10-dichloro-				169	14128	566006
71	Anthracene, 9-benzoyl-10-bromo-				169	14140	82E338
72	Anthracene, 9,10-dichloro-	169	14150	566006	169	14150	66E086
73	Anthracene, 1,4,5,8-tetrachloro-				169	14155	566006
74	Anthracene, 9-benzoyl-10-chloro-				169	14160	82E338
75	Porphine, octaethyl-, zinc(II)				170	14200	71E357
76	Porphine, 2,7,12,17-tetraethyl-3,8,13,18-tetramethyl-, zinc(II)				170	14200	71E357
77	Porphine-2,18-dipropanoic acid, 7,12-diethyl-3,8,13,17-tetramethyl-, dimethyl ester, zinc(II)				171	14260	63E011
78	Anthracene, 9,10-diphenyl-	171	14290	68E131			
79	Quin[2,3-b]acridine-7,14-dione, 5,12-dihydro-2,9-dimethyl-				172	14388	78E878
80	Thiobenzophenone, 4,4'-dimethoxy-	172	14400	726174	176	14700	726174
81	Anthracene, 9-bromo-				173	14400	84B110
82	Anthracene, 9-methyl-	173	14460	57B004	170	14220	66E086
83	Anthracene, 2-methyl-				173	14500	66E086
84	Anthracene, 1,5-dichloro-				174	14568	566006
85	Anthracene, 9,10-dicyano-	175	14600	78E414			
86	Pyronine B cation				175	14600	77E801

Table 2—**Triplet State Energies**—Continued

No.	Compound	$E_T(n)$ kJ/mol	$E_T(n)$ cm^{-1}	Ref.	$E_T(p)$ kJ/mol	$E_T(p)$ cm^{-1}	Ref.
87	Sulforhodamine B, inner salt				175	14600	69E235
88	Anthracene, 9-nitro-	175	14630	57B004			
89	Benzo[a]pyrene	175	14670	57B004	177	14800	779025
90	Anthracene, 9-naphthoyl-				176	14690	82E338
91	2,4-Hexadiene, 2,5-dimethyl-	176[b]	14700[b]	81E270			
92	Thiobenzophenone, 4,4'-bis(dimethylamino)-	176	14700	726174			
93	Anthracene, 9-acetyl-				176	14720	82E338
94	Anthracene, 9-benzoyl-				176	14730	82E338
95	Anthracene, 9-propionyl-				176	14730	82E338
96	Anthracene, 1-chloro-				176	14732	566006
97	Anthracene, 9-cinnamoyl-				176	14740	82E338
98	9-Anthracenecarboxylic acid				177	14760	82E338
99	1,3-Butadiene, 1,4-diphenyl-	177	14800	707199			
100	Fluorescein dianion, 2',4',5',7'-tetrabromo-				177	14800	68Z004
101	Anthracene-9-carboxamide				177	14820	82E338
102	Anthracene	178	14870	57B004	178	14900	779025
103	Benz[g]isoquinoline	178	14870	59E009			
104	Rhodamine 3G cation				178	14870	77E801
105	Rhodamine B, inner salt				178	14890	77E801
106	9-Acridinethione, 10-methyl-	178	14900	83E230			
107	Pyronine cation				178	14900	77E801
108	1-Pyrenecarboxaldehyde	180[b]	15000[b]	83E387			
109	Thioketene, di-tert-butyl-	180[b]	15000[b]	86E675			
110	Benzo[g]quinoline	180	15070	59E009	184	15400	81E444

No.	Compound						
111	Rhodamine 6G cation				181	15100	77E801
112	1-Benzothiopyran-4-thione, 2,6-dimethyl-	181	15110	83E230			
113	9-Xanthione	181	15143	81F218	165	13790	82E338
114	Anthracene-9-carboxaldehyde	182[b]	15200[b]	68B014			
115	2,4,6-Octatrienal	182	15210	60E013			
116	Rhodamine, inner salt, *N,N'*-diethyl				184	15350	77E801
117	Acridine, 9-chloro-				184	15400	86A322
118	Anthracene, 1-amino-				184	15400	69E240
119	Anthracene, 2-amino-				184	15400	69E240
120	Anthracene, 9-amino-				184	15400	69E240
121	Fluorescein dianion, 2',4',5',7'-tetraiodo-				184	15400	69E203
122	Ovalene				184	15400	779025
123	Benz[*a*]anthracene, 7,12-dimethyl-	185	15500	54E002	184	15400	779025
124	Dibenzo[*b,g*]phenanthrene	185	15500	81E192			
125	4-Cholesten-3-thione	186	15500	726174	191	15900	726174
126	Phenazine	186	15561	69E229	187	15600	85B074
127	Indene-1-thione, 4-bromo-2,3-dihydro-2,2,3,3-tetramethyl-	187	15600	83E230			
128	2-Naphthalenethione, 1,1-dimethyl-	187[b]	15600[b]	86A240			
129	Benzo[*a*]phenazinium				187	15650	81E433
130	Naphtho[2,3-*a*]coronene	188	15676	78E314			
131	Acridine, 9-methyl-				188	15690	81E552
132	Acridine, 9-propyl-				188	15690	81E552
133	Colchicine				188	15700	78E359

Table 2—**Triplet State Energies**—Continued

No.	Compound	kJ/mol	$E_T(n)$ cm^{-1}	Ref.	kJ/mol	$E_T(p)$ cm^{-1}	Ref.
134	(E,E)-2,4-Pentadienal, 3-methyl-5-(2,6,6-trimethyl-1-cyclohexen-1-yl)-	188	15700	78E721	189	15800	89E090
135	(E)-Stilbene, 4,4'-dinitro-				188	15740	84E803
136	Dinaphth[1,2-a;1',2'-h]anthracene	190	15852	87E785			
137	Acridine	190	15870	81E147	190	15870	81E552
138	Acridine-d_9				190	15900	779025
139	Benzo[b]chrysene				190	15900	69E235
140	Fluorescein dianion, 4',5'-dibromo-2',7'-dinitro-				190	15900	89E090
141	3-Pentanethione, 2,2,4,4-tetramethyl-	190[b]	15900[b]	85A300			
142	(E)-Stilbene, 4-cyano-4'-methoxy-				191	15930	77E801
143	Indene-1-thione, 2,3-dihydro-2,2,3,3-tetramethyl-	190	15920	83E230			
144	Rhodamine, inner salt				193	16130	78E878
145	Benz[b]acridin-12-one	191	16000	85F172			
146	Benzanthrone				192	16100	779025
147	Fluorescein dianion, 2',7'-dichloro-				192	16100	69E235
148	(E)-Stilbene, 4,4'-dicyano-				192	16100	89E090
149	Acridine, 9-amino-				193	16100	81E552
150	Dibenzo[h,rst]pentaphene				193	16100	779025
151	Benzo[ghi]perylene				194	16180	566005
152	Hexabenzo[a,d,g,j,m,p]coronene	194	16200	76E692			
153	Benz[a]anthracene, 12-methyl-	195	16260	54E002			
154	Benz[j]aceanthrylene, 1,2-dihydro-3-methyl-				192	16100	779025
155	2-Cyclohexenethione, 3,5,5-trimethyl-	195[b]	16300[b]	86A240			
156	Porphine, platinum(II)	195	16300	70E317	195	16300	779025

No.	Compound						
157	(E)-Stilbene, 4-nitro-				208	17400	44E001
158	Dibenz[a,e]aceanthrylene	195	16300	89E090	196	16400	779025
159	(E)-1,3,5-Hexatriene	197	16450	60E013	198	16520	78E761
160	Benz[a]anthracene	197	16500	54E002	197	16500	69E235
161	Fluorescein dianion				197	16500	779025
162	Naphtho[1,2,3,4-def]chrysene				197	16500	89E090
163	(E)-Stilbene, 2-nitro-	198	16500	89E090	198	16500	89E090
164	(E)-Stilbene, 4-benzoyl-				198	16500	89E090
165	(E)-Stilbene, 4-cyano-				199	16600	779025
166	Benz[a]anthracene, 9-methyl-	198	16530	54E002	199	16600	779025
167	Benz[a]anthracene, 10-methyl-	198	16580	54E002	198	16600	76E692
168	Tetrabenzo[a,c,j,l]naphthacene	199	16600	80E113			
169	Indeno[2,1-a]indene				218	18300	89E090
170	Benz[a]anthracene, 2-methyl-	199	16640	54E002	200	16700	779025
171	Benz[a]anthracene, 8-methyl-	199	16650	54E002	201	16800	779025
172	Benz[a]anthracene, 4-methyl-	199	16670	54E002	199	16600	779025
173	Dibenzo[a,j]coronene	199[b]	16700[b]	76E692			
174	Tribenzo[a,d,g]coronene	199[b]	16700[b]	76E692			
175	Benz[b]acridin-12-one, 5-methyl-	200	16700	85F172			
176	(E)-Stilbene, 4-acetyl-	200	16700	89E090	198	16500	89E090
177	(E)-Stilbene, 4-iodo-	200	16700	89E090	202	16900	80E113
178	Benz[a]anthracene, 3-methyl-	200	16720	54E002	201	16800	779025
179	Benz[a]anthracene, 11-methyl-				202	16900	779025
180	1,3-Cyclobutanedithione, 2,2,4,4-tetramethyl-	201	16780	54E002	201[x]	16800[x]	83D207

Table 2—**Triplet State Energies**—Continued

No.	Compound	$E_T(n)$			$E_T(p)$		
		kJ/mol	cm^{-1}	Ref.	kJ/mol	cm^{-1}	Ref.
181	(E)-Stilbene-d_{12}	201	16800	80E113	205	17100	89E090
182	(E)-Stilbene, 4-bromo-				201	16800	89E090
183	Tropolone				201	16800	78E359
184	Benz[a]anthracene, 5-methyl-	201	16810	54E002	202	16900	779025
185	Thebenidine				202	16860	78E761
186	(E)-Stilbene, 4-chloro-				203	16900	89E090
187	Pentaphene				203	16930	566005
188	Pyrene	203	16930	57B004	202	16850	78E761
189	Benz[a]anthracene, 1-methyl-	203	16980	54E002	203	17000	779025
190	Benz[a]anthracene, 6-methyl-	203	16980	54E002	203	17000	779025
191	1-Naphthaleneacrylic acid, methyl ester	203[b]	17000[b]	86A322			
192	(E)-Stilbene, 4,4'-dimethoxy-				203	17000	89E090
193	(E)-Stilbene, 3-methoxy-				203	17000	89E090
194	(E)-Stilbene, 3-nitro-				204	17000	89E090
195	Styrene, β-(3-pyridyl)-, (E)-				204	17100	89E090
196	2,4,6-Cycloheptatrien-1-one, 2-methoxy-				205	17100	78E359
197	(E)-Stilbene, 3,3'-dibromo-				205	17100	80E113
198	(E)-Stilbene, 4-fluoro-				205	17100	89E090
199	(E)-Stilbene, 3-methyl-				205	17100	89E090
200	Styrene, β-(2-pyridyl)-, (E)-				205	17100	89E090
201	Octatetrayne, diphenyl-				205	17150	89E090
202	Acridine Orange				205	17150	566004
203	1,4-Benzoquinone, tetrachloro-	206	17200	69E247	206	17200	69E235
204	Pyridine, 3,3'-(1,2-ethenediyl)bis-, (E)-				206	17200	89E090

No.	Compound		cm⁻¹	Ref.		cm⁻¹	Ref.
205	(E)-Stilbene	206	17200	80E113	206	17200	80E113
206	Benzo[g]pteridine-2,4-dione, 10-methyl-	206	17250	79E564			
207	Dibenzo[g,p]chrysene	207	17300	71E385			
208	(E)-2-Indanthione, hexahydro-	207	17300	83E169			
209	Pyran-4-thione				207	17300	779025
210	Pyridine, 2,2'-(1,2-ethendiyl)bis-, (E)-				207	17300	89E090
211	Thioacetone	207[g]	17300[g]	84E863			
212	Benzo[g]pteridine-2,4-dione, 3,10-dimethyl-	208	17350	79E564			
213	Coumarin, 7-(diethylamino)-3,3'-carbonylbis-	208	17400	89E090	208	17400	82E271
214	Styrene, β-(4-pyridyl)-, (E)-				206	17200	89E090
215	Benz[a]acridine				209	17480	78E761
216	Benzo[g]pteridine-2,4-dione, 7,8,10-trimethyl-				209	17500	69E235
217	Ethene, tetraphenyl-	209[b]	17500[b]	82E204			
218	Riboflavine				209	17500	69E235
219	Triphenylethylene	209[b]	17500[b]	82E204			
220	Diadamantylethanedione	210	17500	83E901			
221	Benzo[a]phenazine	209	17512	85E370			
222	Dibenzo[hi,uv]hexacene	210	17538	79E140	202	16900	81E433
223	Hexabenzo[bc,ef,hi,kl,no,qr]coronene	210[b]	17540[b]	68Z005			
224	(E)-Stilbene, 4-methoxy-	210	17550	79E412	202	16900	89E090
225	Benzo[k]fluoranthene				211	17600	779025
226	2,2'-Bibenzo[b]thiophene				211	17600	85E272
227	Coumarin, 7-(diethylamino)-5',7'-dimethoxy-3,3'-carbonylbis-				211	17600	82E271
228	Dibenzo[c,m]pentaphene				211	17600	566005

Table 2—**Triplet State Energies**—Continued

No.	Compound	kJ/mol	$E_T(n)$ cm^{-1}	Ref.	kJ/mol	$E_T(p)$ cm^{-1}	Ref.
229	9-Fluorenone				211	17600	78E060
230	Indene, 2-phenyl-				211	17600	89E090
231	(E)-Stilbene, α-methyl-	211	17600	82Z015			
232	Benz[c]acridine, 9-methyl-				212	17700	84E803
233	Pyridine, 4,4'-(1,2-ethenediyl)bis-, (E)-	212	17700	89E090	211	17700	89E090
234	Thiocoumarin	212[b]	17700[b]	86A240			
235	Benzo[b]triphenylene	213	17790	57B004	213	17800	779025
236	Benz[c]acridine				213	17800	779025
237	Coumarin, 3-(2-benzofuroyl)-7-diethylamino-				213	17800	82E271
238	Coumarin, 3-benzoyl-7-diethylamino-				213	17800	82E271
239	Coumarin, 3,3'-carbonylbis(7-diethylamino-				213	17800	82E271
240	Pyran-4-thione, 2,6-dimethyl-	213	17800	83E169			
241	Pyrylium, 2,4,6-tris(4-methoxyphenyl)-				213	17800	67F508
242	Acridine, 3,6-diamino-				214	17900	69E235
243	Acridinium, 3,6-diamino-10-methyl-				214	17900	69E235
244	Chrysene, 6-nitro-				215	18000	69Z002
245	Bicyclo[2.2.1]heptane-2,3-dione, 1,7,7-trimethyl-				216	18000	67E114
246	3,4-Hexanedione, 2,2,5,5-tetramethyl-				216	18000	67E114
247	Benzo[a]coronene	216	18000	82E832			
248	Naphthalene, 1,4-dinitro-	215	18003	78E314	215	17955	81E261
249	Ethanedione, dicyclohexyl-	216	18050	81E261			
250	Tricycloquinazoline, 2-bromo-	216	18100	83E901	217	18100	82E257
251	Naphthalene, 1,2-dinitro-				217	18181	776194
252	Benzo[b]carbazole				218	18200	78E761

No.	Name						
253	Coumarin, 7-(diethylamino)-3-thenoyl-				218	18200	82E271
254	Dibenz[a,h]anthracene				218	18200	779025
255	Thiopyrylium, 2,4,6-triphenyl-				218	18200	67F508
256	Coumarin, 3-phenyl-				219	18280	73E372
257	1,3-Cyclohexadiene	219	18300	65E036			
258	1,2-Propanedione, 1-phenyl-	219	18300	83E901			
259	Acridine Yellow				220	18350	83Z077
260	Dinaphtho[2,1-b:1',2'-d]thiophene				220	18350	766267
261	Adamantanethione				220	18400	726174
262	Dibenzo[fg,ij]phenanthro[9,10,1,2,3-pqrst]pentaphene	220	18400	76E692			
263	Fluoranthene	221	18450	57B004	221	18500	779025
264	Dibenzo[a,g]coronene	221	18480	78E314			
265	Benzo[e]pyrene				221	18500	779025
266	Dibenz[a,j]anthracene				221	18500	779025
267	Tricycloquinazoline	221	18500	82E257	224	18700	82E257
268	Coumarin, 7-(diethylamino)-3-(4-dimethylaminobenzoyl)-				222	18500	82E271
269	Pyrylium, 2,4,6-triphenyl-				222	18500	67F508
270	Dibenz[a,j]acridine				223	18600	779025
271	Tribenzo[b,n,pqr]perylene	223[b]	18620[b]	68Z005			
272	Dibenzo[a,c]phenazine	223	18630	69E227	223	18600	779025
273	Benzo[b]naphtho[2,3-d]thiophene				224	18690	68Z005
274	Benzil	223	18700	67E119	227	19000	67E119
275	1,4-Benzoquinone	224	18740	69E247			
276	1,3,5,7-Octatetraynediol	225	18790	566004			

Table 2—**Triplet State Energies**—Continued

No.	Compound	$E_T(n)$ kJ/mol	$E_T(n)$ cm^{-1}	Ref.	$E_T(p)$ kJ/mol	$E_T(p)$ cm^{-1}	Ref.
277	Bicyclo[2.2.1]heptane-2-thione, 1,3,3-trimethyl-				225	18800	726174
278	Dibenz[a,h]acridine	225	18800	706135	229	19100	779025
279	Tricycloquinazoline, 2-methyl-				225	18800	82E257
280	3,5,7,9-Dodecatetrayne				225	18820	566004
281	Naphthalene, 1-[(1-naphthyl)amino]-				225	18850	766421
282	2,3'-Bibenzo[b]thiophene				226	18870	85E272
283	Benzo[g]chrysene				226	18900	779025
284	Benzo[ghi]fluoranthene				226	18900	779025
285	Chrysene, 6-benzoyl-				226	18900	69E232
286	2,3-Pentanedione	226[g]	18900[g]	727134	233	19400	67E114
287	Benzo[c]phenanthrene, 1-methyl-	227	18940	54E002			
288	Cinnoline, 4-methyl-				227	19000	59E008
289	Dibenz[a,c]acridine				227	19000	779025
290	β-Naphthiazoline, 2-benzoyl-N-methyl-				227	19000	78E534
291	(Z)-Stilbene	227[b]	19000[b]	69F410			
292	Naphthalene, 1-anilino-				227	19010	766421
293	Coronene				228	19040	566005
294	Benzo[b]fluoranthene				228	19050	82E059
295	1,3-Diazaazulene	228[b]	19052[b]	73E353			
296	Hexahelicene	228	19100	62E008			
297	1-Naphthalenecarbothioic acid, O-ethyl ester	229[b]	19100[b]	757534	228	19100	779025
298	2,5-Thiophenedione, 3,4-dichloro-	229	19100	73E365			
299	Naphthalene, 1-amino-				229	19150	766421
300	Tetrabenzo[g,lm,uv,a₁,b₁]heptacene	229[b]	19160[b]	68Z005			

No.	Compound			
301	Cinnamic acid, methyl ester	229[b]	19200[b]	86A322
302	Diazene, diethyl-, (E)-	230[g]	19200[g]	767343
303	β-Ionone	230[b]	19200[b]	85E293
304	Naphthalene, 1,5-dinitro-	230	19210	71F587
305	Aniline, N,N-dimethyl-4-nitro-	230	19230	79E415
306	Benzo[def]carbazole	230	19230	78E761
307	Benzo[b]naphtho[2,3-d]furan	230	19230	68Z005
308	Benzo[c]fluorene	231	19300	779025
309	Cyclooctane-1,2-dione, 3,3,8,8,-tetramethyl-	231	19300	82E832
310	Naphthalene, 1,3-dinitro-	231	19300	71F587
311	Naphthalene, 1-nitro-	231	19300	71F587
312	Naphtho[1,2-b]triphenylene	232	19370	566005
313	Hexatriyne, diphenyl-	232	19380	566004
314	Aniline, 4-nitro-	232	19400	566008
315	Benzo[g]pteridine-2,4-dione, 7,8-dimethyl-	232	19400	69E235
316	1,2-Cyclodecanedione	232	19400	67E114
317	Diazene, dimethyl-, (E)-	232[g]	19400[g]	767343
318	Diazene, dipropyl-, (E)-	232	19400	767343
319	1,4-Dioxin, 2,3,5,6-tetraphenyl-	232	19400	79A241
320	2,5-Furandione, 3,4-dichloro-	232	19400	73E365
321	Naphthalene, 1,4-dicyano-	232	19400	767370
322	Oxazole, 2,2'-(1,4-phenylene)bis[5-phenyl-	232	19400	86E128
323	Pyrrole-2,5-dione, 3,4-dichloro-	232	19400	73E365
324	Pentacene-6,13-dione	232	19420	67E115

Table 2—**Triplet State Energies**—Continued

No.	Compound	kJ/mol	$E_T(n)$ cm^{-1}	Ref.	kJ/mol	$E_T(p)$ cm^{-1}	Ref.
325	Anthrone, 1,8-dihydroxy-	233[b]	19500[b]	89E158			
326	Benzo[c]chrysene				233	19500	779025
327	2,2'-Binaphthyl				234	19560	566005
328	Naphtho[1,8-de]-1,3,2-diazaborine, 2,3-dihydro-2-methyl-	234	19562	82E051			
329	Tetrabenzo[a,c,hi,qr]pentacene	234[b]	19570[b]	68Z005			
330	Coumarin, 3-(4-cyanobenzoyl)-5,7-dimethoxy-				234	19600	82E271
331	Coumarin, 5,7-dimethoxy-3,3'-carbonylbis-				234	19600	82E271
332	Naphthalene, octafluoro-	234	19600	68E117			
333	Pyrylium, 2-methyl-4,6-diphenyl-				234	19600	67F508
334	1-Naphthaldehyde	235	19600	68B014	236	19750	60E014
335	Aniline, N-methyl-4-nitro-				235	19610	79E415
336	Dibenzo[c,g]carbazole				235	19610	766267
337	5,12-Tetracenequinone				235	19610	67E115
338	Benzo[c]phenanthrene, 6-methyl-	236	19690	54E002			
339	Naphthalene, 2-[(2-naphthyl)amino]-				236	19690	766421
340	Coumarin, 3,3'-carbonylbis(5,7-dimethoxy-				235	19700	82E271
341	Biacetyl				236[x]	19700[x]	55E008
342	Coumarin, 5,7,7'-trimethoxy-3,3'-carbonylbis-				236	19700	82E271
343	Naphthalene, 1-acetyl-	236	19700	64E021	236	19700	67F523
344	Benzo[c]phenanthrene, 4-methyl-	236	19720	54E002			
345	Benzo[c]phenanthrene, 5-methyl-	236	19720	54E002			
346	Diindeno[1,2-a:2',1'-c]fluorene, 10,15-dihydro-				236	19720	60E009
347	Benzo[c]phenanthrene, 2-methyl-	236	19760	54E002			
348	Benzo[c]phenanthrene, 3-methyl-	236	19760	54E002			

No.	Compound	λ(nm)	cm⁻¹	Ref.	λ(nm)	cm⁻¹	Ref.
349	Diazene, diisopropyl-, (E)-	236	19800	767343	237	19800	69Z002
350	Acenaphthene, 5-nitro-	237	19800	65E036			
351	1,3-Butadiene, 1-methoxy-				237	19800	566005
352	Pentahelicene				239	20000	779025
353	Benzo[c]phenanthrene	237	19840	54E002			
354	1,3-Butadiene, 1-chloro-	238	19900	65E036			
355	Coumarin, 3-benzoyl-5,7-dimethoxy-				238	19900	82E271
356	Naphthalene, 2-nitro-				238	19900	71F587
357	Thiophene, 2,2'-(1,3-phenylene)bis-				238	19900	86A357
358	Naphthalene, 1,5-dibenzoyl-				238	19925	60E014
359	Naphthalene, 2-amino-				239	19960	766421
360	Chrysene				239	20000	779025
361	Coumarin, 3,3'-carbonylbis(7-methoxy-				239	20000	82E271
362	Coumarin, 3-(4-cyanobenzoyl)-7-methoxy-				239	20000	82E271
363	(S)-Dinaphtho[2,1-d:1',2'-f][1,3]dioxepin	239	20000	82E586			
364	Naphthalene, 1,8-dinitro-				239	20000	71F587
365	Coumarin, 5,7-dimethoxy-3-(4-methoxybenzoyl)-				240	20000	82E271
366	Cycloheptane-1,2-dione, 3,3,7,7-tetramethyl-	240	20000	82E832			
367	Dinaphtho[1,2-b:1',2'-d]thiophene				240	20040	766267
368	Naphthalene, 2-anilino-				240	20040	766421
369	Naphtho[1,2-c][1,2,5]thiadiazole	240	20062	72E314			
370	(Z)-Piperylene	240	20070	82Z015			
371	Benzo[b]fluorene				240	20100	779025
372	Coumarin, 5,7-dimethoxy-3-thenoyl-				240	20100	82E271

Table 2—**Triplet State Energies**—Continued

No.	Compound	$E_T(n)$			$E_T(p)$		
		kJ/mol	cm^{-1}	Ref.	kJ/mol	cm^{-1}	Ref.
373	2-Furoic acid, 5-nitro-				240	20100	81A140
374	Naphthalene, 1-cyano-				240	20100	767370
375	Picene				240	20100	779025
376	Benzo[a]fluorene				241	20100	779025
377	1,2-Cycloheptanedione				241	20100	67E114
378	Naphthalene, 1-benzoyl-	241	20100	64E021	240	20100	60E014
379	1-Naphthalenecarboxylic acid				241	20100	67E112
380	Coumarin, 3-acetyl-6-bromo-				241	20120	73E372
381	1,4-Naphthoquinone				241	20160	67E115
382	Naphthalene, 1,5-dihydroxy-				241	20200	67E112
383	9-Acridinone, 2-bromo-				242	20200	78E878
384	Biphenyl, 4,4'-dinitro-				242	20200	44E001
385	Biphenyl, 4,4'-dibenzoyl-	242^x	20222^x	80D196			
386	2,6-Dithiocaffeine				242	20260	76E686
387	Butadiyne, diphenyl-				242	20270	566004
388	Coumarin, 3,3'-carbonylbis-				242	20300	82E271
389	Benzene, nitro-	243	20300	84E530	252	21100	44E001
390	Carbazole, 9-(1-naphthoyl)-				243	20300	69E236
391	Coumarin, 3-benzoyl-7-methoxy-				243	20300	82E271
392	Cyclopentadiene	243	20300	65E036			
393	Ethanedione, dicyclopropyl-	243	20300	83E901			
394	Naphthalene, 1,4-dibromo-				243	20300	67E112
395	Naphthalene, 1-(dimethylamino)-				243	20300	44E001
396	Dibenzo[fg,op]naphthacene				244	20360	566005

No.	Compound	λ	ε	Ref.	λ	ε	Ref.
397	9-Acridinone	244	20370	78E542	252	21050	78E878
398	Coumarin, 3-(4-cyanobenzoyl)-				244	20400	82E271
399	Coumarin, 7-methoxy-3-(4-methoxybenzoyl)-				244	20400	82E271
400	Coumarin, 3-thenoyl-7-methoxy-				244	20400	82E271
401	2,5-Cyclohexadien-1-one, 4,4-di(1-naphthyl)-	244	20400	85F417	244	20400	87E199
402	Imidazole-1-ethanol, 2-methyl-5-nitro-				244	20400	67E112
403	Phenanthrene, 9-acetyl-	244	20400	87F366	244	20400	87F366
404	Quinoline-4-carboxylic acid, ethyl ester				244	20400	67E112
405	p-Terphenyl						
406	9-Acridinone, 10-phenyl-	244	20430	78E542	254	21220	78E542
407	Naphthalene, 1-chloro-	245[b]	20490[b]	78E067	248	20700	44E001
408	Aziridine, 1-(2-naphthoyl)-	245	20500	83F297			
409	Benzoxazole, 4',5-diamino-2-phenyl-				245	20500	88E862
410	Biphenyl, 4-nitro-				245	20500	44E001
411	1,3-Butadiene, 2-chloro-	245	20500	65E036			
412	Carbazole, 9-(2-naphthoyl)-				245	20500	69E236
413	2,4-Hexadiene	245	20500	65E036			
414	Naphthalene, 1-hydroxy-				245	20500	44E001
415	Naphthalene, 1-iodo-				245	20500	63Z003
416	Naphthalene, 2-phenyl-				245	20500	67E112
417	Thioxanthen-9-one-3-carboxylic acid, ethyl ester	245	20500	86E676	245	20500	67E112
418	Phenanthrene, 9-chloro-				246	20530	79E505
419	Tetrabenz[a,c,h,j]anthracene				246	20550	566005
420	Benzoxazole, 4',6-diamino-2-phenyl-				246	20600	88E862

Table 2—**Triplet State Energies**—Continued

No.	Compound	kJ/mol	$E_T(n)$ cm^{-1}	Ref.	kJ/mol	$E_T(p)$ cm^{-1}	Ref.
421	1,3-Butadiene, 2,3-dichloro-	246	20600	65E036			
422	Coumarin, 3-benzoyl-				246	20600	82E271
423	2,5-Cyclohexadien-1-one, 4,4-di(2-naphthyl)-	246	20600	85F417	246	20600	44E001
424	Fluorene, 2-nitro-				246	20600	44E001
425	Naphthalene, 1-[(methylsulfonyl)methyl]-				246	20600	85F371
426	Naphthalene, 1-phenyl-				246	20600	67E112
427	7,8-Benzoflavanone	247	20600	86F287			
428	Flavanone, 6-methoxy-	247[b]	20600[b]	86F287	259	21700	86F287
429	Maleonitrile	247[b]	20600[b]	82E067			
430	Phenanthrene, 3-acetyl-				247	20600	67E112
431	Pyridinium, 2,4,6-tris(4-methoxphenyl)-N-methyl-				247	20600	67F508
432	Styrene, α-phenyl-	247[b]	20600[b]	82E204			
433	Thiophene, 2-nitro-				247	20600	82A153
434	2,2'-Biquinoline	247	20614	82E516			
435	9-Acridinone, 10-methyl-	247	20620	78E542	252	21050	78E542
436	Indolo[3,2-b]carbazole				247	20620	64E025
437	Phenanthrene, 9-bromo-				247	20620	79E505
438	Naphthalene, 1-bromo-	247	20650	57B004	247	20600	63Z003
439	Dibenzo[a,g]carbazole				247	20660	766267
440	Dinaphtho[1,2-b:2',1'-d]thiophene				247	20660	766267
441	Benzo[c]carbazole				247	20700	67E126
442	(E)-Piperylene	247	20700	65E036			
443	Coumarin, 3-(4-methoxybenzoyl)-				248	20700	82E271
444	1-Cyclopentene, 1-phenyl-				248	20700	77F954

No.	Name						
445	Cyclopentene, 1-phenyl-3-acetyl-				248	20700	77F954
446	4-Cyclopentene-1,3-dione, 4,5-dichloro-				248	20700	73E365
447	2-Naphthaldehyde				248	20700	67F523
448	Naphthalene, 2-cyano-				248	20700	44E001
449	Phenanthrene, 1-benzoyl-				248	20700	69E232
450	Dithiouracil				248	20715	78E273
451	Benz[c]acridin-7-one	249	20800	85F172	249	20800	44E001
452	Biphenyl, 2-nitro-				249	20800	82E271
453	Coumarin, 3-thenoyl-						
454	2,4-Hexadiene, 1-hydroxy-	249	20800	65E036			
455	Naphthalene, 2-acetyl-	249	20800	67F523	249	20800	67F523
456	Naphthalene, 2-benzoyl-	249	20800	67F523	249	20800	67F523
457	2-Naphthalenecarboxylic acid				249	20800	67E112
458	Styrene, β-methyl-, (E)-	249	20800	776060			
459	6-Thiocaffeine				249	20800	71E380
460	Benzo[b]naphtho[2,1-d]thiophene				249	20830	68Z005
461	4-Thiouracil, 1,3-dimethyl-				249	20830	76E686
462	Quinoline, 8-chloro-	249[x]	20840[x]	78E067	252	21100	83E417
463	Acenaphthene	250	20872	64E037	248	20700	67E112
464	Psoralen, 4',5'-dihydro-						
465	Benzoxazole, 4'-amino-2-phenyl-				250	20900	87E187
466	Butadiene	250	20900	65E036	250	20900	88E862
467	Naphthalene, 1-methoxy-				250	20900	44E001
468	Benzothiazole, 2-phenyl-	251	20944	736125	250	20870	736125

Table 2—**Triplet State Energies**—Continued

No.	Compound	$E_T(n)$ kJ/mol	$E_T(n)$ cm^{-1}	Ref.	$E_T(p)$ kJ/mol	$E_T(p)$ cm^{-1}	Ref.
469	Naphthalene, 1-fluoro-				251	20970	496001
470	Cambendazole				251	21000	82E624
471	Isoprene	251	21000	65E036			
472	Naphthalene, 2-chloro-				251	21000	44E001
473	Quinoline, 5-chloro				251	21000	83E417
474	Naphthalene, 2-iodo-				252	21040	44E001
475	Tetrabenz[a,c,h,j]acridine				252	21097	84E803
476	Naphthalene, 2-bromo-				252	21100	44E001
477	Naphthalene, 2-cyano-7-methoxy-				252	21100	89E447
478	Naphthalene, 2-hydroxy-				252	21100	44E001
479	2-Naphthalenecarboxylic acid, 7-methoxy-, methyl ester				252	21100	89E447
480	1,3-Butadiene, 2,3-dimethyl-	253	21100	65E036			
481	Phenothiazine	253	21100	757279			
482	Naphthalene	253	21180	57B004	255	21300	44E001
483	Coumarin, 5,7-dimethoxy-				253	21190	73E372
484	Naphthalene, 2,7-dihydroxy-				254	21200	67E112
485	Naphthalene, 1-methyl-				254	21200	44E001
486	Psoralen, 3-carbethoxy-				254	21200	83E324
487	Psoralen, 5-methoxy-				254	21200	83E324
488	Isoquinoline	254	21210	59E009	254	21200	59E008
489	Benzophenone, 4-phenyl-				254	21225	60E014
490	Benzo[f]quinoxaline	254	21262	83E176	251	21000	63E018
491	Benzo[b]naphtho[2,1-d]furan				255	21280	68Z005
492	Coumarin-3-carboxylic acid				255	21280	73E372

No.	Compound	λ	No.	Ref.	λ	No.	Ref.
493	2-Cyclohexen-1-one, 4,4-di(1-naphthyl)-	254	21300	85F416			
494	Naphthalene, 2-methyl-				254	21300	67E112
495	2-Cyclohexen-1-one	255[b]	21300[b]	67F522			
496	Naphthalene, 2-cyano-6-methoxy-				255	21300	89E447
497	2-Naphthalenecarboxylic acid, 7-methoxy-				255	21300	89E447
498	Thioxanthen-9-one-1-carboxylic acid, 3-amino-, ethyl ester	255	21300	86E676			
499	Coumarin, 7-hydroxy-				255	21320	73E372
500	Coumarin, 3-methyl-				255	21320	73E372
501	Quinoxaline				254	21250	59E008
502	Quinoline, 4,6-dichloro-	255	21325	82E355	255	21350	83E417
503	Quinoline, 4-chloro-2-methyl-	255	21340	80E627			
504	Acetophenone, 4'-phenyl-	255	21400	67F523	254	21300	67F523
505	Benzo[a]carbazole				256	21400	779025
506	Indazole, 2-methyl-				256	21400	757167
507	2-Naphthalenecarboxylic acid, 6-methoxy-, methyl ester				256	21400	89E447
508	1-Naphthalenemethanol acetate				256	21400	85F371
509	Phosphoric acid, diethyl 1-naphthalenylmethyl ester				256	21400	85F371
510	Quinoline, 3-chloro				256	21400	83E417
511	6-Phenanthridone, 8-nitro-				256	21415	87E893
512	Quinoline, 4-chloro-	257	21450	80E627	255	21300	83E417
513	9,9'-Biphenanthryl				257	21460	776226
514	Dibenzo[a,i]carbazole				257	21460	766267
515	Psoralen, 5-hydroxy-				257	21460	73E372
516	Naphtho[1,2-c][1,2,5]oxadiazole	257	21499	72E314			

Table 2—**Triplet State Energies**—Continued

No.	Compound	$E_T(n)$ kJ/mol	$E_T(n)$ cm^{-1}	Ref.	$E_T(p)$ kJ/mol	$E_T(p)$ cm^{-1}	Ref.
517	Coumarin, 6-methyl-				257	21500	73E372
518	2-Naphthalenecarboxylic acid, 6-methoxy-				257	21500	89E447
519	Quinoline, 6-methoxy-				257	21500	69Z002
520	2-Cyclohexen-1-one, 4,4-di(2-naphthyl)-	258	21500	85F416			
521	o-Terphenyl				258	21500	67E112
522	Quinoxaline, 2,3-dichloro-	257	21523	82E355			
523	Quinoline, 7-chloro-				258	21550	83E417
524	Quinoline	258	21590	80E627	261	21850	83E417
525	Coumarin	258^b	21600^b	775025	261	21840	73E372
526	Quinoline, 2-methyl-				258	21600	69Z002
527	Quinoline, 4-methyl-				258	21600	69Z002
528	Quinoline-3-carboxylic acid, ethyl ester	258	21600	87F366	259	21700	87F366
529	Styrene	258	21600	776060			
530	Naphthalene, 2-ethoxy-				259	21600	67E112
531	Quinoline, 2,4-dichloro-	259	21640	80E627			
532	Quinoline, 6-chloro-	259	21650	80E627	255	21300	83E417
533	Benzo[1,2-b:5,4-b']dipyran-2,8-dione				259	21670	73E372
534	Flavone	259	21700	86E567			
535	1-Naphthalenemethanminium chloride, N,N,N-trimethyl-				259	21700	85F371
536	Styrene, α-methyl-	260	21700	776060			
537	Thioxanthen-9-one-4-carboxylic acid, ethyl ester	260	21700	86E676			
538	Benzo[h]quinoline	260	21740	59E009	261	21790	59E011
539	9-Bismafluorene, 9-phenyl-	260	21740	81E648			
540	9-Stannafluorene, 9,9-diethyl-	260	21740	81E648			

No.	Compound						
541	Benzoxazole, 2-phenyl-	260	21746	736125	262	21900	88E862
542	1,8-Phenanthroline	260	21755	78E889			
543	Phenanthrene	260	21774	63E024	257	21500	89E090
544	9-Stibafluorene, 9-phenyl-	261	21790	81E648			
545	9,10-Anthraquinone	261	21800	64E021	263	21980	67E115
546	Benzidine, *N,N,N',N'*-tetramethyl-	261	21800	767177			
547	Quinoline, 2,4-dimethyl-				261	21800	69Z002
548	9-Germafluorene, 9,9-diphenyl-	261	21830	81E648			
549	Acetylene, diphenyl-	262	21860	57B004	262	21870	566004
550	Benzo[f]quinoline	262	21880	59E009	262	21865	59E011
551	Phenoxazine	262	21880	72E303	260	21750	72E303
552	Quinoxaline, 2-chloro-3-methyl-	262	21885	82E355			
553	1,9-Phenanthroline	262	21886	78E889			
554	Psoralen, 8-methoxy-				262	21900	73E372
555	Thiophene, 2-(4-cyanobenzoyl)-				262	21900	73E358
556	Psoralen, 8-methyl-				262	21904	73E372
557	Quinazoline	262	21925	78E894	262	21900	59E008
558	6-Phenanthridone, 8-amino-				262	21930	87E893
559	Psoralen				262	21930	73E372
560	6-Thiopurine				263	21980	76E686
561	Biphenyl, 4,4'-dichloro-				263	22000	72E316
562	Biphenyl, 4,4'-dihydroxy-				263	22000	735067
563	Biphenyl, 2-iodo-				263	22000	67E112
564	Cambendazole, 1-amino-				263	22000	82E624

Table 2—**Triplet State Energies**—Continued

No.	Compound	$E_T(n)$ kJ/mol	cm⁻¹	Ref.	$E_T(p)$ kJ/mol	cm⁻¹	Ref.
565	Ethene, 1,2-dicyano-, (E)-	263	22000	82Z015			
566	Phenanthridine				264	22050	78E761
567	Quinoline, 2-chloro-4-methyl-				264	22050	83E417
568	Indene	264[b]	22075[b]	755396			
569	Benz[b]arsindole, 5-phenyl-	264	22080	81E648			
570	Coumarin, 7-hydroxy-4-methyl-				264	22080	73E372
571	9-Phosphafluorene, 9-phenyl-	264	22090	81E648			
572	Angelicin				264	22100	73E372
573	Benzimidazole, 2-(4-thiazolyl)-				264	22100	82E624
574	Biphenyl, 4,4'-dimethoxy-				264	22100	735067
575	1,10-Phenanthroline				264	22100	63E018
576	Phthalazine	264[b]	22100[b]	82E585	275	23000	59E008
577	Thiophene, 2-benzoyl-				265	22100	73E358
578	Thioxanthen-9-one	265	22100	86E676			
579	Thioxanthen-9-one-1-carboxylic acid, ethyl ester	265	22100	86E676			
580	Thioxanthen-9-one-2-carboxylic acid, ethyl ester	265	22100	86E676			
581	Quinoline, 2-chloro-	265	22120	80E627	262	21900	83E417
582	9-Silafluorene, 9,9-diphenyl-	265	22120	81E648			
583	Quinoxaline, 2,3-dimethyl-	265	22149	82E355			
584	4,7-Phenanthroline	265	22150	78E889	266	22200	63E018
585	Phenoxazine, 10-phenyl-	265	22150	72E303	264	22070	72E303
586	1,7-Phenanthroline	265	22154	78E889	265	22150	63E018
587	1,3,5-Hexatriynediol				265	22170	566004
588	2,4,6-Octatriyne	265	22170	60E013	267	22320	566004

No.	Compound						
589	Biphenyl, 4-chloro-	266	22200	72E316			
590	Thiophene, 2-(4-methoxybenzoyl)-	266	22300	73E358			
591	Benzophenone, 4,4'-dithiomethoxy-	267	22300	81C032			
592	Biphenyl, 3-chloro-	267	22300	72E316			
593	Phenoxathiin				267	22300	77E638
594	Benzophenone, 4-thiomethoxy-	268	22400	81C032			
595	Biphenyl, 4-hydroxy-	268	22400	735067			
596	Biphenyl, 3,3',5,5'-tetrachloro-	268	22400	72E316			
597	Psoralen, 4,5',8-trimethyl-	268	22400	73E372			
598	Benzo[2,1-b:3,4-b']bis[1]benzothiophene	268	22420	85E272			
599	Benzene, 1,3,5-triphenyl-	269	22500	67E112			
600	Biphenyl, 4,4'-bis(1,1-dimethylethyl)-	269	22500	735067			
601	Biphenyl, 4,4'-bis(1-methylpropyl)-	269	22500	735067			
602	Biphenyl, 3,3'-dichloro-	269	22500	72E316			
603	Biphenyl, 4,4'-diisopropyl-	269	22500	735067			
604	Biphenyl, 4,4'-dimethyl-	269	22500	735067			
605	m-Terphenyl	269	22500	67E112			
606	Benzo[b]tellurophene	272	22730	89E178	270	22573	89E178
607	Benzene, 1,2,4,5-tetracyano-				271	22650	67E118
608	Benzophenone, 4,4'-dicarbomethoxy-				271	22700	86A166
609	Biphenyl, decafluoro-				272	22700	68E117
610	Fluorene, 2-amino-	272	22700	44E001			
611	Benzophenone, 4,4'-diiodo-				273[b]	22844[b]	81E099
612	Biphenyl	274	22900	735067	274	22871	64E037

Table 2—**Triplet State Energies**—Continued

No.	Compound	kJ/mol	$E_T(n)$ cm^{-1}	Ref.	kJ/mol	$E_T(p)$ cm^{-1}	Ref.
613	Pyrazole, 3-methyl-1,5-diphenyl-				274	22883	83C028
614	Benzophenone, 3,4'-dicarbomethoxy-	274	22900	86A166			
615	Dicumarol				274	22920	73E372
616	Benzophenone, 4,4'-bis(dimethylamino)-	275	23000	68E135	255	21300	79E210
617	Biphenyl, 4,4'-difluoro-				275	23000	735067
618	Styrene, β-methyl-, (Z)-	275	23000	776060			
619	Acetophenone, 2,2,2-trifluoro-4'-methoxy-	276	23000	86A400			
620	Benzophenone, 4,4'-dicyano-				276	23000	80E223
621	Di-2-pyridyl ketone				275[x]	23005[x]	81E099
622	Dibenzo[f,h]quinoxaline	276	23065	69E227	275	23000	63E019
623	Benzophenone, 2,2'-bis(trifluoromethyl)-	276	23100	84F268			
624	Benzophenone, 4,4'-bis(trifluoromethyl)-	276	23100	84F268			
625	Benzophenone, 4-cyano-2',4',6'-triisopropyl-				276	23100	83F055
626	Carbostyril	276	23100	717171			
627	Terephthaldicarboxaldehyde	277[b]	23130[b]	729040			
628	Quinoline-2-carboxylic acid, 4-hydroxy-				277	23188	79B086
629	Acetophenone, 4'-amino-				278	23200	68E135
630	Benzophenone, 4-cyano-	278	23200	64E021	280	23400	80E223
631	1,5-Naphthyridine	278[x]	23215[x]	73E356			
632	6-Phenanthridone, 5-methyl-	278	23251	80E792	282	23585	80E792
633	6-Phenanthridone, 2-amino-				278	23255	87E893
634	Benzo[1,2-b:3,4-b']bis[1]benzothiophene				278	23256	85E272
635	Benzophenone, 4-carbomethoxy-						
636	Benzophenone, 4-cyano-4'-methoxy-	279	23300	81F452	279	23300	80E223

82

No.	Compound	E (kJ/mol)	E (cm⁻¹)	Ref.	E (kJ/mol)	E (cm⁻¹)	Ref.
637	Indazole, 1-methyl-	279	23300	68F291	279	23300	757167
638	Pyridine, 2-benzoyl-	279[x]	23331[x]	746288	273	22800	747390
639	Acetophenone, 4'-thiomethoxy-	279	23364	85E766			
640	[1]Benzopyrano[5,4,3-cde][1]benzopyran-5,10-dione	280	23392	83C028	280	23400	85E766
641	Pyrazole, 3,5-diphenyl-				285	23810	83C028
642	Acetophenone, 2,2,2-trifluoro-3'-methyl-	279	23400	86A400			
643	Benzophenone, 4-cyano-4'-methyl-				279	23400	80E223
644	Benzophenone, 3,3'-bis(trifluoromethyl)-	280	23400	84F268			
645	Benzophenone, 3,3'-dicarbomethoxy-	280	23400	86A166			
646	Triphenylene				280	23400	84E612
647	Pyrido[3,4-b]indole, 1-methyl-	281	23474	87E433	285	23809	87E433
648	Di-4-pyridyl ketone	281	23500	86A400	281[x]	23486[x]	83D200
649	Acetophenone, 4'-acetyl-	281	23500	68E135			
650	Benzophenone, 4-amino-				263	22000	68E135
651	Phthalic anhydride, tetrachloro-	281	23500	67E118			
652	Pyridine, 4-benzoyl-	281	23500	68F291	278	23200	747390
653	Acetophenone, 2,2,2-trifluoro-4'-methyl-	282	23500	86A400			
654	Fluorene, 9,9-dimethyl-	281	23520	81E648			
655	Benzoic acid, 4-(methylamino)-				281	23530	79E415
656	Benzo[b]selenophene	282	23585	89E178	282	23585	89E178
657	Dibenzotellurophene	282	23585	89E178	281	23530	89E178
658	Decafluorobenzophenone	282	23600	89A179	292	24400	89A179
659	Fumaric acid, dimethyl ester	282	23600	82Z015			
660	Acetophenone, 4'-tert-butyl-2,2,2-trifluoro-	283	23600	86A400			

Table 2—**Triplet State Energies**—Continued

No.	Compound	kJ/mol	$E_T(n)$ cm^{-1}	Ref.	kJ/mol	$E_T(p)$ cm^{-1}	Ref.
661	Acetophenone, 4'-chloro-2,2,2-trifluoro-	283	23600	86A400	285	23800	68F291
662	Benzophenone, 4-(trifluoromethyl)-	283	23600	68F291	284	23700	779025
663	Fluorene	282	23601	64E037			
664	Benzoic acid, 4-(diethylamino)-, ethyl ester	283	23640	83E440			
665	Acetophenone, 3',4'-methylenedioxy-	283	23700	67F523	275	23000	67F521
666	Benzo[f][4,7]phenanthroline	284	23700	81E545	285	23800	755398
667	Benzoic acid, 4-(dimethylamino)-, ethyl ester	284	23750	83E440			
668	Indazole	284[b]	23753[b]	755396	284	23700	63E012
669	Benzophenone, 4,4'-dichloro-	285	23800	64E021	286	23900	87F368
670	Benzophenone, 2,4,6-triethyl-				285	23800	83F055
671	Flavanone, 6-methyl-	285	23800	86F287	264	22000	86F287
672	Indole, 3-methyl-				285	23800	84E843
673	Coumarin, 4-hydroxy-				285	23810	73E372
674	Dibenzothiophene	285	23830	81E648	288	24100	766267
675	Dibenzoselenophene	285	23866	89E178	287	23980	89E178
676	Benzophenone, 4'-(4-benzoylbenzyl)-2,4,6-triisopropyl-	286	23900	89F011	287	24000	89F011
677	Benzophenone, 3-chloro-	286	23900	68F291	288	24100	68F291
678	Benzophenone, 4-chloro-	286	23900	68F291	288	24100	68F291
679	Benzophenone, 4-hydroxy-	286	23900	68F291	286	23900	87F368
680	Benzophenone, 3-methoxy-	286	23900	68F291			
681	Benzophenone, 3-(trifluoromethyl)-	286	23900	81F452			
682	Benzophenone, 2,4,6-trimethyl-	286	23900	89A110	284	23700	83F055
683	Pyridine, 3-benzoyl-	286	23900	747390	282	23600	747390
684	6-Phenanthridone	286	23900	68F291	286	23920	80E792

#	Name						
685	Benzo[b]thiophene	287	23970	59E009	288	24040	85E272
686	Benzophenone, 3,3'-dibromo-	287[b]	23975[b]	81E099			
687	Benzophenone	287	24000	67F523	289	24200	80E223
688	Benzophenone, 3-cyano-	287	24000	81F452			
689	Benzophenone, 3,4-dimethyl-	287	24000	67F523	289	24100	67F523
690	Benzophenone, 4-methoxy-	287	24000	68F291	290	24300	80E223
691	Benzophenone, 4-methyl-	287	24000	67F523	290	24300	80E223
692	Benzophenone, 2,3,4,5,6-pentafluoro-	287	24000	89A179			
693	Benzophenone, 2,4,6-triisopropyl-	287	24000	89A110	288	24100	83F055
694	Biphenyl, 4,4'-(trifluoromethyl)-				287	24000	735067
695	Dibenzofuran	287	24000	81E648	293	24515	58E005
696	Indole-3-carboxaldehyde	287	24000	84E843	293	24500	84E843
697	Aniline, N-ethyl-	288	24000	67E116			
698	2,5-Cyclohexadien-1-one, 4,4-diphenyl-				288	24000	67E124
699	Indole-5-carboxylic acid				288	24000	84E843
700	Di-3-pyridyl ketone				287[x]	24030[x]	83D200
701	Benzoic acid, 4-amino-				288	24040	79E415
702	Benzonitrile, 4-(diethylamino)-	288	24060	83E440	288	24100	79E415
703	Pyrazole, 1,3,5-triphenyl-				288	24096	83C028
704	Acetophenone, 2',3',4',5',6'-pentafluoro-	288	24100	89A179			
705	Aniline, N-methyl-	288	24100	67E116			
706	Benzophenone, 3-carbomethoxy-	288	24100	81F452			
707	Benzophenone, 4,4'-dibromo-				288	24100	66E094
708	Benzophenone, 2,5-dimethyl-	288	24100	67F523	290	24300	67F523

Table 2—**Triplet State Energies**—Continued

No.	Compound	$E_T(n)$ kJ/mol	$E_T(n)$ cm^{-1}	Ref.	$E_T(p)$ kJ/mol	$E_T(p)$ cm^{-1}	Ref.
709	Benzophenone, 4,4′-dimethyl-	288	24100	67F523	290	24300	80E223
710	Benzophenone, 4-fluoro-	288	24100	68F291	292	24400	81C032
711	Benzophenone, 2-(trifluoromethyl)-	288	24100	68F291			
712	1,4-Naphthoquinone, 5-hydroxy-	288	24100	86E628	297	24900	86E628
713	Benzophenone, 2-carbomethoxy-	289	24100	81F452			
714	Benzophenone, 2,4-dimethyl-	289	24100	67F523	291	24300	67F523
715	Acetophenone, 2-fluoro-	289	24200	86A401			
716	Benzaldehyde, 2,3,5,6-tetramethyl-	289[b]	24200[b]	775025			
717	Benzophenone, 2-benzyl-	289	24200	67F523	292	24400	67F523
718	Indole, 1-methyl-	289	24200	771021	292	24400	771021
719	Acetophenone, 4′-cyano-	290	24200	86A400	291	24300	68F291
720	Benzophenone, 4,4′-di-*tert*-butyl-	290	24200	86A400			
721	Benzophenone, 2-methyl-	290	24200	68F291			
722	Benzonitrile, 4-(dimethylamino)-	290	24210	83E440	288	24100	79E415
723	Pyridine, 4-acetyl-	290	24235	83E630	295	24654	83E630
724	Benzaldehyde, 2-chloro-	290	24240	72E315			
725	Isoindole-1,3-dione, 2-[(dimethylamino)methyl]-	290	24300	78E088	292	24400	78E088
726	Aniline, *N,N*-diphenyl-				291	24300	69Z002
727	4-Pyridinecarboxaldehyde	291[x]	24300[x]	79D171			
728	Acridan				291	24330	73E359
729	Benzoic acid, 4-amino-, methyl ester				292	24380	79E415
730	Acetophenone, 2,2,2-trifluoro-	292	24400	86A400			
731	Indole, 3-acetyl-	292	24400	84E843	294	24600	84E843
732	Indole, 5-fluoro-	292	24400	84E843	291	24300	84E843

No.	Compound						
733	Pyrido[3,4-b]indole, 7-methoxy-1-methyl-				292	24400	84E102
734	1,3,5-Triazine, 2,4-diphenyl-	292	24400	66D173			
735	Carbazole, N-methyl-	292	24450	81E648	292	24450	68Z005
736	Benzophenone, 4,4'-dimethoxy-	293[b]	24470[b]	81E099	292	24400	80E223
737	Acetophenone, 2-chloro-	293	24500	68F291			
738	Acetophenone, 2,2,2,4-tetrafluoro-	293	24500	86A400			
739	Pyridine, 2-acetyl-	294	24538	83E630	298	24899	83E630
740	Carbazole				294	24540	58E005
741	Benzonitrile, 2-amino-	294[b]	24552[b]	755396			
742	Benzene, 1,4-dicyano-	294	24560	62E018	295	24700	70E318
743	Benzophenone, 4,4'-difluoro-	294	24600	86A400	294	24600	81C032
744	Benzophenone, 4-methoxy-2',4',6'-triisopropyl-				294	24600	83F055
745	Carbazole, N-phenyl-				294	24600	68Z005
746	1,3,5-Triazine, 2,4,6-triphenyl-	294	24600	66D173			
747	Acetophenone, 3'-cyano-	295	24600	68F291	307	25600	68F291
748	Benzophenone, 4-methyl-2',4',6'-triisopropyl-				295	24600	83F055
749	Benzonitrile, 4-(methylamino)-	295	24660	83E440			
750	Benzaldehyde, 2-fluoro-	295	24670	72E315			
751	Benzoic acid, 4-amino-, ethyl ester	295	24690	83E440			
752	Acetophenone, 4'-hydroxy-	295	24700	68B014	303	25300	66E094
753	Acetophenone, 2,2,2-trifluoro-4'-(trifluoromethyl)-	295	24700	86A400			
754	Benzotriazole				295	24700	63E012
755	Limonene	295	24700	86F465			
756	Acetophenone, 2,2,2-triphenyl-	296	24700	67F523	300	25000	67F523

Table 2—**Triplet State Energies**—Continued

No.	Compound	$E_T(n)$ kJ/mol	$E_T(n)$ cm^{-1}	Ref.	$E_T(p)$ kJ/mol	$E_T(p)$ cm^{-1}	Ref.
757	1,4-Naphthoquinone, 5,8-dihydroxy-	296	24700	86E628	282	23600	86E628
758	2-Pyridinecarboxaldehyde	296^x	24760^x	79D171			
759	Dibenz[b,f]azepine, 10,11-dihydro-				296	24770	73E359
760	Acetophenone, 2,2-difluoro-	297	24800	86A401			
761	Aniline	297	24800	67E116	321	26800	44E001
762	Dimethyl phthalate				297	24800	72D317
763	3-Pyridinecarboxaldehyde	297^x	24810^x	79D171			
764	Pyridazine	297	24850	67E117	290^x	24251^x	67B017
765	Pyridine, 3-acetyl-	298	24899	83E630	307	25633	83E630
766	Acetophenone, 4'-bromo-	297	24900	67F523	299	25000	67F523
767	2-Cyclohexen-1-one, 4,4-diphenyl-				298	24900	67E124
768	2-Thiouracil				298	24932	78E273
769	Benzaldehyde, 3-chloro-	298	24940	72E315			
770	Benzonitrile, 4-amino-	298	24940	83E440			
771	Acetophenone, 3',5'-dimethyl-	299	25000	67F523	293	24500	756176
772	Benzaldehyde, 3-fluoro-	299	25000	72E315	298	24900	67F521
773	1,2,3-Triazole, 4-benzoyl-5-methyl-				299	25000	717346
774	Benzaldehyde, 4-chloro-	300	25090	72E315			
775	Acetophenone, 4'-methoxy-	300	25100	67F523	299	25000	67F521
776	Acetophenone, 4'-(trifluoromethyl)-	300	25100	67F523	301	25200	67F523
777	Benzaldehyde, 4-methoxy-	300	25100	67F523	295	24700	69E227
778	Anthrone, 10,10-dimethyl-				301	25100	89A345
779	Benzene, 1,1'-sulfonylbis[4-chloro-						
780	Aniline, N-phenyl-	301	25100	77E662	301	25140	73E359

No.	Compound						
781	Anthrone	301	25157	89E178	301	25150	706018
782	Benzofuran	301	25160	72E315	301	25130	89E178
783	Benzaldehyde, 2-methyl-	301	25190	717384			
784	Acetophenone, 3',4'-dimethyl-	301	25190	59E009	300	25060	717384
785	Acetylene, phenyl-	301	25190	756304			
786	Benzonitrile, 3-methoxy-	301	25200	72E315			
787	Benzaldehyde				298	24950	706018
788	Benzoic acid, 4-cyano-, methyl ester				301	25200	72D317
789	Indole				296	24800	84E843
790	Methanone, (1,2-dimethyl-2-cyclopenten-1-yl)phenyl-	301			301	25200	767471
791	Acetophenone, 4'-chloro-	301		84E843	301	25100	68F291
792	Acetophenone, 2',4',6'-trimethyl-	302	25200	68F291	303	25300	67F523
793	Benzaldehyde, 3-methyl-	302	25200	89A110			
794	Acetophenone, 4'-acetyl-2,2,2-trifluoro-	302	25200	72E315			
795	Acetophenone, 3'-methoxy-	303	25280	86A400	303	25300	67F521
796	Benzene, 1,1'-sulfonylbis[4-methoxy-	303	25300	67F523			
797	Caffeine	303	25300	77E662	303	25300	71E380
798	Formaldehyde	303[g]	25316[g]	80Z097	303	25320	79E415
799	Phenol, 3-cyano-	303	25320	86E058	304	25400	67F523
800	α-Tetralone	303	25340	62E018	305	25500	70E318
801	Benzene, 1,2-dicyano-	303	25400	67F523	304	25400	67F523
802	Acetophenone, 3'-bromo-	304	25400	86A400			
803	Acetophenone, 4'-tert-butyl-	304	25400	86A400			
804	Acetophenone, 3'-chloro-		25400				

Table 2—**Triplet State Energies**—Continued

No.	Compound	$E_T(n)$ kJ/mol	$E_T(n)$ cm^{-1}	Ref.	$E_T(p)$ kJ/mol	$E_T(p)$ cm^{-1}	Ref.
805	Acetophenone, 3'-(trifluoromethyl)-	304	25400	67F523	307	25600	67F523
806	Dibenzo[b,e][1,4]dioxin	304	25400	77E638			
807	4-Chromanone	304	25445	717384	304	25445	717384
808	Benzaldehyde, 4-methyl-	305	25490	72E315	298	24900	67F523
809	Acetophenone, 4'-methyl-	305	25500	72E294	305	25500	67F521
810	Benzene, 1,1'-sulfonylbis[4-methyl-	305	25500	77E662			
811	Dimethyl terephthalate						
812	Flavanone	305	25500	86F287	305	25500	72D317
813	Benzaldehyde, 4-fluoro-	305	25520	72E315			
814	2-Thiouracil, 1-methyl-	306	25562	81D022			
815	Acetophenone, 3'-methyl-	306	25600	67F523			
816	2-Δ²-Thiazoline, (4'-chlorobenzoyl)amino-	306	25600	83F476	303	25400	67F521
817	Benzene, hexachloro-	307	25600	67E112			
818	4-Cholesten-3-one	307	25600	717222			
819	Testosterone acetate	307	25600	68E132			
820	Benzoic acid, 2-methoxy-						
821	Benzoic acid, 2-cyano-, methyl ester	308[b]	25777[b]	78D082	307	25640	79E415
822	Benzonitrile, 4-bromo-	309	25800	77E662	307	25700	72D317
823	Benzene, 1,1'-sulfonylbis-	309	25800	67E118	308	25770	79E415
824	Phthalic anhydride	309	25800	83F476			
825	2-Δ²-Thiazoline, (4'-methylbenzoyl)amino-	309[x]	25840[x]	78D082			
826	Benzonitrile, 4-chloro-	310	25900	86E058			
827	1-Indanone, 2,2-dimethyl-				313	26136	85E829
828	Methanone, phenyl(1,2,3-trimethyl-2-cyclopenten-1-yl)-				310	25900	767471

No.	Compound						
829	Progesterone	310	25900	717222	310	25905	717449
830	Xanthone	310	25906	717449	311	26034	81E561
831	Acetophenone	310[b]	25933	81E561			
832	Benzene, pentachloro-	310[b]	25944[b]	84E054	311	25970	79E415
833	Benzonitrile, 2-bromo-	311	25970	756304			
834	Benzonitrile, 2-methoxy-	311	26000	67F523	311	26000	67F523
835	Acetophenone, 2-allyl-	311	26000	67F523	311	26000	67F523
836	Acetophenone, 2-(2-phenylethyl)-				311	26000	67F523
837	Acetophenone, 2-(phenylmethyl)-	311	26000	67F523			
838	Acetophenone, 2-propyl-				312	26100	67F523
839	Benzoic acid, 3,5,-dimethyl-				311	26000	767546
840	Propiophenone	312	26080	86E058	313	26150	706018
841	Acetophenone, 2-ethyl-				313	26100	680600
842	Benzonitrile, 2-(trifluoromethyl)-				313	26100	81F452
843	Pyrazine-d_4	313[b]	26146[b]	82E367			
844	Aniline, 4-chloro-	313[x]	26154[x]	78D082			
845	Borine, triphenyl-	313	26171	83E638			
846	2-Δ^2-Thiazoline, (2'-chlorobenzoyl)amino-	313	26200	83F476			
847	Benzene, 1,4-dimethoxy-	314[b]	26250[b]	72D316			
848	Benzoic acid, 4-methoxy-, ethyl ester	314	26250	83E440			
849	1-Indanone	314	26250	86E058	317	26500	66E097
850	Benzonitrile, 3-methyl-	314	26270	62E018	315	26400	767546
851	Benzoyl chloride	314	26280	59E009			
852	Benzonitrile, 2-chloro-	315	26291	85E829	315	26317	85E829

Table 2—**Triplet State Energies**—Continued

No.	Compound	$E_T(n)$ kJ/mol	$E_T(n)$ cm^{-1}	Ref.	$E_T(p)$ kJ/mol	$E_T(p)$ cm^{-1}	Ref.
853	Benzonitrile, 3-chloro-	315	26291	85E829	316	26395	85E829
854	Benzoic acid, 3,4-dimethyl-				315	26300	767546
855	Deoxythymidine 5'-monophosphate				315	26300	673066
856	2-Δ²-Thiazoline, benzoylamino-	315	26300	83F476			
857	2-Δ²-Thiazoline, (2'-methylbenzoyl)amino-	315	26300	83F476			
858	Pyrazine, 2,5-dimethyl-				315	26318	58E006
859	Benzoic acid, 3-methoxy-				315	26320	79E415
860	Benzonitrile, 3-bromo-				315	26320	79E415
861	Benzonitrile, 4-methoxy-	315	26320	83E440	315	26300	756304
862	Phenol, 2-cyano-				315	26320	79E415
863	Phenol, 4-cyano-				315	26320	79E415
864	Borane, dichlorophenyl-	315	26347	85D003			
865	Pyrazine	315	26361	78E312	311	25991	58E006
866	Chromone	316	26385	717384	313	26180	717384
867	Benzoic acid, 4-methoxy-, methyl ester				315	26400	767546
868	Benzene, 1,3-dicyano-				316	26400	70E318
869	1,3,5-Triazine				316	26400	61E009
870	Benzonitrile, 4-methyl-	316	26410	62E018	317	26500	62E018
871	Purine	316[b]	26434[b]	755396			
872	Benzonitrile, 2-methyl-	316	26450	62E018	318	26550	62E018
873	Benzoic acid, 4-methoxy-				317	26460	79E415
874	2,4-Hexadiyne, 1,6-dichloro-	317	26460	60E013			
875	Pyridine, 3-cyano-	317[b]	26460[b]	72D316	322	26946	69E226
876	Aniline, N,N-dimethyl-				317	26500	61E013

No.	Compound						
877	Benzoic acid, 3-methyl-, methyl ester				318	26500	767546
878	Pyridine, 2-cyano-				317[x]	26519[x]	84E054
879	Pyrazine, 2-methyl-				316	26400	767546
880	Benzene, 1,2,3,4-tetrachloro-				318	26600	756028
881	Benzoic acid, 3-methyl-				318	26600	81F452
882	Aniline, 4-bromo-				319	26650	69E226
883	Benzonitrile, 3-fluoro-				319	26700	673066
884	Benzonitrile, 4-(trifluoromethyl)-				319	26700	72D317
885	Borinane, 1-phenyl-				319[x]	26700[x]	84E477
886	Pyridine, 4-cyano-				320	26700	767546
887	Adenosine 5'-monophosphate				320[x]	26710[x]	84E054
888	Benzoic acid, 3-cyano-, methyl ester				320	26738	77E581
889	Pyrazine, tetramethyl-				323	27000	767546
890	Benzene, 1,2,4,5-tetrabromo-	317[b]	26510[b]	72D316	320	26800	81F452
891	Benzoic acid, 4-methyl-, methyl ester	317	26511	78E312	320	26800	72D317
892	Benzene, 1,2,3,5-tetrachloro-	318	26580	776222	318	26600	82E624
893	Pyridine, 2-methoxy-	318[x]	26600[x]	78D082			
894	Pyrazine, 2,6-dimethyl-	319	26630	80E641			
895	Benzonitrile	320[x]	26700[x]	67E112			
896	Aniline, 4-methyl-	320	26758	78E312			
897	Benzene, 1,2,4,5-tetrachloro-	320	26780	62E018			
898	Benzonitrile, 3-(trifluoromethyl)-	321[b]	26795[b]	78D082			
899	Trimethyl 1,3,5-benzenetricarboxylate	320[x]	26800[x]	67E112			
900	Benzimidazole	321[x]	26810[x]	755396			

Table 2—**Triplet State Energies**—Continued

No.	Compound	$E_T(n)$			$E_T(p)$		
		kJ/mol	cm^{-1}	Ref.	kJ/mol	cm^{-1}	Ref.
901	Aniline, 4-fluoro-	321[b]	26838[b]	78D082			
902	Benzonitrile, 2-fluoro-				321	26850	756028
903	Triphenylene, dodecahydro-	321	26850	82E344			
904	Benzamide, 3-methyl-				321	26870	79E415
905	Benzoic acid, 4-methyl-	322	26880	776222	320	26800	767546
906	Dimethyl isophthalate				322	26900	72D317
907	Anthracene, 1,2,3,4,5,6,7,8-octahydro-	322	26950	82E344			
908	Mesitylpentamethyldisilane				322	26950	84E082
909	Aniline, N,N-diethyl-				323	27000	61E013
910	Phenanthrene, 1,2,3,4,5,6,7,8-octahydro-	323	27000	82E344			
911	Silane, tris(trimethylsilyl)mesityl-				323	27000	84E082
912	Benzamide, 4-methyl-				323	27010	79E415
913	Benzoic acid, ethyl ester	324	27100	83E440			
914	Benzoic acid, 2-methyl-	324	27100	776222			
915	Disilane, (2,5-dimethylphenyl)pentamethyl-						
916	Benzoic acid	324	27110	63E014	324	27100	84E082
917	1,3-Benzodioxole	325[b]	27189[b]	755396	326	27200	767546
918	Guanosine 5'-monophosphate						
919	Pyrazole, 3,5-dimethyl-1-phenyl-				325	27200	673066
920	Benzonitrile, 4-fluoro-	326[b]	27229[b]	78D082	325	27210	83C028
921	Benzoic acid, methyl ester	326	27240	63E014	325	27200	756028
922	Benzene, 1-methoxy-4-methyl-	326[b]	27280[b]	78D082	326	27200	72D317
923	1,2,3-Triazole, 4-acetyl-5-methyl-						
924	Butadiyne	327	27300	60E013	326	27300	717346

#	Compound						
925	Benzene, 1-bromo-4-methoxy-	327[b]	27350[b]	78D082			
926	(Z)-2-Butene	327[g]	27360[g]	71E391			
927	Benzene, pentamethyldisilyl-				327	27370	84E082
928	Benzene, 1-chloro-4-methoxy-	327[b]	27377[b]	78D082			
929	2,4-Hexadiyne, 1,6-dihydroxy-	328	27380	60E013			
930	Uridine 5′-monophosphate				328	27400	737541
931	1,3-Octadiyne	328	27420	60E013			
932	Benzene, hexamethyl-	328[x]	27423[x]	64E022			
933	Ethylene	330	27550	82Z015			
934	Benzene, 1-methoxy-4-fluoro-	330[b]	27563[b]	78D082			
935	Benzene, 1,3,5-tribromo-	330	27600	67E112			
936	Benzene, 1,3,5-trichloro-				331	27600	67E112
937	Benzamide, 2-methyl-				331	27690	79E415
938	Benzene, 1,2,3,4-tetramethyl-	331	27700	82E344			
939	Xanthene				331	27700	88E763
940	Pyridine	332[b]	27770[b]	84E679			
941	Benzamide	332[b]	27785[b]	63E014	330	27550	63E014
942	Acetone				332	27800	66E095
943	Benzene, 1,4-dibromo-	332[x]	27800[x]	67E112			
944	Tetrazole, 5-phenyl-, ion(1-)				332	27800	69F401
945	Ethene, 1,2-dichloro-, (E)-	333	27800	82Z015			
946	5,7-Dodecadiyne	333	27820	60E013			
947	Benzene, 1-bromo-4-chloro-				333	27900	67E112
948	Pyridine, 2,4,6-trimethyl-				333	27900	81E297

Table 2—Triplet State Energies—Continued

No.	Compound	$E_T(n)$			$E_T(p)$		
		kJ/mol	cm^{-1}	Ref.	kJ/mol	cm^{-1}	Ref.
949	Cytidine 5'-monophosphate				334	27900	673066
950	2,4-Hexadiyne	334	27910	60E013			
951	Benzene, 1,2,4,5-tetramethyl-	334	27920	82E344	335	28000	67F523
952	Benzene, 1,4-dichloro-	335	28000	67E112			
953	Benzene, 1,3,5-trimethyl-	335	28010	61E014	336	28075	64E022
954	p-Xylene	337	28135	61E014	337	28145	60E016
955	Pyridine, 2,6-dimethyl-				337	28160	69E226
956	Benzene, methoxy-	338	28200	67F523	338	28200	67F523
957	Pyrimidine	338	28214	70E310			
958	1,3,2-Benzothiazaborolidine, 2-methyl-	338	28215	82E051			
959	Benzene, cyclopropyl-	339	28300	59E009			
960	Naphthalene, 1,2,3,4-tetrahydro-	339	28300	82E344			
961	Phenyl ether				339	28320	84E090
962	m-Xylene	339	28325	61E014	336	28120	60E016
963	Benzene, 1-fluoro-4-methyl-	341^g	28500^g	747355			
964	Phenol				342	28563	84E090
965	Benzene, chloro-	342^g	28570^g	59E009			
966	Pyridinium, 2,6-dimethyl-				342	28600	81E297
967	1,3,2-Benzodiazaborolidine, 2-methyl-	343	28678	82E051			
968	o-Xylene	343	28705	61E014	344	28760	60E016
969	Indan	344	28750	74E517			
970	Benzylamine				345	28800	69Z002
971	Pyrimidine, 2-chloro-	345	28800	70E310			
972	2-Indanone				345^x	28853^x	85E384

No.	Name						
973	Anthracene, 9,10-dihydro-	346	28920	61E014	346	28900	88E763
974	Toluene	347[g]	29000[g]	747355	347	29000	67F523
975	Pyridine, 4-amino-	347[g]	29000[g]	747355	347	28985	73E338
976	Benzene, 1,4-difluoro-	347[g]	29000[g]	747355			
977	Benzene, 1-fluoro-2-methyl-	348[g]	29100[g]	747355			
978	Benzene, 1,2,4,5-tetrafluoro-	348[g]	29100[g]	747355			
979	Benzene, 1-fluoro-3-methyl-	349[g]	29150[g]	59E009			
980	Benzene, hexafluoro-	349[g]	29200[g]	747355			
981	Benzene, (trifluoromethyl)-	349[g]	29200[g]	747355			
982	Benzene, pentafluoro-	351[g]	29300[g]	747355			
983	Benzene, 1,2,3,4-tetrafluoro-	351[g]	29300[g]	747355			
984	Octanoic acid, 8-phenyl-				350	29200	706023
985	Benzene, 1,2-difluoro-				352	29450	71E391
986	Benzene, 1,2,4-trifluoro-				351	29300	706023
987	Butanoic acid, 4-phenyl-				351	29300	706023
988	Phenylacetic acid				351	29300	706023
989	3-Phenylpropionic acid				353	29500	67F523
990	Benzene	353	29500	67F523			
991	Benzene, 1,3-difluoro-	353[g]	29500[g]	747355			
992	Benzene, fluoro-	353[b]	29500[b]	57B004			
993	Benzene, 1,2,3-trifluoro-	353	29500	71E391	349	29150	71E391
994	Benzene, 1,2,3,5-tetrafluoro-	354[g]	29600[g]	747355			
995	Pyridinium				355	29700	81E297
996	Benzene, 1,3,5-trifluoro-	358[g]	29900[g]	747355	358	29890	71E391

Table 2—**Triplet State Energies**—Continued

No.	Compound	$E_T(n)$ kJ/mol	$E_T(n)$ cm^{-1}	Ref.	$E_T(p)$ kJ/mol	$E_T(p)$ cm^{-1}	Ref.
997	Methyl methacrylate	358[g]	29900[g]	89M184			
998	Pyridine, 4-hydroxy-				358	29940	76E691
999	Methyl acrylate	372[g]	31100[g]	89M184			
1000	Pyridinium, 2-amino-				392	32800	81E297
1001	Acetylene	502[g]	41900[g]	78M177			
1002	1-Butyne	502[g]	41900[g]	78M177			

[b] Aromatic solvent, benzene-like; [g] Gas-phase measurement; [x] Crystalline medium

Section 3

Flash Photolysis: Designing Experiments

Table 3 is intended to bring together information pertinent to designing triplet-triplet energy transfer experiments in flash photolysis. These experiments are central to triplet photosensitized photochemistry, measurements of triplet quenching rates, and estimations of many triplet-triplet extinction coefficients. Table 3 can also be useful for the design of steady-state photochemistry, but triplet extinction coefficients and corresponding wavelengths are included which are needed to quickly monitor time-resolved experiments.

The information presented in Table 3 is shown pictorially in Fig. 3-1. The data for four of the columns is already presented in Table 1 in SI units, although the criteria used for the selection of compounds in Section 1 differ from that adopted here. For convenience, the triplet energy, E_T, in Table 3 is given in units of kcal/mol, and the singlet energy is given as an excitation wavelength, λ_S^{0-0}. The other two items from Table 1, namely the triplet quantum yields, ϕ_T, and triplet lifetimes, τ_T, are carried along unmodified.

The central new item of Table 3 is the information on triplet-triplet absorption. Whenever possible, this absorption is characterized by both a wavelength, λ_T, and a molar absorption coefficient, ε_T, at that wavelength. When an absorption maximum exists, the wavelength refers to that maximum. In cases where molar absorption coefficients are not available, but the triplets are important donors or acceptors, only a wavelength is given, and this isolated wavelength refers to an absorption maximum. Molar absorption coefficients are often called decadic extinction coefficients. A review of the primary literature was presented previously, [86Z026] and an extensive statistical analysis of this data was also performed, [87Z100] leading to the publication of a set of recommended values for, and procedures for the determination of the molar absorptivity of transients. [91B003]

In Table 3, the solvent type is a dual one. Benzene and some aromatic solvents have a large influence on triplet-triplet absorption. There is a strong tendency to broaden the peaks such that the extinction coefficient at the maximum is lowered, but the oscillator strength (related to the total area) in the transition remains constant. To distinguish aromatic solvents for the (λ_T, ε_T) pair of a given compound from nonaromatic solvents, "b" for "benzene-like" is used for aromatic solvents, and "nb" is used for other solvents. One or the other of these two symbols follow the "/" in the solvent column and refer only to the (λ_T, ε_T) column. The symbol preceding the "/" in the solvent column is a nonpolar/polar classification (discussed in the introduction of Section 1) for the four properties, λ_S^{0-0}, E_T, ϕ_T, and τ_T.

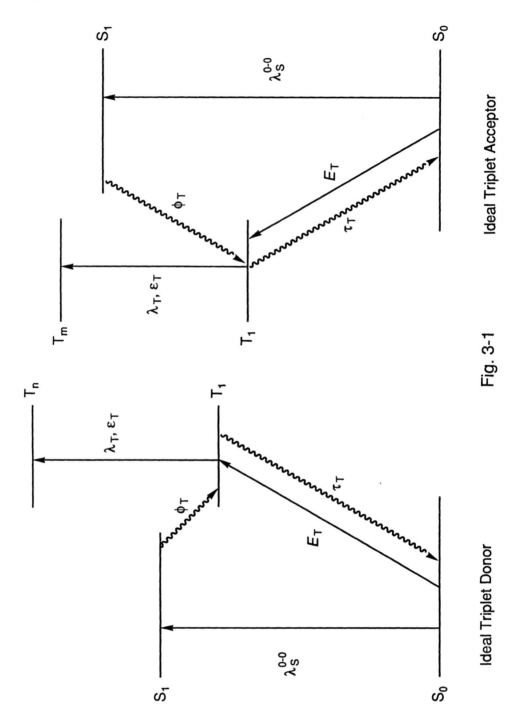

Fig. 3-1

Table 3

Flash Photolysis Parameters

No.	Solv	λ_S^{0-0} (nm)	Ref.	E_T (kcal/mol)	Ref.	ϕ_T	Ref.	λ_T (nm), ε_T (M^{-1} cm^{-1})	Ref.	τ_T (µs)	Ref.
1	**Acenaphthene**										
	n/b	319	71Z001	59.7	64E037	0.46	69E202	422	82A292		
	p/nb	318	68Z003	59.2	67E112	0.58	68E098	430, 6000	87Z100	3300	68Z003
2	**Acetone**										
	n					0.90	707357			6.3	84B051
	p/nb			79.4	66E095	1	66F206	300, 600	87Z100	47	717489
3	**Acetophenone**										
	n	363	64E028	74.1	81E561	1	85A268			0.23	717179
	p/nb	354	81E561	74.4	81E561	1	85A268	330, 7160	87Z100	0.14	717179
4	**Acetophenone, 4'-methoxy-**										
	n	352	66E094	71.8	67F523	1	85A268				
	p/nb			71.5	67F521	1	85A268	360, 8840	87Z100		
5	**Acetophenone, 4'-methyl-**										
	n	352	68F298	72.9	72E294	1	85A268				
	p/nb			72.8	67F521	1	85A268	331, 11400	737198		
6	**Acetophenone, 4'-phenyl-**										
	n			61.1	67F523						
	p/nb			60.8	67F523			435, 130000	87Z100		
7	**Acetophenone, 4'-(trifluoromethyl)-**										
	n	360	68F298	71.7	67F523	1	85A268				
	p/nb	360	697155	72.0	67F523	1	85A268	455, 2290	87Z100		

Table 3—**Flash Photolysis Parameters**—Continued

No.	Solv	λ_S^{0-0} (nm)	Ref.	E_T (kcal/mol)	Ref.	ϕ_T	Ref.	λ_T (nm), ε_T (M^{-1} cm^{-1})	Ref.	τ_T (µs)	Ref.
8	**Acridine**										
	n/b			45.4	81E147	0.5	78E222	440, 24300	71E360	10000	677498
	p/nb	380	779025	45.0	84E803	0.82^s	85E408	432.5, 31500	71E360	14	78E394
9	**Acridine, 3,6-diamino-**										
	p/nb	485	69E235	51.1	69E235			550, 9510	87Z100		
10	**Acridine Orange, conjugate monoacid**										
	p/nb	524	71Z014	45.7	68Z003	0.10	727073	540, 9570	87Z100	105	79E219
11	**9-Acridinethione, 10-methyl-**										
	n/b	633	83E230	42.6	83E230	0.95	84E342	520, 9300	84E342	2.6	84E342
	p/nb					0.90	84E342	520, 8790	87Z100	2.3	84E342
12	**Acridinium, 3,6-diamino-**										
	p/nb	478	71Z014	48.9	68Z003	0.22	727073	550, 8270	87Z100	20	80F373
13	**Acridinium, 3,6-diamino-10-methyl-**										
	p/nb	485	69E235	51.1	69E235			620, 8600	87Z100		
14	**9-Acridinone**										
	n/b	393	78E542	58.2	78E542	0.99	766377	620, 37800	766377	20	766377
	p/nb	413	78E542	60.2	78E878			620, 41400	87Z100	9.2	89A120
15	**Angelicin**										
	n					0.009	78E157				
	p/nb	362	73E372	63.2	73E372	0.031	84E794	450, 4330	87Z100	2.6	84E794

No.	Compound / row										
16	**Aniline**										
	n	300	82B140	71.0	67E116	0.75	80E253			0.72	87B098
	p/nb	312	80E253	76.6	44E001	0.90	80E253	320	69E215		
17	**Aniline, *N,N*-dimethyl-**										
	n	313	71Z001								
	p/nb	319	71Z001								
18	**Aniline, *N*-phenyl-**										
	n			75.8	61E013	0.32	706243	460, 4000	85E653	0.3	706243
	p/nb	322	73E359	71.9	73E359	0.47	706243	530, 10400	71E360	0.5	706243
19	**Anthracene**										
	n/b	376	85E096	42.5	57B004	0.71	757282	432.5, 45500	71E360	670	85E555
	p/nb	375	779025	42.5	779025	0.66	85E408	425, 64700	71E360	3300	747049
20	**Anthracene, 9-bromo-**										
	n/b	390	79E108					430, 48000	690087	43	62E009
	p/nb			41.3	84B110	0.98	89B155	430, 47500	87Z100	19.5	84B110
21	**Anthracene, 9-cyano-**										
	n/b	403	81E183			0.040	737140	435, 10300	690087	600	737140
	p/nb	417	737140			0.021	737140	435, 8490	87Z100	1800	737140
22	**Anthracene, 9,10-dibromo-**										
	n/b	406	84E393	40.2	566006	0.70	84E393	427.5, 48000	84E393	36	62E009
	p/nb	405	756125			0.82	84E785	425, 46300	87Z100	11	78E394

Table 3—Flash Photolysis Parameters—Continued

No.	Solv	λ_S^{0-0} (nm)	Ref.	E_T (kcal/mol)	Ref.	ϕ_T	Ref.	λ_T (nm), ε_T (M^{-1} cm^{-1})	Ref.	τ_T (µs)	Ref.
23	**Anthracene, 9,10-dichloro-**										
	n/b	401	71Z001	40.5	566006	0.29	84E393	425, 46000	84E393	100	62E009
	p/nb	403	66E086	40.5	66E086	0.45	706054	425, 42500	87Z100		
24	**Anthracene, 9,10-dicyano-**										
	n/b	421	85A009	41.7	78E414	0.23	84E393	440, 9000	84E393	100	78E414
	p/nb	428	85A009					440, 9180	87Z100		
25	**Anthracene, 9,10-dimethyl-**										
	n/b	398	71Z014			0.02	68F286	435, 35300	87Z100		
	p/nb	403	68Z003			0.032	68Z003	435, 35300	87Z100	8000	67E129
26	**Anthracene, 9,10-diphenyl-**										
	n/b	393	85E096	40.9	68E131	0.02	83E281	445, 14500	87Z100	2500	746270
	p/nb	392	71Z014			0.02	83E281	445, 15600	87Z100	3000	746270
27	**Anthracene, 9-methyl-**										
	n/b	387	85E096	41.3	57B004	0.48	756505	430, 42000	690087		
	p/nb	391	68Z003	40.7	66E086	0.67	67E129	430, 45900	87Z100	10000	67E129
28	**Anthracene, 9-phenyl-**										
	n	387	71Z001			0.37	65F031				
	p/nb	394	68Z003			0.51	67E109	428, 14600	87Z100	15000	67E129
29	**1-Anthracenesulfonate ion**										
	p/nb					0.98	85E452	440, 20000	85E452	103	85E452

#											
30 2-Anthracenesulfonate ion											
	p/nb					0.65	85E452	425, 30000	85E452	83	85E452
31 9,10-Anthraquinone											
	n/b	421	85E712	62.4	64E021	0.90	65F030	390, 10300	720392	0.11	81F130
	p			62.8	67E115						
32 Anthrone											
	n										
	p/nb			71.9	706018			341, 74000	87Z100	0.170	766464
33 Azulene											
	n/b	704	71Z014	39	757247			360, 4000	81F275	11	84E491
	p/nb							360, 4140	87Z100	3	81F275
34 Benz[b]acridin-12-one											
	n	488	78E878	45.7	85F172			590, 53600	87Z100		
	p/nb			46.1	78E878						
35 Benz[a]anthracene											
	n/b	385	71Z014	47.2	54E002	0.79	69E208	490, 20500	71E360	100	64E015
	p/nb	390	78E761	47.2	78E761	0.79	727047	480, 28800	71E360	9400	64E026
36 Benzene											
	n	260	71Z014	84.3	67F523	0.25	69E202	235, 11000	87Z100		
	p/nb	261	506002	84.3	67F523	0.15	736048				

Table 3—**Flash Photolysis Parameters**—Continued

No.	Solv	λ_S^{0-0} (nm)	Ref.	E_T (kcal/mol)	Ref.	ϕ_T	Ref.	λ_T (nm), ε_T (M^{-1} cm^{-1})	Ref.	τ_T (μs)	Ref.
37	**Benzene, chloro-**										
	n	272	71Z014	81.7	59E009	0.6	83F178			1.6	84E529
	p/nb					0.7	85E488	300, 6150	87Z100	0.715	85E488
38	**Benzene, 1,4-dichloro-**										
	n	282	82B140	80.1	67E112	0.8	83F178			430	85E406
	p/nb	280	79B163			0.95	85E406	310, 3800	85E406	330	85E406
39	**Benzene, methoxy-**										
	n	274	496002	80.7	67F523	0.64	82E060	252	757161	3.3	757161
	p/nb	278	82B140	80.7	67F523						
40	**Benzidine, N,N,N',N'-tetramethyl-**										
	n			62.3	767177	0.52	84E405				
	p/nb					0.41	84E405	475, 38700	87Z100	5	84F074
41	**Benzil**										
	n	485	69E223	53.4	67E119	0.92	65F030			150	717447
	p/nb			54.3	67E119			480	79E690	1500	68E128
42	**Benzo[a]coronene**										
	n	432	59E013	51.5	78E314						
	p/nb					0.55	68E104	570, 22300	87Z100		
43	**Benzoic acid**										
	n			77.5	63E014						
	p/nb	279	71Z014	77.9	767546			320, 1000	87Z100		

No.	Compound		λ	Ref		Ref		Ref		Ref		Ref
44	**Benzo[rst]pentaphene**	n/b	433	64Z007	40.2	62E015			490	71E361	170	71E361
		p/nb	433	779025	40.4	779025			490	71E361		71E361
45	**Benzo[ghi]perylene**	n/b	408	696020	46.3	566005	0.53	696020	470	71E361	150	71E361
		p/nb	407	779025					465, 39300	87Z100		
46	**Benzo[c]phenanthrene**	n	370	71Z014	56.7	54E002			517, 4800	58E001		
		p/nb	373	779025	57.1	779025						
47	**Benzophenone**	n/b	379	64E028	68.6	67F523	1.0	69E202	530, 7220	83B067	6.9	776016
		p/nb	384	63Z003	69.2	80E223	1	85A268	525, 6250	87Z100	50	766276
48	**Benzophenone, 4,4′-bis(dimethylamino)-**	n			65.8	68E135	0.91	777004			25	777603
		p/nb	405	79E210	60.8	79E210	0.47	85A268	500	777603	20	777603
49	**Benzophenone, 4-chloro-**	n			68.3	68F291	1	85A268	320, 12800	84B033		
		p/nb			68.8	68F291	1	85A268				
50	**Benzophenone, 4,4′-dichloro-**	n			68.0	64E021	1	85A268	320, 23300	87Z100	2.2	776016
		p/nb			68.4	87F368	1	85A268				

Table 3—Flash Photolysis Parameters—Continued

No.	Solv	λ_S^{0-0} (nm)	Ref.	E_T (kcal/mol)	Ref.	ϕ_T	Ref.	λ_T (nm), ε_T (M^{-1} cm^{-1})	Ref.	τ_T (μs)	Ref.
51	**Benzophenone, 4,4'-dimethoxy-**										
	n	365	66E094	70.0	81E099	1	85A268			3.6	776016
	p/nb			69.8	80E223	1	85A268	350, 14100	87Z100		
52	**Benzophenone, 4-fluoro-**										
	n			68.9	68F291	1	85A268				
	p/nb			69.8	81C032	1	85A268	315, 21900	87Z100		
53	**Benzophenone, 4-methoxy-**										
	n			68.6	68F291	1	85A268				
	p/nb			69.4	80E223	1	85A268	335, 10100	87Z100	7.2	82A082
54	**Benzophenone, 4-methyl-**										
	n	370	697155	68.6	67F523	1	85A268				
	p/nb	366	697155	69.4	80E223	1	85A268	315, 21100	87Z100		
55	**Benzophenone, 4-(trifluoromethyl)-**										
	n	373	697155	67.6	68F291	1	85A268				
	p/nb	368	697155	68.1	68F291	1	85A268	320, 22400	87Z100		
56	**Benzo[g]pteridine-2,4-dione, 7,8,10-trimethyl-**										
	n	500	69E235	50.0	69E235	0.30	777617	650, 6090	87Z100	30	68E100
	p/nb									320	70E295
57	**Benzo[a]pyrene**										
	n/b	405	71Z014	41.9	57B004			475	71E361	8700	78A345
	p/nb	403	779025	42.3	779025			465	78A345	8800	64E026

No.	Compound / State									
58	**Benzo[e]pyrene**									
	n/b	368	71Z014				560	71E361	120	71E361
	p/nb	366	779025	52.8			555, 17000	87Z100		
59	**1,4-Benzoquinone**									
	n	457	69E247	53.6			450	79B007	0.53	80B112
	p/nb									
60	**1,4-Benzoquinone, tetrachloro-**									
	n/b	450	69E247	49.2			516	727069	2.0	697272
	p/nb				0.98	79B061	510, 6990	87Z100	1.2	697272
61	**1,4-Benzoquinone, tetramethyl-**									
	n/b				1.0	767144	490, 6950	71E360	21	767144
	p/nb				1.0	767144	490, 5330	71E360	15	767144
62	**Benzo[b]triphenylene**									
	n/b	375	71Z014	50.9			450	71E361	90	71E361
	p/nb	374	779025	50.9			450, 29900	87Z100		
63	**Benzoxazole, 2,2'-(1,4-phenylene)bis-**									
	n				1.0	60E012	480, 18600	87Z100	0.48	82E632
	p/nb									
64	**Biacetyl**									
	n/b						315, 5160	71E360	638	726211
	p/nb						315, 4580	87Z100		

Table 3—**Flash Photolysis Parameters**—Continued

No.	Solv	λ_S^{0-0} (nm)	Ref.	E_T (kcal/mol)	Ref.	ϕ_T	Ref.	λ_T (nm), ε_T (M^{-1} cm^{-1})	Ref.	τ_T (μs)	Ref.
65	**1,1′-Binaphthyl**										
	n	325	71Z001					615, 12400	87Z100	14	771048
	p/nb	328	776226								
66	**2,2′-Binaphthyl**										
	n	333	71Z001	55.9				450, 24800	87Z100		
	p/nb				566005						
67	**Biphenyl**										
	n/b	286	71Z001	65.4	64E037	0.84	757282	367.3, 27100	71E360	130	69E208
	p/nb	306	776226	65.5	735067			361.3, 42800	71E360		
68	**Biphenyl, 4,4′-diamino-**										
	p/nb	346	71Z001					460, 35500	87Z100		
69	**Biphenylene**										
	n	392	64Z007					350, 10500	87Z100	100	720464
	p/nb										
70	**1,3-Butadiene, 1,4-diphenyl-**										
	n/b			42.3	707199	0.020	82E365	390, 45000	87Z100	1.6	82E365
	p/nb	358	80B135			≤0.002	82E365	390, 54500	87Z100	5.0	84E319
71	**C$_{60}$**										
	n/b	620	91E003	36.0	91E368	1	92E260	740, 12000	92E205	250	92E205
	p/nb							740, 14000	92E260		

#	Compound								nm, ε			
72	C₇₀	n/b	648	91E594	35.3	91D034	0.97	92E142	490, 12000	92E205	250	92E205
		p/nb							470, 19000	92E260		
73	Carbazole	n/b	332	71Z001			0.36	65F030	418	84F248	170	77A178
		p/nb	345	78E761	70.2	58E005			425, 14000	87Z100		
74	2,2'-Carbocyanine, 1,1'-diethyl-	p/nb					0.0029	79E243	635, 58000	87Z100	190	79E243
75	4,4'-Carbocyanine, 1,1'-diethyl-	p/nb					$<6\times10^{-4}$	736051	778, 35600	87Z100	1100	736051
76	β-apo-8'-Carotenal	n	518	78E431			0.003	78E721				
		p/nb							520, 223000	87Z100	10	733001
77	β-apo-14'-Carotenal	n/b	457	78E432			0.54	79E546	480, 114000	87Z100	8	83E026
		p/nb	465	78E432			−0.033	79E546	480, 116000	87Z100	10.3	79E546
78	β-Carotene	n/b	524	71Z014	21	757247	<0.001	776412	520	767094	70	66E089
		p/nb	704	92N199	20.3	92N199			515, 187000	87Z100	9	81B115
79	Chlorophyll a	n/b	677	81F121					460	55E003	1500	58R001

Table 3—**Flash Photolysis Parameters**—Continued

No.	Solv	λ_S^{0-0} (nm)	Ref.	E_T (kcal/mol)	Ref.	ϕ_T	Ref.	λ_T (nm), ε_T (M^{-1} cm^{-1})	Ref.	τ_T (μs)	Ref.
79	**Chlorophyll *a*—Continued**										
	p/nb	671	79E838	29.8	79E838	0.53	86R013	460, 48000	79B037	800	68Z003
80	**Chlorophyll *b***										
	n/b	660	79E838	31.1	79E838			316, 36500	59B002	2500	59B002
	p/nb	667	68Z003	32.6	68Z003	0.81	86R013	450, 24300	87Z100	1500	68Z003
81	**Chrysene**										
	n/b	361	71Z014			0.85	757282	575	761024	710	69E208
	p/nb	360	779025	57.1	779025	0.85	68E098	580, 29800	87Z100		
82	**Coronene**										
	n/b	428	64Z007					480	761024		
	p/nb	429	779025	54.4	566005	0.56	68E098	480, 15000	87Z100		
83	**Coumarin**										
	n/b			61.7	775025			400, 11000	79E282	3.8	79E282
	p/nb	341	73E372	62.4	73E372	0.054	79E282	400, 10100	87Z100	1.3	79E282
84	**Coumarin, 5,7-dimethoxy-**										
	n									10	79E282
	p/nb	352	73E372	60.6	73E372	0.072	79E282	450, 10500	87Z100		
85	**1,3-Cyclohexadiene**										
	n/b	292	399001	52.4	65E036			310	84A458	30	80B021
	p/nb							303, 2300	87Z100		

No.	Compound	Type	λ	ref		ref	Φ	ref	λ, ε	ref		ref
86	Deoxythymidine 5′-monophosphate	p/nb	293	673066	75.2	673066	0.055	79B087	370, 4000	79B087	25	79B087
87	Dibenz[a,h]anthracene	n/b	395	71Z014		727047	0.90	727047	584, 13000	83F075		
		p/nb	395	779025	52.1	779025	0.9	83F075	580, 25100	87Z100		
88	2,2′-Dicarbocyanine, 1,1′-diethyl-	p/nb					$<3 \times 10^{-4}$	736051	780, 69300	87Z100	480	736051
89	Fluorene	n	301	71Z014	67.5	64E037	0.22	706049			150	69E208
		p/nb	301	779025	67.9	779025	0.32	68E098	380, 22700	87Z100		
90	9-Fluorenone	n/b	450	706018	50.3	78E060	0.94	757282	430, 5900	757282	500	78E495
		p/nb					0.48	78E495	425, 6040	87Z100	100	78E495
91	Fluorescein dianion	p/nb	520	69E235	47.2	69E235	0.02	82E660	535, 8700	60A001	20000	60A001
92	Fluorescein dianion, 2′,4′,5′,7′-tetrabromo-	p/nb	571	68Z004	42.3	68Z004	0.33	82E660	580, 10200	87Z100	30	84E216
93	Fluorescein dianion, 2′,4′,5′,7′-tetraiodo-	p/nb	565	68Z004	44.0	69E203	0.83	82E660	526, 26000	67E031	630	64E016
94	(E,E,E)-2,4,6-Heptatrienal, 5-methyl-7-(2,6,6-trimethyl-1-cyclohexen-1-yl)-	n/b	400	78E431	35.9	84E180	0.66	79E546	430, 63000	79E546	6.2	79E546

Table 3—**Flash Photolysis Parameters**—Continued

No.	Solv	λ_S^{0-0} (nm)	Ref.	E_T (kcal/mol)	Ref.	ϕ_T	Ref.	λ_T (nm), ε_T (M^{-1} cm^{-1})	Ref.	τ_T (μs)	Ref.
94	(*E,E,E*)-2,4,6-Heptatrienal, 5-methyl-7-(2,6,6-trimethyl-1-cyclohexen-1-yl)—Continued										
	p/nb					0.41	79E546	440, 52000	87Z100	10.9	79E546
95	1,3,5-Hexatriene, 1,6-diphenyl-										
	n/b			35.6	72B007	0.029	82E365	420, 105000	87Z100	20	82E365
	p/nb	399	78B145			0.020	761088	420, 114000	87Z100	30	82E365
96	Indole										
	n/b	288	51Z002	72.0	84E843	0.43	81E082	430	771021	16	777037
	p/nb	299	63E012	70.8	84E843	0.23	81E082	430, 4260	87Z100	11.6	757163
97	β-Ionone										
	n/b	405	85E293	55	85E293	0.49	85E293	345	85E293	0.16	78E721
	p/nb							330, 85300	87Z100		
98	Isoquinoline										
	n			60.6	59E009	0.21	87E642				
	p/nb	320	59E008	60.6	59E008			418, 11900	87Z100		
99	Methylene Blue cation										
	p/nb	664	63E027	33	67F524	0.52	69E203	420, 14400	87Z100	450	756162
100	Naphthalene										
	n/b	311	71Z014	60.6	57B004	0.75	757282	425, 13200	71E360	175	62E009
	p/nb	311	506002	60.9	44E001	0.80	68E098	415, 24500	71E360	1800	747049
101	Naphthalene, 1-bromo-										
	n	321	71Z014	59.0	57B004					270	62E009

No.	Compound								
101	**Naphthalene, 1-bromo-—Continued**								
	p/nb	59.0	63Z003			425, 11500	87Z100	830	61E005
102	**Naphthalene, 1-chloro-**								
	n			58.6	78E067	0.79	757282	420, 29500	87Z100	280	62E009
	p/nb	319	79B163	59.2	44E001				
103	**Naphthalene, 1,4-dicyano-**								
	n	336	84F449	55.5	767370	0.19	84B066	455, 6730	87Z100	40	84B066
	p/nb								
104	**Naphthalene, 1,4-diphenyl-**								
	n	336	71Z001					444, 32500	87Z100
	p/nb								
105	**Naphthalene, 1-hydroxy-**								
	n	321	71Z001	58.6	44E001	>0.27	65F030	430, 9000	85A406
	p/nb	323	71Z014						
106	**Naphthalene, 2-hydroxy-**								
	n	330	71Z001	60.3	44E001			435, 6680	87Z100	67	737113
	p/nb	330	71Z014						
107	**Naphthalene, 1-methoxy-**								
	n			330	757282	0.45	757282	440, 9980	87Z100	5500	68Z003
	p/nb	320	79B163	59.8	44E001	0.50	68E098		

Table 3—**Flash Photolysis Parameters**—Continued

No.	Solv	λ_S^{0-0} (nm)	Ref.	E_T (kcal/mol)	Ref.	ϕ_T	Ref.	λ_T (nm), ε_T (M^{-1} cm^{-1})	Ref.	τ_T (μs)	Ref.
108	**Naphthalene, 2-methoxy-**										
	n					0.50	757282			50	77E663
	p/nb							435, 21400	87Z100		
109	**Naphthalene, 1-methyl-**										
	n/b	317	71Z001			0.58	757282	425	84E092	25	767159
	p/nb	317	71Z014	60.6	44E001			420, 20200	87Z100		
110	**Naphthalene, 2-methyl-**										
	n	319	71Z001			0.56	757282				
	p/nb	319	71Z014	60.8	67E112			420, 30600	87Z100		
111	**Naphthalene, 1-nitro-**										
	n	382	71Z014	55.2	71F587	0.63	687061			0.93	81B064
	p/nb							525	81B064	4.9	81B064
112	**Naphthalene, 2-nitro-**										
	n	380	71Z014			0.83	71F587			0.53	767269
	p/nb			56.9	71F587			360, 3600	87Z100	1.70	767269
113	**Naphthalene, 1-phenyl-**										
	n	315	71Z001			0.52	757282				
	p/nb			58.8	67E112			490, 21700	87Z100		
114	**Naphthalene, 2-phenyl-**										
	n	325	71Z001			0.43	757282				
	p/nb			58.7	67E112			430, 43000	87Z100		

No.	Compound / Row	λ	Ref		Ref	Φ	Ref	λ, ε	Ref		Ref
115	**1,3,5,7-Octatetraene, 1,8-diphenyl-**										
	n/b	443	72E330	31.6	72B007	0.005	82E365	440, 178000	87Z100	40	82E365
	p/nb					0.006	761088	440, 198000	87Z100	34	82E365
116	**1,3,4-Oxadiazole, 2,5-diphenyl-**										
	n	311	71Z001					425, 980	87Z100	0.300	777265
	p/nb	310	71Z001								
117	**2,2'-Oxadicarbocyanine, 3,3'-diethyl-**										
	p/nb					<0.005	726156	650, 81400	87Z100	5000	726156
118	**Oxazole, 2,5-bis(4-biphenylyl)-**										
	n	375	71Z001					560, 110000	87Z100	0.285	777265
	p/nb										
119	**Oxazole, 2,5-diphenyl-**										
	n	335	71Z001			0.12	80E439	500, 14800	87Z100	1700	80E439
	p/nb	336	71Z001							2500	747049
120	**Oxazole, 2,2'-(1,4-phenylene)bis[5-phenyl-**										
	n/b	379	71Z001			0.054	86E128	550, 37600	86E128	1750	86E128
	p	385	71Z001	55.5	86E128						
121	**Pentacene**										
	n/b	585	71Z014	18.0	716279	0.16	727073	505, 120000	727348	110	61E005
	p/nb							305, 595000	87Z100		

Table 3—**Flash Photolysis Parameters**—Continued

No.	Solv	λ_S^{0-0} (nm)	Ref.	E_T (kcal/mol)	Ref.	ϕ_T	Ref.	λ_T (nm), ε_T (M^{-1} cm^{-1})	Ref.	τ_T (μs)	Ref.
122	(*E,E*)-2,4-Pentadienal, 3-methyl-5-(2,6,6-trimethyl-1-cyclohexen-1-yl)-										
	n	358	78E431	45.0	78E721	0.20	84E036			0.1	78E721
	p/nb					0.45	79E546	385, 32300	87Z100	0.19	79E546
123	**Pentaphene**										
	p/nb	424	64Z006	48.4	566005			493, 45900	84E390		
124	**Perylene**										
	n/b	435	71Z014	35.4	69E238	0.014	66E101	490, 14300	71E360		
	p/nb	439	68Z003	36.0	68Z003	0.0088	66E101	485, 13400	87Z100	5000	68Z003
125	**Phenanthrene**										
	n/b	346	71Z014	62.3	63E024	0.73	757282	492.5, 15700	71E360	145	62E009
	p/nb	347	79E505	61.4	89E090	0.85	68E098	482.5, 25200	71E360	910	61E005
126	**1,10-Phenanthroline**										
	n/b	342	83E180					440	82A259	26	777201
	p/ab	339	63E018	63.2	63E018			445	777201	35	777201
127	**Phenazine**										
	n/b			44.5	69E229	0.21	716169	440	717154	42	716169
	p/nb	438	79E967	44.6	85B074	0.45	85B074	355, 37700	87Z100	770	85B074
128	**Phenazinium, 3,7-diamino-2,8-dimethyl-5-phenyl-**										
	p/nb					0.50	82E660	420, 10000	87Z100	67	677322
129	**Phenazinium, 3,7-diamino-5-phenyl-**										
	p/nb					0.10	89A343	440, 29000	89A343	25	89A343

No.	Compound											
130	Phenol	n	278	71Z014	81.7	84E090	0.32	82E060	250	757161	3.3	757161
		p/nb	283	84E090								
131	Phenothiazine	n										
		p/nb	370	71Z014	60.4	757279	0.54	83E835	460, 27000	87Z100		
132	Phenoxazine	n/b	362	72E303	62.6	72E303			465	707186	32	72E303
		p/nb	366	72E303	62.2	72E303			465	707186	44	707186
133	Phenoxazinium, 3,7-diamino-, conjugate monoacid	p/nb					≤0.003	767661	650, 16000	767246	55	767246
134	*p*-Phenylenediamine, *N,N,N',N'*-tetramethyl-	n/b	358	71Z001					620, 12200	87Z100	1.4	82E474
		p/nb	364	71Z001					605, 12200	71E360	0.5	84B061
135	Pheophytin *a*	n	670	79E838	31.0	79E838	0.95	70E296	407, 62800	87Z100	750	70E296
		p/nb	667	70E296								
136	Pheophytin *b*	n	658	79E838	32.1	79E838	0.75	70E296	423, 71200	87Z100	1050	70E296
		p/nb	654	70E296								

Table 3—**Flash Photolysis Parameters**—Continued

No.	Solv	λ_S^{0-0} (nm)	Ref.	E_T (kcal/mol)	Ref.	ϕ_T	Ref.	λ_T (nm), ε_T (M^{-1} cm^{-1})	Ref.	τ_T (μs)	Ref.
137	**Phthalazine**										
	n/b			63.1	82E585	0.29	82E203	396	747093	2.7	87E642
	p/nb	387	59E008	65.8	59E008	0.44	87E642	421, 4450	87Z100	21.3	757309
138	**Phthalocyanine**										
	n	704	69E231	28.7	78A378	0.14	78A378			130	78A378
	p/nb							480, 29900	87Z100		
139	**Phthalocyanine, magnesium(II)**										
	n	687	69E231							100	65A001
	p/nb					0.23	70E319	480, 32300	87Z100	430	86E784
140	**Phthalocyanine, zinc(II)**										
	n/b	683	71E386	26.2	71E386	0.65	81E457	480, 51000	87Z100		
	p/nb					0.04	86E784	480, 28900	87Z100	270	86E784
141	**Picene**										
	n/b	376	64Z006					560	71E361	160	71E361
	p/nb	376	779025	57.4	779025			630, 45500	87Z100		
142	**Porphine**										
	n	615	753056	36.0	753056	0.90	82F161	419, 98600	87Z100		
	p/nb	610	743135	36.4	743135						
143	**Porphine, 2,7,12,17-tetraethyl-3,8,13,18-tetramethyl-, zinc(II)**										
	n	573	69E231								
	p/nb	576	71E357	40.6	71E357			440, 99000	87Z100		

#	Compound										
144	**Porphine, tetrakis(4-sulfonatophenyl)-**										
	p/nb					0.78	82E622	790, 3400	84E203	420	82E622
145	**Porphine, tetrakis(4-sulfonatophenyl)-, zinc(II)**										
	p/nb					0.84	82E622	840, 6000	84E203	80	82N068
146	**Porphine, tetrakis(4-trimethylammoniophenyl)-**										
	p/nb					0.80	83E462	800, 3200	83E462	540	83E462
147	**Porphine, tetrakis(4-trimethylammoniophenyl)-, zinc(II)**										
	p/nb					0.82	83E462	840, 5000	83E462	1200	83E462
148	**Porphine, tetraphenyl-**										
	n/b	667	82Z053	33.0	82Z053	0.82	82F161	790, 6000	84E203	1500	84E203
	p/nb	645	743135	33.5	743135	0.88	83F182	785, 6000	87Z100		
149	**Porphine, tetraphenyl-, magnesium(II)**										
	n	611	82Z053	34.1	82Z053						
	p/nb	620	71E357	34.2	71E357	0.85	83F182	485, 72000	87Z100	1400	81E271
150	**Porphine, tetraphenyl-, zinc(II)**										
	n/b	605	82Z053	36.6	756229	0.88	82F161	845, 8200	84E203	1200	81E271
	p/nb	602	71E357	36.6	71E357	0.90	83F182	470, 71000	81E271		
151	**Porphine, zinc(II)**										
	p/nb	569	71E357	39.6	71E357			840, 7000	80N087		
152	**Porphine-2,18-dipropanoic acid, 7,12-diethenyl-3,8,13,17-tetramethyl-, dimethyl ester**										
	n/b					0.80	82F161	710, 9000	771078	550	80B017

Table 3—Flash Photolysis Parameters—Continued

No.	Solv	λ_S^{0-0} (nm)	Ref.	E_T (kcal/mol)	Ref.	ϕ_T	Ref.	λ_T (nm), ε_T (M^{-1} cm^{-1})	Ref.	τ_T (µs)	Ref.
152	**Porphine-2,18-dipropanoic acid, 7,12-diethenyl-3,8,13,17-tetramethyl-, dimethyl ester—Continued**										
	p	627	743135	36.0	743135						
153	**Porphine-2,18-dipropanoic acid, 3,7,12,17-tetramethyl-, dimethyl ester**										
	n					0.63	80E200			210	80E200
	p/nb							440, 23900	87Z100		
154	**Porphycene**										
	n/b					0.42	86E633	380, 66000	86E633	200	86E633
155	**Psoralen**										
	n/b	365	73E372			0.034	78E157	450, 8100	78E157		
	p/nb			62.7	73E372	0.06	79E678	450, 11200	87Z100	5	79E678
156	**Psoralen, 5-methoxy-**										
	n/b	375	73E372			0.067	78E157	450, 10200	78E157		
	p/nb			60.7	83E324	0.1	83E324	450, 9450	87Z100	4.2	83E324
157	**Psoralen, 8-methoxy-**										
	n/b	387	73E372			0.011	78E157	480, 10000	78E157	1.1	78E157
	p/nb			62.6	73E372	0.04	83E324	370, 17700	87Z100	10	79E678
158	**Psoralen, 4,5',8-trimethyl-**										
	p/nb	372	73E372	64.0	73E372	0.093	79B042	470, 25700	87Z100	7.1	79B042
159	**Purine**										
	n	286	71Z014	75.6	755396						
	p/nb							390, 4100	87Z100		

No.	Compound											
160	**Pyranthrene**											
		n					0.55	83F075	500, 20600	87Z100		757309
		p/nb					0.52	83F075				
161	**Pyrazine**											
		n	327	83E031	75.4	78E312	0.33	67E117	260, 3600	87Z100		
		p/nb			74.3	58E006	0.87	757309			4.5	757309
162	**Pyrene**											
		n/b	372	71Z014	48.4	57B004	0.37	82E042	420, 20900	71E360	180	70E295
		p/nb	373	78E761	48.2	78E761	0.38	68E098	412.5, 30400	71E360	11000	68Z003
163	**1-Pyrenecarboxaldehyde**											
		n/b			43	83E387	0.78	83E387	440, 20100	83E387	50	83E387
		p/nb					0.65	83E387	440, 18400	87Z100	38	83E387
164	**Quinoline**											
		n	313	80E627	61.7	80E627	0.31	65F030				
		p/nb	314	80E627	62.5	83E417			425, 6750	87Z100		
165	**Quinoxaline**											
		n	381	82E355	61.0	82E355	0.99	82E203				
		p/nb	375	59E008	60.8	59E008	0.90	85E408	425, 8100	707240	29.4	757309
166	**(*all-E*)-Retinal**											
		n/b	426	717003	29.4	84E180	0.43^{w}	75E529	450, 58400	87Z100	9.3	82A288
		p/nb					0.12^{w}	78E467	450, 69300	87Z100	18	62E007

Table 3—**Flash Photolysis Parameters**—Continued

No.	Solv	λ_S^{0-0} (nm)	Ref.	E_T (kcal/mol)	Ref.	ϕ_T	Ref.	λ_T (nm), ε_T (M^{-1} cm^{-1})	Ref.	τ_T (μs)	Ref.
167	**(*all-E*)-Retinol**										
	n	373	69E251	33.5	84E180	0.017	776412			25	716113
	p/nb	366	69E251			~0.003	85E190	405, 80000	776412		
168	**Rhodamine, inner salt, *N,N'*-diethyl**										
	p/nb	534	77E801	43.9	77E801	0.005	777041	615, 12000	78A304		
169	**Rhodamine 6G cation**										
	p/nb	546	77E801	43.2	77E801	0.0021	747050	620, 16000	87Z100	3500	747050
170	**Riboflavine, conjugate monoacid**										
	p/nb	455	71Z014			0.40	777617	415, 7560	87Z100	19	777617
171	**Rubrene**										
	n/b	542	81E716	26.3	81E346	0.0092	86E782	495, 26000	81E716	120	86E782
	p/nb					0.023	89B155	495, 26500	87Z100	80	68E103
172	**(*E*)-Stilbene**										
	n/b	334	62E012	49.3	80E113			360	720447	14	680379
	p/nb			49.3	80E113			378, 34000	87Z100	62	81E214
173	**Styrene**										
	n	288	776060	61.8	776060	0.40	82E181	325, 2210		0.025	82E181
	p/nb								87Z100		
174	**p-Terphenyl**										
	n/b	310	71Z001			0.11	69E208	460, 90000	71E360	450	69E208
	p/nb			58.3	67E112			460, 72700	87Z100		

#											
175	**Tetracene**										
	n/b	472	71Z014	29.3	64E038	0.62	757282	465, 31200	71E360	400	85E555
	p/nb	471	779025			0.66	717459	465, 57900	87Z100		
176	**Thiobenzophenone**										
	n/b	627	756061	39.3	756061	1.0^w	84A221	400, 4800	84A221	1.7	84A221
	p/nb	626	726174	40.6	726174			400, 4950	87Z100		
177	**Thiobenzophenone, 4,4′-bis(dimethylamino)-**										
	n/b	583	726174	42.2	726174	1.0	84A221	335, 14400	84A221	1.3	84A221
	p/nb							335, 14900	87Z100		
178	**Thiobenzophenone, 4,4′-dimethoxy-**										
	n			41.2	726174	1.0^w	84A221	295, 20500	87Z100	1.4	84A221
	p/nb	606	726174	42.1	726174						
179	**Thionine cation**										
	p/nb	612	697141	39	697141	0.62	69E203	770, 10900	87Z100	72	697141
180	**Thionine cation, conjugate monoacid**										
	p/nb							650, 18000	87Z100	16	777315
181	**Thioxanthen-9-one**										
	n/b			63.3	86E676			650, 30000	79E099	95	81A294
	p/nb							650, 26200	87Z100	73	737190
182	**Thioxanthione**										
	n/b			39.6	82E214	1.0^w	84A221	505, 2500	84A221	0.83	84A221

Table 3—**Flash Photolysis Parameters**—Continued

No.	Solv	λ_S^{0-0} (nm)	Ref.	E_T (kcal/mol)	Ref.	ϕ_T	Ref.	λ_T (nm), ε_T (M^{-1} cm^{-1})	Ref.	τ_T (μs)	Ref.
182	**Thioxanthione**—Continued										
	p/nb							505, 2580	87Z100		
183	**Thymidine**										
	n	284	80E025								
	p/nb	287	80E025			0.069	79B087	370, 2320	87Z100	25	79B087
184	**Thymine**										
	n	283	80E025								
	p/nb	283	80E025			0.06	757510	340, 4000	757510	10	757510
185	**1,3,5-Triazine**										
	n	327	71Z014								
	p/nb			75.5	61E009			245, 6000	87Z100	0.91	757066
186	**Triphenylene**										
	n/b	343	71Z001			0.86	69E202	430, 6190	87Z100	55	61E005
	p/nb	340	779025	67.0	84E612	0.89	66E098	430, 13500	87Z100	1000	61E005
187	**Tryptophan**										
	p/nb	300	62Z002			0.18	777432	460, 5000	87Z100	14.3	757163
188	**Uracil**										
	n	278	80E025								
	p/nb	278	80E025			0.1	79B087	350, 1730	87Z100	2	757510
189	**Uracil, 1,3-dimethyl-**										
	n	283	80E025								

189 Uracil, 1,3-dimethyl—Continued											
p/nb	283	80E025				0.02	81E042	380, 8000	87Z100		
190 Uridine											
p/nb						0.078	79B087	370, 4130	87Z100	20	79B087
191 Uridine 5'-monophosphate											
p/nb	287	673066	78.3	737541		0.044	79B087	390, 5810	87Z100	33	79B087
192 9-Xanthione											
n/b	621	81F218	43.3	81F218		0.8^w	84A221	345, 15400	84A221	1.8	84A221
p/nb								345, 15900	87Z100		
193 Xanthone											
n/b	370	717449	74.1	717449				610, 5300	767171	0.02	767171
p/nb		717449	74.1	717449				605, 6480	87Z100	17.9	767171

[s] Very solvent-dependent; [w] Wavelength dependent

Section 4

Low Temperature Photophysics of Organic Molecules

The same three excited states that dominate the photophysics of organic molecules at room temperature are also dominant at low temperature. However, by immobilizing a molecule in glasses, crystalline media, or polymer matrices, the radiative processes of the triplet state (phosphorescence) can not only compete with nonradiative decay channels but also with deactivation due to extrinsic agents such as impurities, residual oxygen, etc. Some information from low temperature that is relevant for room temperature photochemical experiments is listed below.

First, as discussed in the introduction to Section 2, triplet energies are routinely estimated from phosphorescence spectra. Triplet energies have several uses. One of the main uses is for assessing potential photosensitizers and quenchers. Triplet energies, whether from low temperature or not, are tabulated in Sections 1 and 2.

Second, when no triplet quantum yields, ϕ_T, measurements have been done, the phosphorescence quantum yield, ϕ_{ph}, gives a lower bound to the ϕ_T because

$$\phi_{ph} = \phi_T \, k_r^T \, \tau_{ph} \tag{4-1}$$

where k_r^T is the natural radiative rate constant of the triplet state and τ_{ph} is the measured phosphorescence lifetime given by

$$\frac{1}{\tau_{ph}} = k_r^T + k_{isc}^T + k_{pc}^T \tag{4-2}$$

corresponding to all decay channels out of the lowest triplet state. See Fig. 4-1 for a schematic definition of the rate constants in terms of the transitions underlying them.

Third, the nature of the triplet can often be guessed from the natural lifetime of the phosphorescence. For a rule-of-thumb n, π^* triplets have natural phosphorescence lifetimes $(1/k_r^T)$ that are 0.1 to 0.001 times less than those of π, π^* triplets. The natural phosphorescence lifetimes can be computed from Eq. (4-1) when the quantities, ϕ_T, ϕ_{ph}, and τ_{ph} are known. These latter three quantities are among the five parameters presented in Table 4.

For completeness, low-temperature fluorescence data is also presented in Table 4. Singlet states, which are not easy to study at room temperature because of temperature activated radiationless transitions, can often be investigated at low temperature. Even excited singlet states which fluoresce at room temperature can often be analyzed more conveniently at low temperature where the fluorescence lifetimes are usually longer and thus easier to measure.

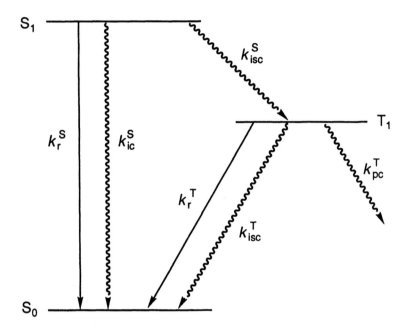

Fig. 4-1

Table 4

Photophysical Parameters of Organic Molecules at Low Temperature

No.	Solv	ϕ_T	Ref.	τ_{fl} (ns)	Ref.	ϕ_{fl}	Ref.	τ_{ph} (s)	Ref.	ϕ_{ph}	Ref.	E_T (cm^{-1})	Ref.
1	**Acenaphthene**												
	n											20872	64E037
	p	0.23	737055			0.57	79E793	2.6	746103	0.05	79E793	20700	67E112
2	**Acetanilide**												
	p							3.6	62E019	0.05	62E019		
3	**Acetone**												
	p							0.0006	496001	0.042	766343	27800	66E095
4	**Acetophenone**												
	n	0.90	83E630					0.0021	72E294	0.65	83E630	25933	81E561
	p					0	62E025	0.0050	81E561	0.6	83E630	26034	81E561
5	**Acetophenone, 4′-amino-**												
	p					0.11	78E891	0.72	78E891	0.14	78E891	23200	68E135
6	**Acetophenone, 4′-hydroxy-**												
	n											24700	68B014
	p							0.26	70E309	0.55	70E309	25300	66E094
7	**Acetophenone, 4′-methoxy-**												
	n							0.35	67F523			25100	67F523
	p							0.26	67F521	0.72	70E309	25000	67F521
8	**Acetophenone, 4′-methyl-**												
	n							0.027	72E294			25500	72E294
	p							0.084	67F521	0.61	67F521	25500	67F521

No.	Compound	State						
9	Acridan	p		12 73E359	0.38 73E359	3.4 73E359	0.51 73E359	24330 73E359
10	Acridine	n	0.15 757014			0.155 779025		15870 81E147
		p						15740 84E803
11	9-Acridinone	n						20370 78E542
		p		12.9 78E878	0.78 78E878	2.50 78E878	0.014 78E878	21050 78E878
12	9-Acridinone, 10-phenyl-	n						20430 78E542
		p		7.3 78E878	0.78 78E878	2.58 78E878	0.021 78E878	21220 78E542
13	Adenine	p	0.17 86D067		0.008 90E280	2.0 90E280	0.011 90E280	
14	Adenine, conjugate acid	p			0.010 90E280	3.35 66E102	0.038 90E280	
15	Adenine ion(1-)	p			0.003 90E280	2.8 66E102	0.004 90E280	
16	Adenosine	p			0.005 90E280	2.8 90E280	0.035 90E280	
17	Adenosine, conjugate acid	p			0.002 90E280		0.008 90E280	

Table 4—Low Temperature Data—Continued

No.	Solv	ϕ_T	Ref.	τ_{fl} (ns)	Ref.	ϕ_{fl}	Ref.	τ_{ph} (s)	Ref.	ϕ_{ph}	Ref.	E_T (cm^{-1})	Ref.
18	**Adenosine ion(1-)**												
	p					0.004	90E280			0.032	90E280		
19	**Adenosine 5′-monophosphate**												
	p	0.02	673066	2.8	737541	0.006	90E280	2.3	90E280	0.025	90E280	26700	673066
20	**Adenosine 5′-monophosphate, conjugate acid**												
	p					<0.001	66E102			0.005	90E280		
21	**Adenosine 5′-monophosphate ion(1-)**												
	p					0.008	90E280			0.028	90E280		
22	**Aniline**												
	n					0.17	80E253	4.2	80E253	0.65	80E253	24800	67E116
	p					0.21	84E090	4.0	84E090	0.67	84E090	26800	44E001
23	**Aniline, N-phenyl-**												
	p	0.87	706243	2.6	73E370	0.12	73E370	1.9	73E370	0.74	73E370	25140	73E359
24	**Anthracene**												
	n			6.5	65E038	0.34	79E793	−0.04	79E793	0.0003	79E793	14870	57B004
	p			5.7	65E038	0.32	79E793	0.12	79E793	0.0007	79E793	14900	779025
25	**Anthracene-d_{10}**												
	p	0.53	68E105										
26	**Anthracene, 9,10-diphenyl-**												
	n			7.90	766100	1.0	746270						
	p			7.95	766100							14290	68E131

No.	Compound		Col 1	Col 2	Col 3	Col 4	Col 5
27	**9,10-Anthraquinone**	n			0.0033 81E594		21800 64E021
		p			0.0033 81E594	0.29 87B054	21980 67E115
28	**Anthrone**	p			0.0015 706018	0.51 706018	25150 706018
29	**Benzaldehyde**	n		0 62E025	0.0015 67F523		25200 72E315
		p			0.0023 67F523	0.49 706018	24950 706018
30	**Benzaldehyde, 4-hydroxy-**	p			0.10 70E309	0.32 70E309	
31	**Benzaldehyde, 4-methoxy-**	n			0.094 67F523		25100 67F523
		p			0.10 72E293	0.32 72E293	24700 69E227
32	**Benz[a]anthracene**	n	0.67 68E104		0.50 78E761		16500 54E002
		p	60 65E038				16520 78E761
33	**Benzene**	n		0.17 746445	4.50 746445	0.15 746445	29500 67F523
		p		0.19 72E293	6.3 72E293	0.18 72E293	29500 67F523
34	**Benzene, 1,4-dicyano-**	n					24560 62E018

Table 4—**Low Temperature Data**—Continued

No.	Solv	ϕ_T	Ref.	τ_{fl} (ns)	Ref.	ϕ_{fl}	Ref.	τ_{ph} (s)	Ref.	ϕ_{ph}	Ref.	E_T (cm^{-1})	Ref.
34	**Benzene, 1,4-dicyano-—Continued**												
	p					0.42	86E917	2.1	70E318	0.25	86E917	24700	70E318
35	**Benzil**												
	n							0.0051	67E119			18700	67E119
	p							0.0056	67E119	0.67	62E019	19000	67E119
36	**Benzo[*a*]coronene**												
	n	0.64	68E104										
	p					0.39	69E216	5.65^a	68E104	0.07	69E216	18003	78E314
37	**Benzofuran**												
	n	0.37	89E178									25157	89E178
	p					0.63	89E178	2.35	89E178	0.24	89E178	25130	89E178
38	**Benzoic acid**												
	n							3.06	756077			27110	63E014
	p							2.1	706023	0.27	62E019	27200	767546
39	**Benzoic acid, 4-(dimethylamino)-, ethyl ester**												
	n			2.3	83E440	0.58	83E440	2.3	83E440	0.26	83E440	23750	83E440
40	**Benzoic acid, 4-methyl-**												
	n							2.41	776222	0.90	776222	26880	776222
	p							1.7	79E415			26800	767546
41	**Benzonitrile**												
	n											26780	62E018

#	Compound											
41	Benzonitrile—Continued	0.59	68E105									
		p			0.30	86E917	3.10	86E917	0.43	86E917	27000	767546
42	Benzonitrile, 4-amino-	3.3	756176									
		n			0.15	756176	2.1	83E440	0.64	83E440	24940	83E440
		p					2.45	756176	0.11	756176	24500	756176
43	Benzonitrile, 4-chloro-											
		n			0.004	85E829	0.15	79E415	0.10	85E829	25840	78D082
		p			0.003	85E829			0.11	85E829	26136	85E829
44	Benzonitrile, 4-methoxy-	7	756304									
		n			0.20	83E440	1.9	83E440	0.73	83E440	26320	83E440
		p			0.22	756304	1.6	756304	0.24	756304	26300	756304
45	Benzonitrile, 4-methyl-											
		n					3.40	84E864			26410	62E018
		p			0.36	84E864	3.80	84E864	0.33	84E864	26500	62E018
46	Benzo[ghi]perylene	240	696020									
		p			0.35	696020	0.438	779025			16180	566005
47	Benzo[c]phenanthrene	96	87E785									
		n			0.097	87E785	3.1	87E785	0.092	87E785	19840	54E002
		p					3.34	779025			20000	779025
48	Benzo[f][4,7]phenanthroline											
		n			0.11	81E545	4.95	81E545	0.43	81E545	23700	81E545

Table 4—Low Temperature Data—Continued

No.	Solv	ϕ_T	Ref.	τ_{fl} (ns)	Ref.	ϕ_{fl}	Ref.	τ_{ph} (s)	Ref.	ϕ_{ph}	Ref.	E_T (cm^{-1})	Ref.
48	**Benzo[f][4,7]phenanthroline—Continued**												
	p					0.15	81E545	6.3	81E545	0.50	81E545	23800	755398
49	**Benzo[a]phenazine**												
	n					0.063	81E433	0.110	81E433	0.02	81E433	17512	85E370
	p											16900	81E433
50	**Benzophenone**												
	n							~0.001[ab]	78E538			24000	67F523
	p					0	62E025	0.0060	67F523	0.84	67F523	24200	80E223
51	**Benzophenone, 4,4′-dimethyl-**												
	n							0.0056	67F523			24100	67F523
	p							0.0064	67F523	0.86	67F523	24300	80E223
52	**Benzo[g]pteridine-2,4-dione, 3,10-dimethyl-**												
	n			10.12	79E564	0.62	79E564	0.168	79E564	0.0048	79E564	17350	79E564
53	**Benzo[g]pteridine-2,4-dione, 10-methyl-**												
	n			10.4	79E564	0.70	79E564	0.134	79E564	0.0025	79E564	17250	79E564
54	**Benzo[b]thiophene**												
	n	0.98	89E178									23970	59E009
	p					0.035	85E272	0.35	85E272	0.42	89E178	24040	85E272
55	**Biacetyl**												
	p							0.00225	496001	0.23	63E009	19700	55E008

No.	Compound		val	ref	val	ref	val	ref	val	ref	val	ref
56	**1,1′-Binaphthyl**	p			0.70	776226	2.6	776226	0.11	776226		
57	**Biphenyl**	n					2.00	63E021			22871	64E037
		p			0.14	746103	4.6	72E293	0.24	72E293	22900	735067
58	**Biphenyl, 4,4′-dichloro-**	p					0.28	72E316	0.63	72E316	22000	72E316
59	**C_{70}**	n	15	73E359			53000[b]	91D034	0.0013[b]	91D034	12300	91D034
60	**Carbazole**	p	0.23	737055	0.44	73E359	6.8	78E761	0.24	73E359	24540	58E005
61	**Carbazole, *N*-methyl-**	n			0.50	81E648	7.0	81E648	0.10	81E648	24450	81E648
		p					8.0	766474			24450	68Z005
62	**Chlorophyll *a***	n	0.65	80F701	0.35	80F701	0.002[a]	82E129				
		p			0.54	79E838	0.0014	79E838	1×10^{-5}	79E838	10400	79E838
63	**Chlorophyll *b***	n					0.0028	79E838			10900	79E838
		p			0.14	79E838			3×10^{-5}	79E838	11400	68Z003

Table 4—Low Temperature Data—Continued

No.	Solv	ϕ_T	Ref.	τ_{fl} (ns)	Ref.	ϕ_{fl}	Ref.	τ_{ph} (s)	Ref.	ϕ_{ph}	Ref.	E_T (cm^{-1})	Ref.
64 Chrysene													
	n							2.20	63E021				
	p	0.70	757014	55	746103	0.23	72E293	2.54	779025	0.050	746103	20000	779025
65 Chrysene-d_{12}													
	p			56	746103	0.23	746103	13.4	746103	0.20	746103		
66 Coronene													
	n							8.45	63E021				
	p	0.68	69E216	320	69E216	0.27	69E216	9.5	779025	0.12	69E216	19040	566005
67 Coronene-d_{12}													
	p	0.67	69E216	355	69E216	0.28	69E216	35	69E216	0.40	69E216		
68 Coumarin													
	n					0.009	73E372	0.45	73E372	0.055	73E372	21600	775025
	p					0.002	90E280	0.66	90E280	0.005	90E280	21840	73E372
69 Cytidine													
	p					0.018	90E280			0.005	90E280		
70 Cytidine 5'-monophosphate													
	p	0.03	737541			0.04	90E280	0.4	90E280	0.015	90E280	27900	673066
71 Cytidine 5'-monophosphate, conjugate acid													
	p									0.005	90E280		
72 Cytidine 5'-monophosphate ion(1-)													
	p					0.005	90E280			0.018	90E280		

#	Compound		Col 1	Col 2	Col 3	Col 4	Col 5	Col 6
73	**Cytosine**	p			0.002 90E280	0.8 90E280	0.006 90E280	
74	**Cytosine ion(1-)**	p			0.10 66E102		0.03 66E102	
75	**Deoxythymidine 5′-monophosphate**	p	<0.003 82Z025	3.2 737541	0.16 737541	0.35 673067	~0 737541	26300 673066
76	**2′-Deoxyuridine**	p			0.0004 90E280		0.002 90E280	
77	**Dibenz[*a,h*]acridine**	n		7.0 706135				18800 706135
		p				0.84 706135 / 2.31 779025		19100 779025
78	**Dibenz[*a,j*]acridine**	n		15 706135				
		p				1.6 706135 / 1.04 779025		18600 779025
79	**Dibenz[*a,h*]anthracene**	p	0.98 68E105	45.5 746103	0.19 746103	1.60 779025	0.021 746103	18200 779025
80	**Dibenzofuran**	n			0.31 81E648	4.1 81E648	0.31 81E648	24000 81E648
		p	0.60 89E178		0.40 89E178	5.6 89E178		24515 58E005
81	**Dibenzothiophene**	n			0.027 81E648	1.3 81E648	0.97 81E648	23830 81E648

Table 4—**Low Temperature Data**—Continued

No.	Solv	ϕ_T	Ref.	τ_{fl} (ns)	Ref.	ϕ_{fl}	Ref.	τ_{ph} (s)	Ref.	ϕ_{ph}	Ref.	E_T (cm^{-1})	Ref.
81	**Dibenzothiophene**—Continued												
	p	0.97	89E178			0.025	89E178	1.5	766267	0.47	89E178	24100	766267
82	**Diphenylmethane**												
	n			38.0	71E394	0.18	71E394	7.8	71E394	0.41	71E394		
83	**Ethene, tetraphenyl-**												
	n					0.9	82E204					17500	82E204
	p	<0.01	82E204			0.95	82E204						
84	**Fluoranthene**												
	n							0.84	63E021			18450	57B004
	p			58	766401			0.990	779025			18500	779025
85	**Fluorene**												
	n					0.83	726112	5.10	63E021			23601	64E037
	p					0.8	726112	5.70	63E021	0.07	62E019	23700	779025
86	**9-Fluorenone**												
	p			20	706018	0.09	706018	0.0019	78E060	3×10^{-5}	78E060	17600	78E060
87	**Fluorescein dianion**												
	p	0.0078	80F701			0.97	80F701	0.3	77E801	0.0003	77E801	16500	69E235
88	**Fluorescein dianion, 2′,4′,5′,7′-tetrabromo-**												
	p	0.022	61E011					0.0093	61E011	0.022	61E011	14800	68Z004
89	**Guanine**												
	p					0.06	66E102	1.42	66E102	0.07	66E102		

No.	Compound		val	ref	val	ref	val	ref	val	ref	val	ref
90	Guanine ion(1-)	p			0.20	66E102	1.35	66E102	0.06	66E102		
91	Guanosine	p			0.006	90E280	1.2	90E280	0.042	90E280		
92	Guanosine, conjugate acid	p			0.50	90E280			≤0.03	90E280		
93	Guanosine ion(1-)	p			0.04	90E280	1.3	66E102	0.025	90E280		
94	Guanosine 5'-monophosphate	p	0.15	673066	0.028	90E280	1.2	90E280	0.095	90E280	27200	673066
95	Guanosine 5'-monophosphate, conjugate acid	p			0.040	90E280			≤0.02	90E280		
96	Guanosine 5'-monophosphate ion(1-)	p			0.04	90E280	1.25	66E102	0.06	90E280		
97	Indan	n					7.6	74E517			28750	74E517
		p			0.43	746251	8.0	746251	0.16	746251		
98	Indole	n	0.34	757014	0.6	69E244	2.4	84E843			25200	84E843
		p									24800	84E843

Table 4—Low Temperature Data—Continued

No.	Solv	ϕ_T	Ref.	τ_{fl} (ns)	Ref.	ϕ_{fl}	Ref.	τ_{ph} (s)	Ref.	ϕ_{ph}	Ref.	E_T (cm^{-1})	Ref.
99	**Isoorotic acid**												
	n			2.4	83E064	0.44	83E064			≤0.0005	83E064		
	p					0.04	90E280	~0.2	90E280	0.04	90E280		
100	**Isoquinoline**												
	n	0.19	717235					0.71	776085			21210	59E009
	p	0.24	717235	9.0	766314	0.61	766314	1.0	766314	0.038	766314	21200	59E008
101	**Naphthalene**												
	n			262	84F060	0.45	746103	2.30	746103	0.033	746103	21180	57B004
	p	0.29	707232	273	68E129	0.45	766054	2.6	79E793	0.039	766054	21300	44E001
102	**Naphthalene-d_8**												
	n			304	706053	0.50	746103	18.5	746103	0.16	746103		
	p	0.25	69E234	250	706053	0.41	766054	20.4	766054	0.34	766054		
103	**Naphthalene, 1-amino-**												
	n			10.9	70E321								
	p			15.1	70E321			1.5	496001			19150	766421
104	**Naphthalene, 2-amino-**												
	n			20.6	70E321								
	p			20.5	70E321	0.080	82E663	1.2	81E644	0.0024	82E663	19960	766421
105	**Naphthalene, 1-chloro-**												
	n											20490	78E067
	p			12.9	68E129	0.03	62E025	0.305	83E417	0.16	62E025	20700	44E001

No.	Compound	Type											
106	**Naphthalene, 1,4-dichloro-**	p					0.04	766054	0.11	766054	0.25	766054	
107	**Orotic acid**	n					0.0015	83E064	0.110	83E064	0.017	83E064	
		p					≤0.008	90E280	0.3	90E280				
108	**Phenanthrene**	n											21774	63E024
		p	0.46	737055	63	746103	0.20	79E793	3.6	78E761	0.16	79E793	21500	89E090
109	**Phenanthrene-d_{10}**	p	0.45	69E234	65	746103	0.20	79E793	15.4	79E793	0.66	79E793	
110	**6-Phenanthridone**	p					0.31	87E893	4.9	87E893	0.51	80E792	23920	80E792
111	**Phenol**	n					0.44	82E060			0	82E060	
		p					0.45	84E090	2.5	84E090	0.40	84E090	28563	84E090
112	**Phenothiazine**	n											
		p							0.049	80E109	0.34	80E109	21100	757279
113	**Phenoxazine**	n					0.064	72E303	2.60	72E303	0.29	72E303	21880	72E303
		p					0.11	72E303	2.76	72E303	0.41	72E303	21750	72E303

Table 4—**Low Temperature Data**—Continued

No.	Solv	ϕ_T	Ref.	τ_{fl} (ns)	Ref.	ϕ_{fl}	Ref.	τ_{ph} (s)	Ref.	ϕ_{ph}	Ref.	E_T (cm^{-1})	Ref.
114	Phenoxazine, 10-phenyl-												
	n					0.082	72E303	2.31	72E303	0.94	72E303	22150	72E303
	p					0.12	72E303	2.65	72E303	0.72	72E303	22070	72E303
115	Phenylacetic acid												
	p					0.17	746251	5.0	746251	0.35	746251	29300	706023
116	Phenyl ether												
	p					0.14	84E090	1.0	84E090	0.67	84E090	28320	84E090
117	Pheophytin *a*												
	n	0.79	80F701			0.21	80F701	0.0010	79E838	3×10^{-5}	79E838	10800	79E838
118	Pheophytin *b*												
	n					0.20	79E838			5×10^{-5}	79E838	11200	79E838
119	Phthalazine												
	n											22100	82E585
	p	0.49	72E323			0	726177	0.42	726177			23000	59E008
120	Phthalocyanine, zinc(II)												
	n					0.3^b	71E386	0.0011^b	71E386	0.0001^b	71E386	9150	71E386
121	Picene												
	p	0.36	68E105					2.7	779025			20100	779025
122	Porphine												
	n					0.055^b	753056	0.011^b	753056	0.0001^b	753056	12600	753056
	p							0.009	78E504	5×10^{-5}	78E504	12700	743135

No.	Compound										
123	**Porphine, magnesium(II)**										
		n	0.07	753056	0.10	753056	0.0006	753056	13700	753056
		p	0.07	753056	0.10	753056	0.0006	753056	13700	753056
124	**Porphine, octaethyl-**										
		n	0.16	753056	0.016	753056	0.0006	753056	13000	753056
		p							13000	743135
125	**Porphine, octaethyl-, zinc(II)**										
		p			0.057	71E357	0.065	71E357	14200	71E357
126	**Porphine, 2,7,12,17-tetraethyl-3,8,13,18-tetramethyl-**										
		n	0.17[b]	753056	0.015[b]	753056	0.0006[b]	753056	13000	753056
		p	0.17	753056	0.020	753056	0.0003	753056	13000	753056
127	**Porphine, 2,7,12,17-tetraethyl-3,8,13,18-tetramethyl-, magnesium(II)**										
		n	0.25	753056	0.085	753056	0.006	753056	13600	753056
		p	0.25	753056			0.0051	753056		
128	**Porphine, 2,7,12,17-tetraethyl-3,8,13,18-tetramethyl-, zinc(II)**										
		p			0.057	71E357	0.070	71E357	14200	71E357
129	**Porphine, tetraphenyl-**										
		n			0.006	80E540	4×10^{-5}	80E540	11500	82Z053
		p			0.0066	743135	2×10^{-5}	743135	11700	743135
130	**Porphine, tetraphenyl-, magnesium(II)**										
		n			0.045	81E271	0.015	81E271	11900	82Z053

Table 4—**Low Temperature Data**—Continued

No.	Solv	ϕ_T	Ref.	τ_{fl} (ns)	Ref.	ϕ_{fl}	Ref.	τ_{ph} (s)	Ref.	ϕ_{ph}	Ref.	E_T (cm^{-1})	Ref.
130	Porphine, tetraphenyl-, magnesium(II)—Continued												
	p							0.046	71E357	0.015	71E357	12000	71E357
131	Porphine, tetraphenyl-, zinc(II)												
	n							0.026	81E271	0.012	81E271	12800	756229
	p							0.025	71E357	0.014	71E357	12800	71E357
132	Porphine, zinc(II)												
	p							0.0425	78E504	0.01	71E357	13900	71E357
133	Porphine-2,18-dipropanoic acid, 7,12-diethenyl-3,8,13,17-tetramethyl-, dimethyl ester												
	p							0.0046	743135	5×10^{-5}	743135	12600	743135
134	Propiophenone												
	n							0.0063	746085			26080	86E058
	p							0.0038	706018	0.70	706018	26150	706018
135	Psoralen												
	p					0.019	73E372	0.66	73E372	0.13	73E372	21930	73E372
136	Psoralen, 5-methoxy-												
	p					0.019	73E372	1.42	83E324	0.22	73E372	21200	83E324
137	Psoralen, 8-methoxy-												
	p					0.013	73E372	0.72	73E372	0.17	73E372	21900	73E372
138	Purine												
	n	0.51	757014									26434	755396
	p	0.35	82E207					3.6[a]	80B077				

No.	Compound													
139	**Pyrazine**	n					0.0006	67E117	0.0185	78E312	0.30	67E117	26361	78E312
		p							0.020	496001			25991	58E006
140	**Pyrene**	n	0.22	696019	515	696019	0.88	746103	0.55	746103	0.0021	746103	16930	57B004
		p		696019			0.92	746103	0.58	746103	0.0022	746103	16850	78E761
141	**Pyrene-d_{10}**	n	0.15	696019	460	746103	0.80	746103	3.35	746103	0.009	746103		
		p					0.87	79E793	3.95	79E793	0.012	79E793		
142	**Pyridazine**	n	0.66	88E230			0.01	67E117			$<10^{-5}$	67E117	24850	67E117
		p	0.24	72E323									24251	67B017
143	**Pyridine**	n					0.005	84E679	0.7	84E679	0.004	84E679	27770	84E679
		p					0.025	84E679	2.4	84E679	0.015	84E679		
144	**Pyridine, 2-amino-**	p							1.9	73E338	0.068	73E338		
145	**Pyridine, 4,4'-(1,2-ethenediyl)bis-, (E)-**	n					0.03	89E090	0.012	89E090	$\leq$0.0001	89E090	17700	89E090
		p											17700	89E090

Table 4—**Low Temperature Data**—Continued

No.	Solv	ϕ_T	Ref.	τ_{fl} (ns)	Ref.	ϕ_{fl}	Ref.	τ_{ph} (s)	Ref.	ϕ_{ph}	Ref.	E_T (cm^{-1})	Ref.
146	**Pyridine, 2-hydroxy-**												
	p							0.10	77E581	0.009	77E581		
147	**Pyridinium, 2-amino-**												
	p					0.07	66E102			0.33	66E102		
148	**Pyrimidine**												
	n					0.0058	67E117			0.14	67E117	28214	70E310
149	**Pyrimidine, 2-amino-**												
	p					0.15	66E102			0.22	66E102		
150	**4-Pyrimidone**												
	p					0.02	66E102			0.005	66E102		
151	**4-Pyrimidone, conjugate monoacid**												
	p					0.01	66E102			0.01	66E102		
152	**4-Pyrimidone ion(1-)**												
	p					0.03	66E102			0.008	66E102		
153	**Pyronine cation**												
	p							1.6	77E801	4×10^{-5}	77E801	14900	77E801
154	**Quinazoline**												
	n							0.56	78E894			21925	78E894
	p							0.68	78E894			21900	59E008
155	**Quinoline**												
	n	0.43	717235			$<10^{-5}$	776085	1.04	776085			21590	80E627

No.	Compound							
155	**Quinoline—Continued**							
	p	0.25 737055		0.026 83E417	1.30 80E627	0.09 83E417	21850 83E417
156	**Quinoline, 4-chloro-**							
	n				0.30 80E627		21450 80E627
	p			0.014 83E417	0.40 80E627	0.20 83E417	21300 83E417
157	**Quinoxaline**							
	n	0.27 717235			0.25 68E116		21325 82E355
	p	0.18 707232			0.27 72E293	0.42 72E293	21250 59E008
158	**Reserpic acid**							
	p			0.95 83E489	3.7 83E489	0.14 83E489	
159	**Reserpine**							
	p			0.33 83E489	3.2 83E489	0.05 83E489	
160	**(all-E)-Retinoic acid**							
	n		1.7 80E137	0.44 80E137			
	p		7.5 80E137	0.48 80E137			
161	**Rhodamine, inner salt**							
	p			1.0 65E042	2.2 77E801	0.00006 77E801	15930 77E801
162	**Rhodamine, inner salt, N,N'-diethyl**							
	p	0.0015 83Z077			2.2 77E801	6×10^{-5} 77E801	15350 77E801
163	**Rhodamine B, inner salt**							
	p	0.003 83Z077		1.0 65E042	1.6 77E801	5×10^{-5} 77E801	14890 77E801

Table 4—Low Temperature Data—Continued

No.	Solv	ϕ_T	Ref.	τ_{fl} (ns)	Ref.	ϕ_{fl}	Ref.	τ_{ph} (s)	Ref.	ϕ_{ph}	Ref.	E_T (cm^{-1})	Ref.
164	**Rhodamine 6G cation**												
	p							1.7	755120	8×10^{-5}	77E801	15100	77E801
165	**Riboflavine**												
	p	0.7	776328			0.35	776328			0.71	776328	17500	69E235
166	**(E)-Stilbene**												
	n							0.1^x	85E575			17200	80E113
	p					0.90	79E640	0.022^a	707199			17200	80E113
167	**(E)-Stilbene, 4-iodo-**												
	n							0.0001	89E090			16700	89E090
	p					0.02	89E090	0.00008	89E090	<0.0001	89E090	16900	80E113
168	**(E)-Stilbene, 4-nitro-**												
	n					0.004	89E090	0.009	89E090	0.0003	89E090	16300	89E090
	p					0.003	89E090	0.010	89E090	0.0003	89E090	17400	44E001
169	**Styrene**												
	n			18.8	79E265	0.46	79E265			<0.001	79E265	21600	776060
170	**p-Terphenyl**												
	p			1.3	65E038			2.6	64E036			20400	67E112
171	**Thiobenzophenone**												
	n							2.5×10^{-5}	756061	0.021	756061	13760	756061
	p							4.4×10^{-5}	726174			14200	726174

No.	Compound						
172	Thiofluorenone	n			1.5×10^{-5} 756061	0.035 756061	13300 756061
173	Thioxanthene	n			0.025 77E638	0.46 77E638	
174	Thioxanthen-9-one	n			8.8×10^{-6} 756199	0.03 756199	22100 86E676
175	Thymidine	n	1.0 80E025	0.003 80E025		0.040 80E025	
		p		0.033 90E280	≤0.5 90E280	0.006 90E280	
176	Thymidine 5'-monophosphate	p		0.14 90E280	≤0.4 90E280	≤0.01 90E280	
177	Thymidine 5'-monophosphate, conjugate acid	p		0.11 90E280		≤0.02 90E280	
178	Thymidine 5'-monophosphate ion(1-)	p		0.10 90E280	≤0.02 90E280	≤0.02 90E280	
179	Thymine	n	1.0 80E025	0.005 80E025	0.075 80E025	0.016 80E025	
		p	<0.5 83E064	0.012 90E280	≤0.6 90E280	0.003 90E280	
180	Thymine, 1,3-dimethyl-	n		0.032 80E025			
		p		0.006 90E280		0.001 90E280	

Table 4—**Low Temperature Data**—Continued

No.	Solv	ϕ_T	Ref.	τ_{fl} (ns)	Ref.	ϕ_{fl}	Ref.	τ_{ph} (s)	Ref.	ϕ_{ph}	Ref.	E_T (cm^{-1})	Ref.
181	**Thymine, 1-methyl-**												
	p					0.009	90E280			0.002	90E280		
182	**Thymine, 1-methyl-, ion(1-)**												
	p					0.02	66E102			0.002	66E102		
183	**Toluene**												
	n			53.3	71E394	0.29	82E344	7.73	82E344	0.26	82E344	28920	61E014
	p					0.27	746251	7.7	746251	0.26	746251	29000	67F523
184	**Triphenylene**												
	p	0.54	69E234			0.07	72E293	15.2	779025	0.41	72E293	23400	84E612
185	**Triphenylene-d_{12}**												
	p	0.88	69E234			0.08	72E293	21.4	72E293	0.53	72E293		
186	**Triphenylethylene**												
	n					0.9	82E204						
	p	<0.01	82E204			0.9	82E204					17500	82E204
187	**Triphenylmethane**												
	n			35.5	71E394	0.15	71E394	7.2	71E394	0.53	71E394		
	p							5.1	677472	0.011	677472		
188	**Uracil**												
	n					≤0.0005	83E064			0.0014	80E025		
	p					0.0002	90E280			0.001	90E280		

No.	Compound	Form						
189	Uracil, 1,3-dimethyl-	n	0.0001	80E025				
		p	0.0003	90E280			0.002	90E280
190	Uracil, 1,3-dimethyl-5-styryl-, (E)-	n	0.43	87E138	0.41	87E138		
		p	0.67	87E138	0.38	87E138		
191	Uracil, 1-methyl-	p	0.0002	90E280			0.001	90E280
192	Uracil, 1-methyl-, ion(1-)	p	0.01	66E102			0.002	66E102
193	Uracil, 3-methyl-	p	0.0002	90E280			0.001	90E280
194	Uracil, 6-methyl-	p	0.0002	90E280			0.001	90E280
195	Uracil ion(1-)	p	0.0013	66E102			0.0006	66E102
196	Uridine	p	0.0005	90E280	≤0.5	90E280	0.002	90E280
197	Uridine ion(1-)	p	0.002	66E102			0.001	66E102

Table 4—**Low Temperature Data**—Continued

No.	Solv	ϕ_T	Ref.	τ_{fl} (ns)	Ref.	ϕ_{fl}	Ref.	τ_{ph} (s)	Ref.	ϕ_{ph}	Ref.	E_T (cm^{-1})	Ref.
198	**Uridine 5′-monophosphate**												
	p	<0.003	82Z025			≤0.002	90E280	≤0.6	90E280	0.006	90E280	27400	737541
199	**Uridine 5′-monophosphate, conjugate acid**												
	p					≤0.002	90E280			<0.006	90E280		
200	**Uridine 5′-monophosphate ion(1-)**												
	p					≤0.002	90E280			0.006	90E280		
201	**9-Xanthione**												
	n							4.3×10^{-5}	74E514	0.11	74E514	15143	81F218
	p							4.8×10^{-5}	74E514	0.15	74E514		
202	**Xanthone**												
	n											25906	717449
	p							0.020	706018	0.44	706018	25905	717449

No.	Compound					
203	**m-Xylene**					
	n		0.28 72E313		0.25 72E313	28325 61E014
	p			8.1 51E003		28120 60E016
204	**o-Xylene**					
	n	62.3 82E344	0.33 82E344	7.46 82E344	0.45 82E344	28705 61E014
	p			5.2 51E002		28760 60E016
205	**p-Xylene**					
	n		0.45 72E313	4.60 72E313	0.49 72E313	28135 61E014
	p			6.2 51E002		28145 60E016
206	**Zinc(II) chlorophyll a**					
	p		0.23 79E838	0.0010 79E838	4×10^{-5} 79E838	11000 79E838
207	**Zinc(II) chlorophyll b**					
	p		0.08 79E838	0.0026 79E838	0.0002 79E838	11700 79E838

[a] Decay of triplet-triplet absorption; [b] Aromatic solvent, benzene-like; [x] Crystalline medium

Section 5

Ground-State Absorption Spectra

In this section, absorption spectra of 35 common sensitizers and/or quenchers are reproduced from the literature. The compounds correspond roughly to those in Section 1 of the *First Edition*, where molar absorption coefficients were given in Table 1-1 at three important mercury emission wavelengths. This change in format in the *Second Edition* was made in order to supply comparable information to photochemists who now have an arsenal of lasers available with a variety of excitation wavelengths.

Molar absorption coefficients (decadic extinction coefficients) are given by

$$\varepsilon(\lambda_{ex}) = \frac{A(\lambda_{ex})}{c\,l} \tag{5-1}$$

where c is the concentration of the absorbing species in moles per liter, l is the optical pathlength in cm, λ_{ex} is the excitation wavelength, and A is the absorbance given by

$$A(\lambda_{ex}) = \log_{10} \frac{I_0(\lambda_{ex})}{I(\lambda_{ex})} \; . \tag{5-2}$$

I_0 is a measure of the intensity of the light probe before entering the sample, and I is the intensity of the light beam transmitted through the sample of pathlength l. The units of ε are liters / (mole cm). The units of intensity are usually that of a flux, namely photons / (cm^2 s). Note that the logarithm in Eq. (5-2) is base 10.

The absorption spectra are presented on semilog plots to accommodate the wide dynamic range of molar absorption coefficients. The wavelength scales are in common, with the exception of azulene. Wavelengths λ are in nanometers. The ground-state absorption spectra of the compounds in both polar and nonpolar solvents are presented on the same plot when both are available.

Acetone

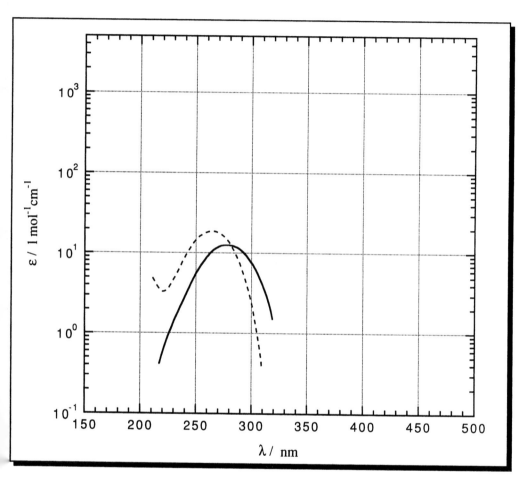

Figure 5-1   Absorption spectrum of acetone in water (---) and isooctane.
Adapted from 82B140, with permission.

Acetophenone

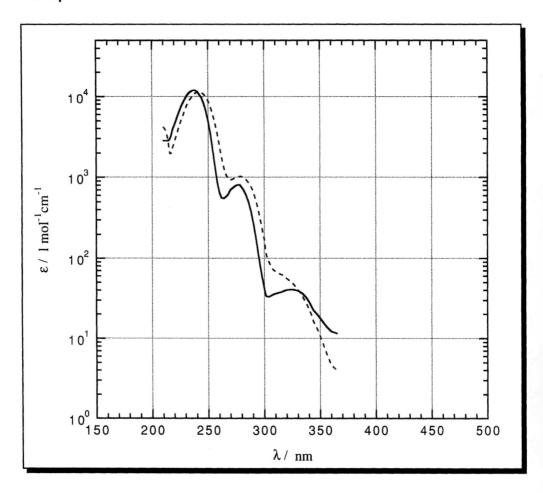

Fig. 5-2   Absorption spectrum of acetophenone in ethanol (---) and hexane.
Adapted from 61Z003, with permission.

Anthracene

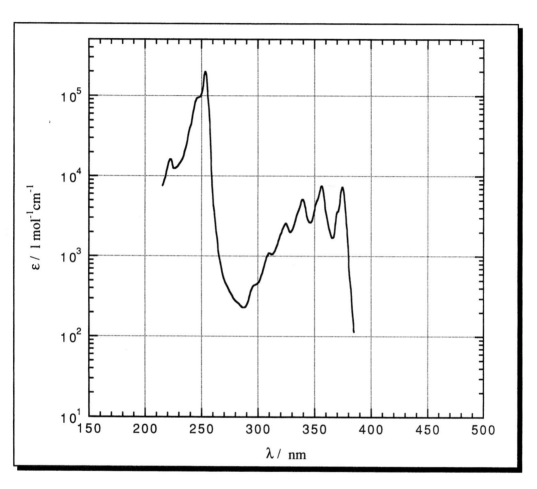

Fig. 5-3   Absorption spectrum of anthracene in cyclohexane.
Adapted from 51Z002, with permission.

Anthracene, 9-cyano-

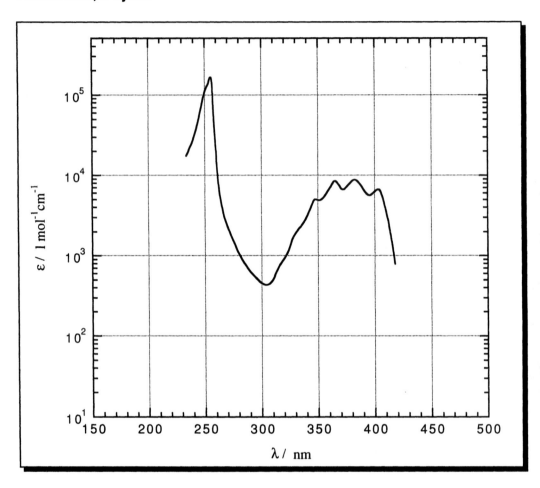

Fig. 5-4   Absorption spectrum of 9-cyanoanthracene in ethanol.
Adapted from 51Z002, with permission.

Anthracene, 9,10-dicyano-

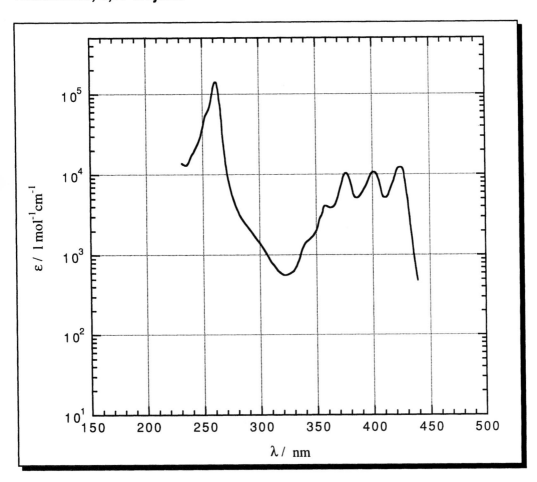

Fig. 5-5   Absorption spectrum of 9,10-dicyanoanthracene in 2-methyltetrahydrofuran.
Adapted from 71Z014, with permission.

Anthracene, 9-10-diphenyl-

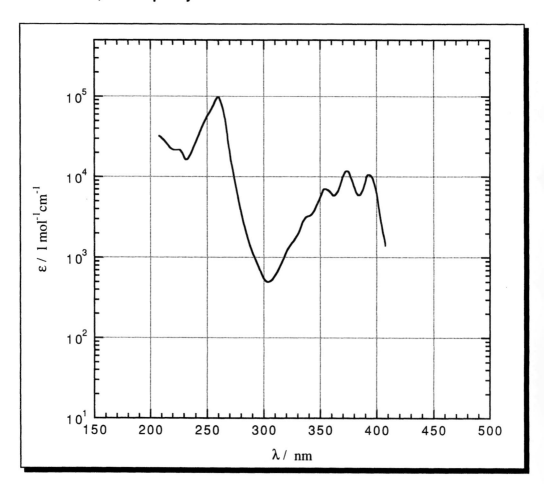

Fig. 5-6   Absorption spectrum of 9,10-diphenylanthracene in ethanol.
Adapted from 71Z014, with permission.

Azulene

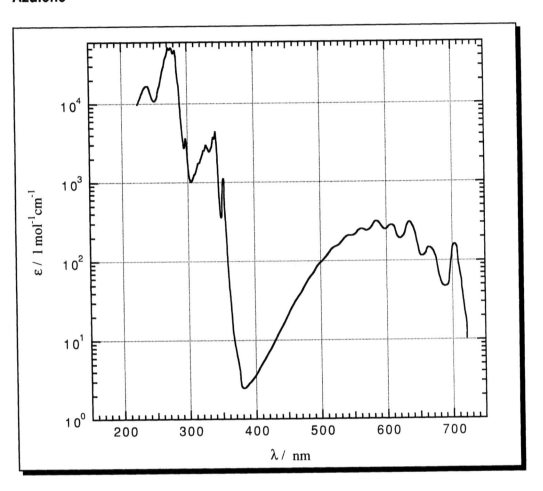

Fig. 5-7   Absorption spectrum of azulene in heptane.
Adapted from 71Z014, with permission.

Benzaldehyde

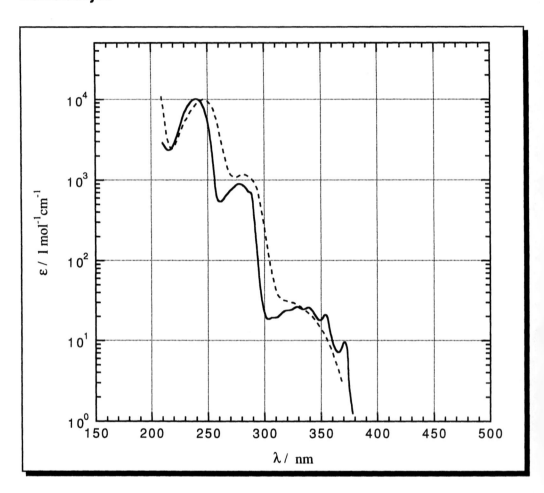

Fig. 5-8   Absorption spectrum of benzaldehyde in ethanol (---) and hexane.
Adapted from 61Z003, with permission.

Benzene

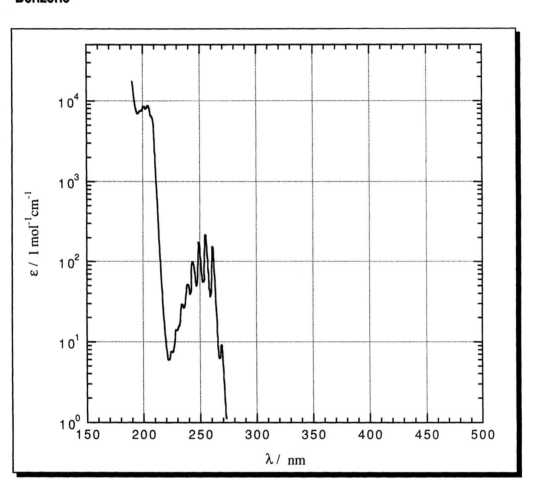

Fig. 5-9   Absorption spectrum of benzene in hexane.
Adapted from 71Z014, with permission.

Benzene, 1,2,4,5-tetramethyl-

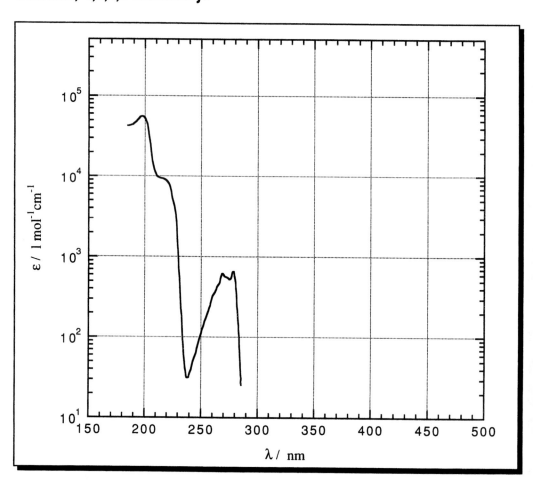

Fig. 5-10   Absorption spectrum of durene in heptane.
Adapted from 71Z014, with permission.

Benzene, 1,3,5-trimethyl-

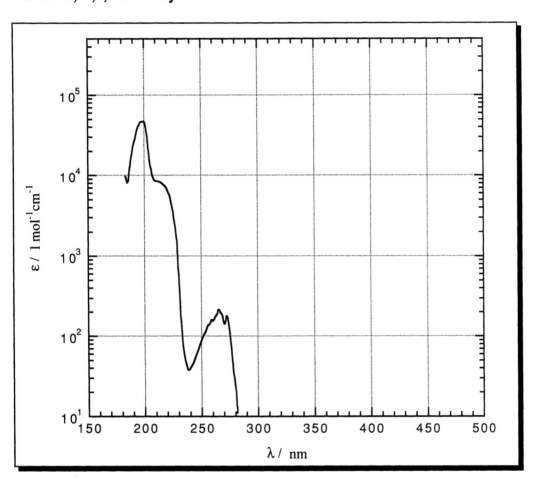

Fig. 5-11   Absorption spectrum of mesitylene in ethanol.
Adapted from 71Z014, with permission.

Benzil

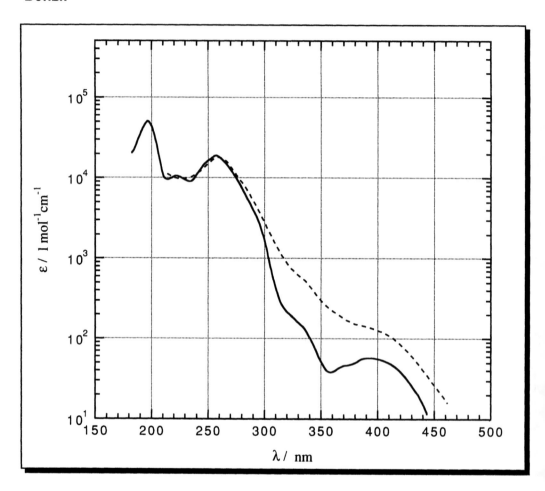

Fig. 5-12   Absorption spectrum of benzil in ethanol (---) and heptane.
Adapted from 61Z003 and 71Z014 respectively, with permission.

Benzophenone

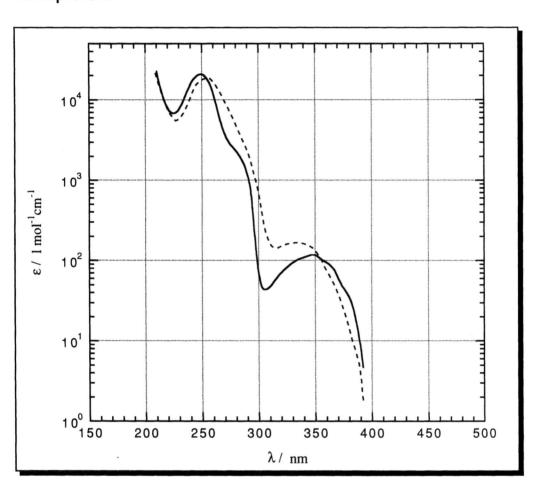

Fig. 5-13   Absorption spectrum of benzophenone in ethanol (---) and hexane.
Adapted from 61Z003, with permission.

Benzophenone, 4-phenyl-

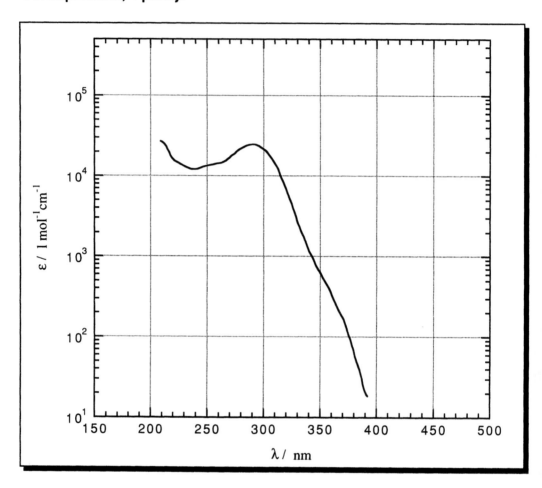

Fig. 5-14   Absorption spectrum of 4-benzoylbiphenyl in methanol.
            Adapted from 61Z003, with permission.

Biacetyl

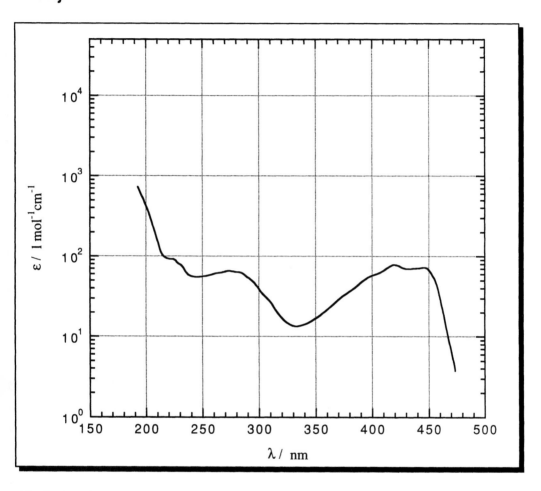

Fig. 5-15   Absorption spectrum of biacetyl in hexane.
Adapted from 71Z014, with permission.

Biphenyl

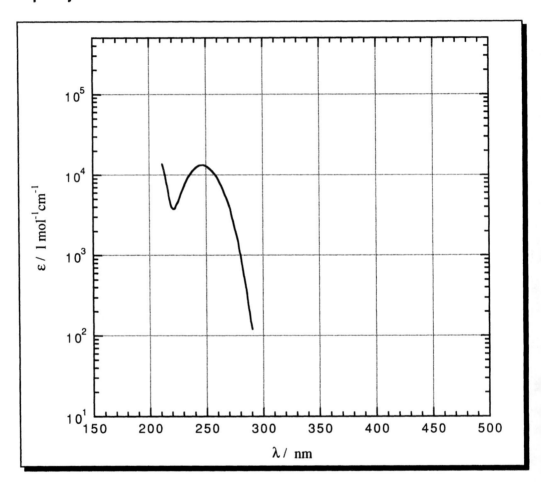

Fig. 5-16   Absorption spectrum of biphenyl in ethanol.
Adapted from 61Z003, with permission.

Fluorene

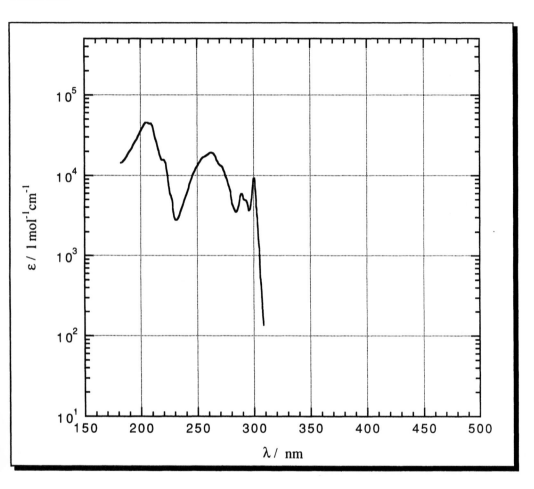

Fig. 5-17   Absorption spectrum of fluorene in heptane.
Adapted from 71Z014, with permission.

9-Fluorenone

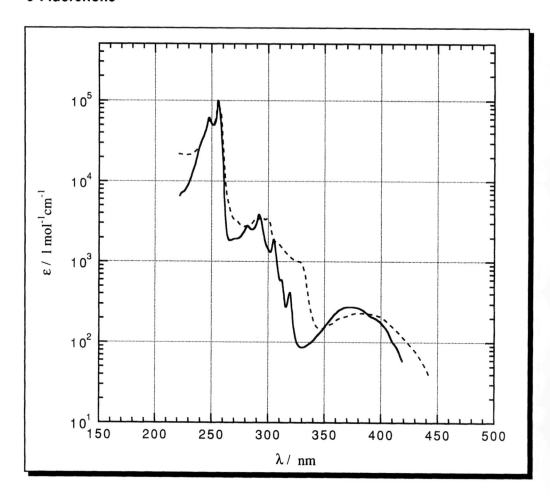

Fig. 5-18   Absorption spectrum of 9-fluorenone in ethanol (---) and cyclohexane.
Adapted from 71Z014 and 51Z002 respectively, with permission.

1,5-Hexadiene, 2,5-dimethyl-

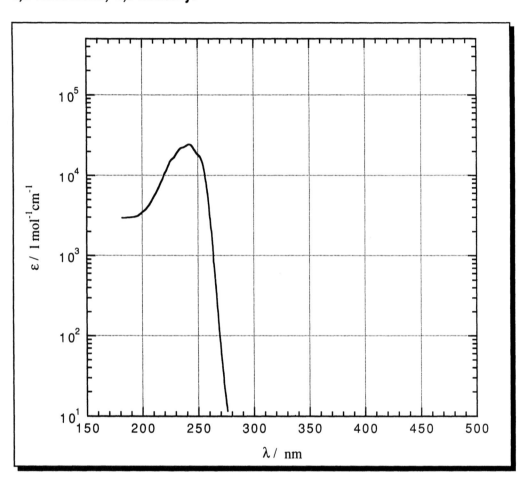

Fig. 5-19   Absorption spectrum of 2,5-dimethyl-1,5-hexadiene in heptane.
Adapted from 71Z014, with permission.

Naphthalene

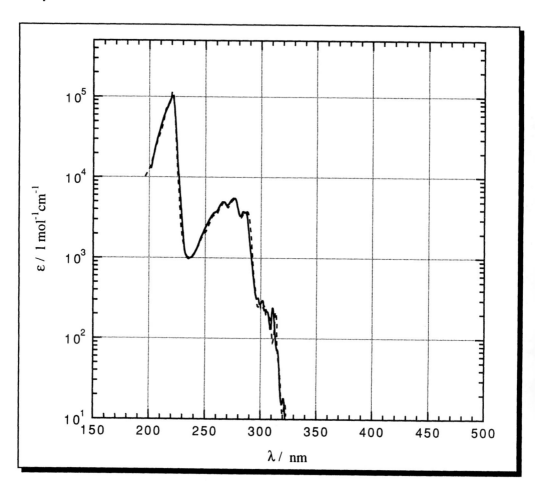

Fig. 5-20   Absorption spectrum of naphthalene in ethanol (---) and hexane.
Adapted from 61Z003 and 71Z014 respectively, with permission.

Naphthalene, 2-acetyl-

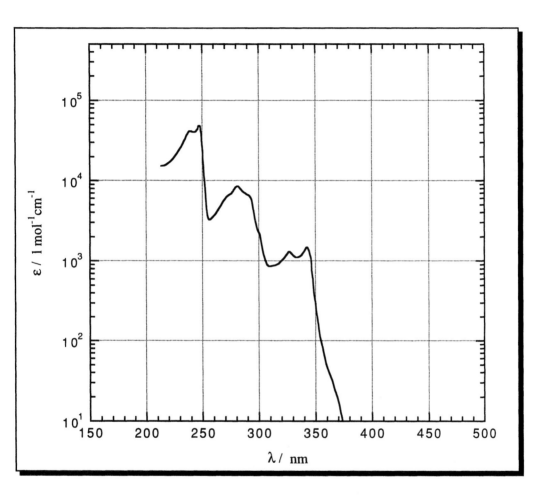

Fig. 5-21   Absorption spectrum of 2'-acetonaphthone in cyclohexane.
Adapted from 61Z003, with permission.

Naphthalene, 1-benzoyl-

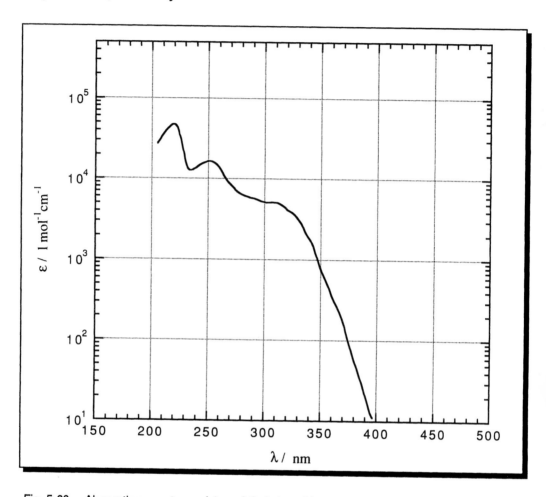

Fig. 5-22   Absorption spectrum of 1-naphthyl phenyl ketone in methanol.
Adapted from 61Z003, with permission.

Naphthalene, 2-benzoyl-

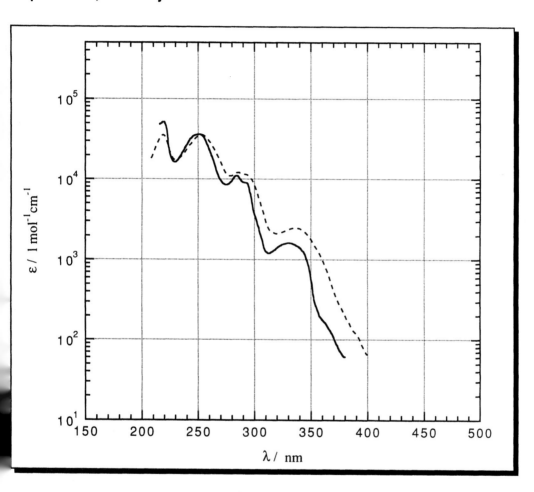

Fig. 5-23 Absorption spectrum of 2-naphthyl phenyl ketone in ethanol (---) and cyclohexane. Adapted from 51Z002, with permission.

Naphthalene, 1-cyano-

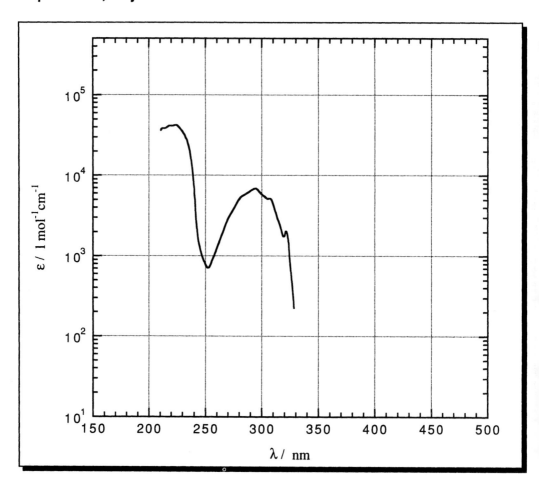

Fig. 5-24   Absorption spectrum of 1-cyanonaphthalene in ethanol.
Adapted from 71Z014, with permission.

Naphthalene, 1-methyl-

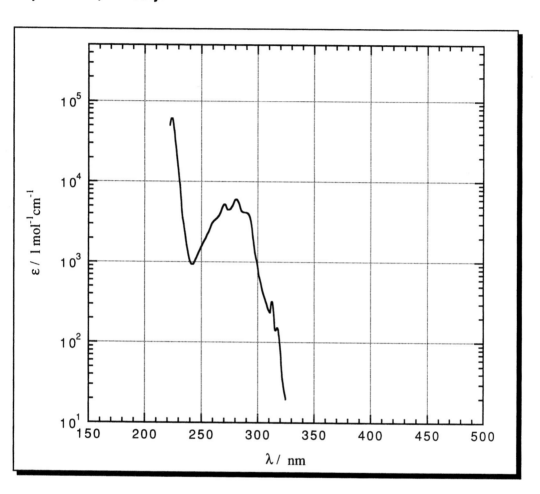

Fig. 5-25   Absorption spectrum of 1-methylnaphthalene in ethanol.
Adapted from 71Z014, with permission.

Phenanthrene

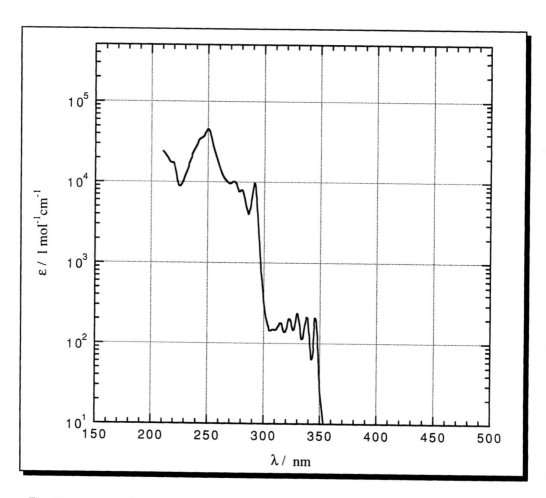

Fig. 5-26 Absorption spectrum of phenanthrene in isooctane.
Adapted from 61Z003, with permission.

Piperylene

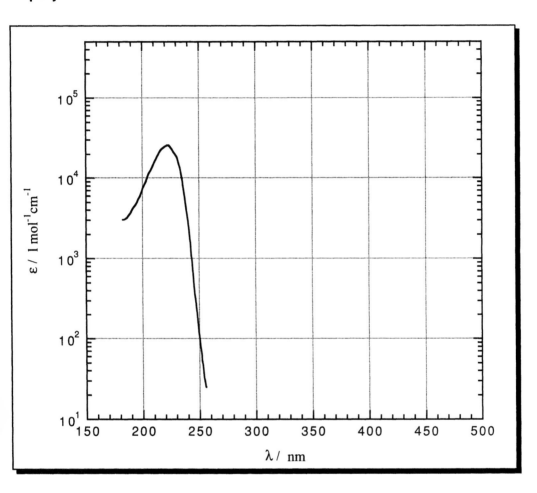

Fig. 5-27   Absorption spectrum of 1,3-pentadiene in heptane.
Adapted from 71Z014, with permission.

Pyrene

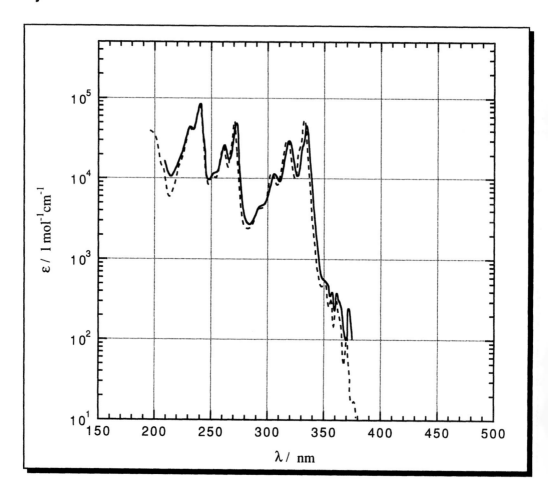

Fig. 5-28   Absorption spectrum of pyrene in ethanol (---) and light petroleum.
Adapted from 51Z002 and 71Z014 respectively, with permission.

p-Quaterphenyl

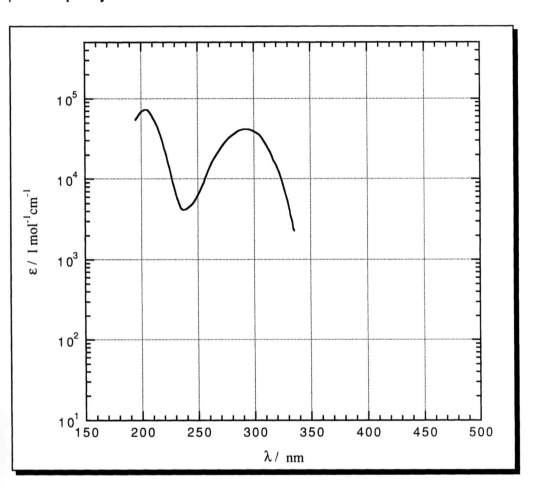

Fig. 5-29   Absorption spectrum of *p*-quaterphenyl in hexane.
Adapted from 71Z014, with permission.

(E)-**Stilbene**

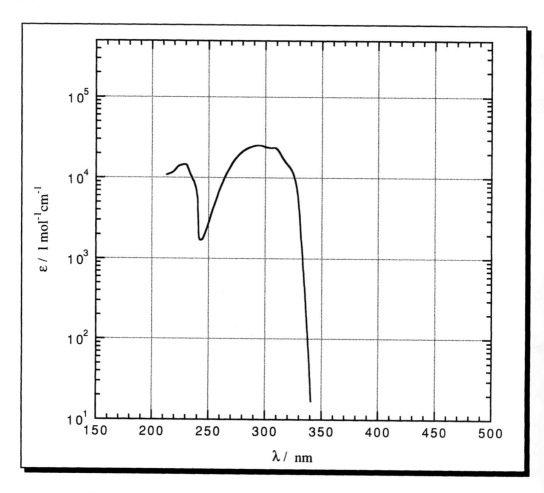

Fig. 5-30   Absorption spectrum of *(E)*-stilbene in cyclohexane.
                Adapted from 61Z003, with permission.

(Z)-Stilbene

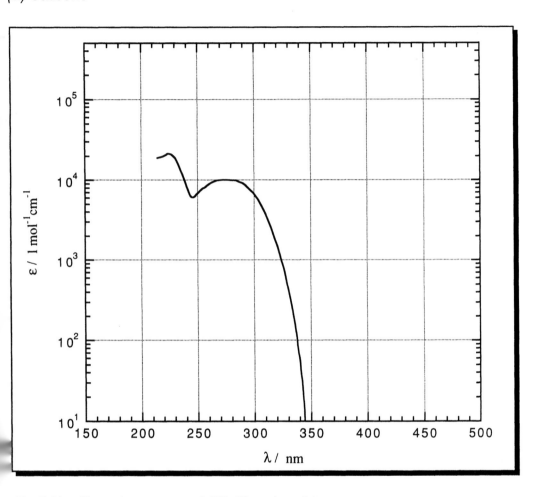

Fig. 5-31   Absorption spectrum of *(Z)*-stilbene in cyclohexane.
Adapted from 61Z003, with permission.

p-Terphenyl

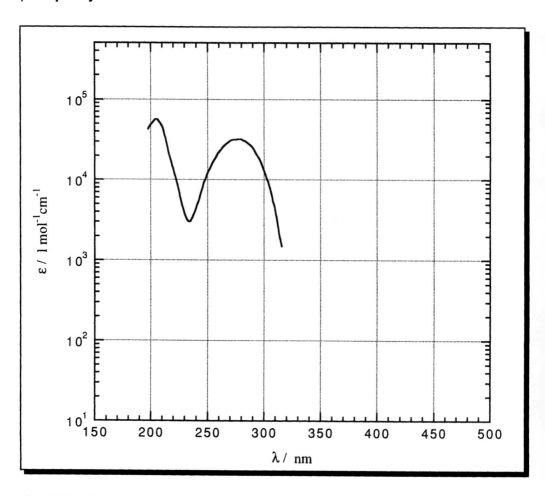

Fig. 5-32   Absorption spectrum of p-terphenyl in ethanol.
             Adapted from 71Z014, with permission.

Toluene

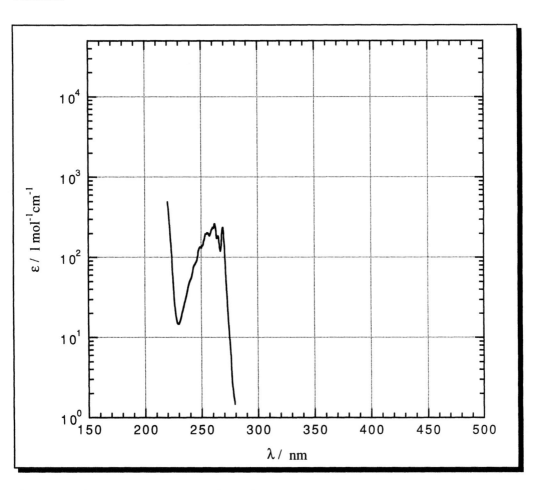

Fig. 5-33   Absorption spectrum of toluene in cyclohexane.
Adapted from 51Z002, with permission.

Triphenylene

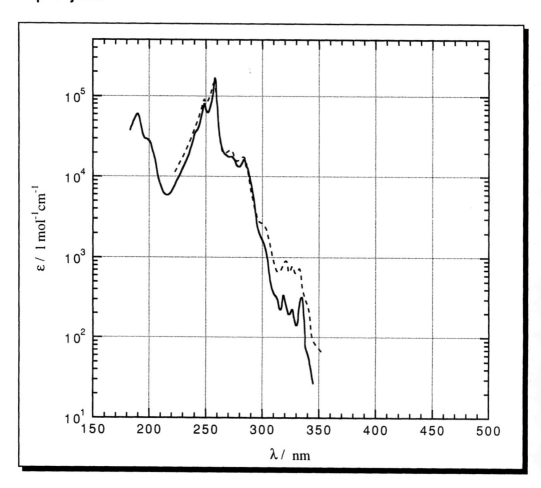

Fig. 5-34   Absorption spectrum of triphenylene in ethanol (---) and heptane.
Adapted from 51Z002 and 71Z014 respectively, with permission.

Xanthone

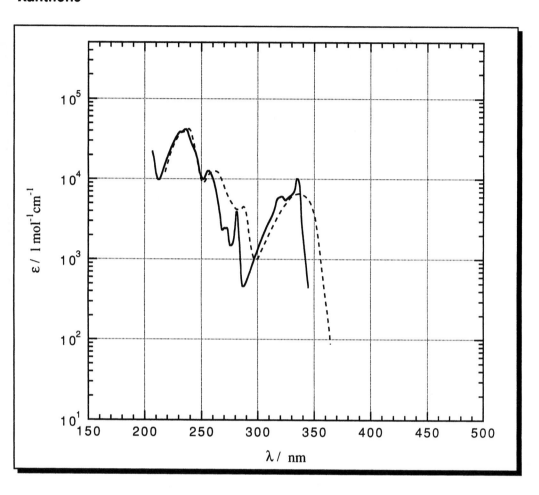

Fig. 5-35   Absorption spectrum of xanthone in ethanol (---) and cyclohexane.
Adapted from 51Z002 and 71Z014 respectively, with permission.

Section 6

ESR and ODMR Parameters of the Triplet State

For many organic molecules the g-factor of the excited triplet state is almost isotropic and close to that of the free electron. However a large splitting of the first-order ($\Delta M_S = \pm 1$) esr spectrum is common. This doublet, due to dipolar interactions between the unpaired electron spins, is strongly anisotropic leading to very broad esr transitions in nonoriented samples. The splitting is present even in the absence of an applied external magnetic field.

The traceless (spin) dipole-dipole interaction tensor is often described by two terms which are linear combinations of the principal axis components, (D_q ; $q = x,y,z$) as

$$D = \tfrac{3}{2} D_z \ \text{ and } \ E = \tfrac{1}{2}(D_x - D_y).$$

Determination of the absolute signs of the so-called zero-field splitting parameters, D and E, is difficult and often only magnitudes are reported. These magnitudes depend on the choice of axis system used to characterize the molecular environment. Care should thus be taken to compare measurements between several laboratories, and it is recommended that the original references be consulted in cases of conflict.

Second order ($\Delta M_S = \pm 2$) transitions are also detected as a single line, for a given orientation, at roughly half the field (or frequency) value of the $\Delta M_S = \pm 1$ transition. Though intrinsically weaker in intensity, these half-field resonances generally show much less anisotropy and have proven extremely useful for investigations of randomly oriented triplets. Often only the signal derived from the transitions occuring at minimum magnetic fields are measured and a combination of the magnitudes of D and E is reported as

$$D^* = (D^2 + 3E^2)^{1/2}.$$

Many modern techniques couple optical detection schemes, for example phosphorescence, with microwave irradiation to conveniently characterize the excited state parameters.

Table 6

ESR Parameters of Excited Triplet States

No.	Host	D/cm^{-1}	E/cm^{-1}	$D*$/cm^{-1}	Ref.
1	**Acenaphthene**				
	EtOH (77 K)	0.0966	−0.1040		68D217
2	**Acetaldehyde**				
	MCH (77 K)	0.148	0.028		90D074
3	**Acetone**				
	MCH (77 K)	0.142	0.058		90D074
4	**Acetophenone**				
	1,4-Dibromobenzene	−0.3350	0.0492		74E515
5	**Acetophenone, 4′-chloro**				
	1,4-Dibromobenzene	−0.1981	0.0617		74E515
6	**Acetophenone, 4′-methoxy-**				
	1,4-Dimethoxybenzene	−0.1051	0.0611		79E156
7	**Acetophenone, 2′-methyl-**				
	Toluene/EtOH (77 K)	±0.060	±0.0025		87D053
8	**Acetophenone, 4′-phenyl-**				
	EtOH (77 K)	0.0954	−0.0082		775062
9	**Acetylene, diphenyl-**				
	EPA			0.152	72D313
10	**Acridine**				
	MeOH/H$_2$O (90 K)			0.069	745194
11	**Acridine, 3-amino-**				
	MeOH/H$_2$O (90 K)			0.062	745194
12	**Acridine, 9-amino-**				
	MeOH/H$_2$O (90 K)			0.065	745194
13	**Acridine, 3,6-bis(diethylamino)-**				
	MeOH/H$_2$O (90 K)			0.069	745194
14	**Acridine, 3,6-diamino-**				
	MeOH/H$_2$O (90 K)			0.076	745194
15	**Acridine, 3,9-diamino-**				
	MeOH/H$_2$O (90 K)			0.069	745194
16	**Acridine Orange, free base**				
	EtOH (77 K)	0.061	0.015	0.069	755120

Table 6—**Triplet State ESR Parameters**—Continued

No.	Host	D/cm^{-1}	E/cm^{-1}	D^*/cm^{-1}	Ref.
17	**Acridine Red cation**				
	EtOH (77 K)			0.068	755120
18	**Acriflavine cation**				
	MeOH/H$_2$O (90 K)			0.076	745194
19	**Adenosine 5′-monophosphate**				
	PEG/H$_2$O (77 K)	0.119	0.027	0.128	663075
20	**Aniline**				
	EPA (77 K)			0.1317	625016
	1,4-Dibromobenzene (1.8 K)	0.1160	−0.0597		86D051
21	**Aniline, *N,N*-diphenyl-**				
	Lucite (77 K)			0.0801	645022
22	**Aniline, 4-methyl-**				
	1,4-Dibromobenzene (1.8 K)	0.1119	−0.0533		86D051
23	**Aniline, *N*-phenyl-**				
	EPA (77 K)			0.0994	625016
24	**Anthracene**				
	EPA	0.072	0.007		64Y005
25	**9,10-Anthraquinone, 1,8-dichloro-**				
	EtOH (10 K)	−0.018	0.0003		87D086
26	**9,10-Anthraquinone**				
	EPA (77 K)	−0.351	0.005		84D056
	Octane (77 K)	−0.2894	0.0041		84D056
27	**7-Azaindole**				
	1-PrOH (77 K)	0.0919	0.0465	0.123	80E163
	3-MP (77 K)			0.123	80E163
28	**Bacteriochlorophyll *a***				
	MTHF (100 K)	0.0232	0.0058		78D004
29	**Bacteriopheophytin *a***				
	MTHF (100 K)	0.0256	0.0054		78D004
30	**Benzaldehyde**				
	1,4-Dibromobenzene	−0.3350	0.0492		74E515
31	**Benzaldehyde, 4-chloro-**				
	1,4-Dibromobenzene	−0.1944	0.0587		74E515

Table 6—**Triplet State ESR Parameters**—Continued

No.	Host	D/cm^{-1}	E/cm^{-1}	D^*/cm^{-1}	Ref.
32	**Benzaldehyde, 4-(dimethylamino)-**				
	Durene	−0.0796	0.0460		74E515
33	**Benzaldehyde, 2-hydroxy-**				
	MCH (77 K)	0.0697	0.014		88D039
34	**Benzaldehyde, 4-methoxy-**				
	1,4-Dimethoxybenzene	−0.188	0.0525		74E506
35	**Benzaldehyde, 4-methyl-**				
	1,4-Dibromobenzene	−0.1547	0.0649		74E515
36	**Benz[*a*]anthracene**				
	2-MTHF	0.079	±0.014	0.083	66D175
37	**Benzene**				
	EtOH (77 K)	0.1581	0.0044	0.1583	705030
	Cyclohexane (1.2 K)	0.1593	±0.0093		756441
38	**Benzene, 1,4-dibromo-**				
	Xylene	0.268	±0.0597		80D074
39	**Benzene, 1,4-dichloro-**				
	Xylene	0.144	−0.0228		73E352
40	**Benzene, 1,4-dicyano-**				
	1,4-Dibromobenzene (77 K)	0.128	0.018		72D316
41	**Benzene, 1,4-dihydroxy-**				
	EPA (77 K)			0.1321	625016
42	**Benzene, 1,4-dimethoxy-**				
	1,4-Dibromobenzene (77 K)	0.131	0.027		72D316
43	**Benzene, 1,2,4,5-tetrachloro-**				
	Durene (4.2 K)	0.1518	0.02908		72D318
44	**Benzene, 1,2,4,5-tetracyano-**				
	Hexane (1.25 K)	0.1243	0.0136	0.1265	78E306
45	**Benzene, 1,3,5-triphenyl-**				
	3-MP (77 K)	0.111	<<0.001	0.113	66D173
46	**Benzil**				
	Benzil (77 K)	±0.119	±0.0464		755071
47	**Benzimidazole**				
	Benzoic acid (4.2 K)	0.1123	±0.0270		84D226

Table 6—**Triplet State ESR Parameters**—Continued

No.	Host	D/cm^{-1}	E/cm^{-1}	$D*/cm^{-1}$	Ref.
47	**Benzimidazole**—Continued				
	1,4-Dibromobenzene	0.1133	−0.0409		755396
48	**Benzofuran**				
	1,4-Dibromobenzene	0.1076	−0.0530		755396
49	**Benzoic acid**				
	EPA (77 K)			0.1385	625016
50	**Benzoic acid, methyl ester**				
	EtOH/MeOH (4:1) (77 K)	0.133	0.011	0.134	72D317
51	**Benzonitrile**				
	1,4-Dibromobenzene (77 K)	0.1370	0.0072		72D312
	EtOH (90 K)	0.1363	0.0062	0.1389	765078
52	**Benzonitrile, 4-bromo-**				
	Xylene	0.1681	±0.0129		78D082
53	**Benzonitrile, 4-chloro-**				
	Xylene (1.8 K)	0.1349	±0.0045		78D082
54	**Benzonitrile, 4-fluoro-**				
	EtOH (90 K)	0.1412	0.0032	0.1423	765078
55	**Benzonitrile, 4-methoxy-**				
	EtOH (77 K)	±0.1260	±0.0216	0.1329	91D221
56	**Benzonitrile, 4-methyl-**				
	1,4-Dibromobenzene (77 K)	0.136	0.005		72D316
57	**Benzophenone**				
	Benzophenone	±0.152	∓0.021		69D288
58	**Benzophenone, 4-phenyl-**				
	EtOH (77 K)	0.0946	−0.0087		775062
59	**Benzo[a]pyrene**				
	EPA (77 K)			0.0758	625016
60	**Benzo[e]pyrene**				
	2-MTHF (77 K)	0.090	±0.023	0.098	66D175
61	**Benzo[h]quinoline**				
	EtOH (77 K)	0.1011	−0.0468		68D217
62	**1,4-Benzoquinone**				
	EPA (77 K)	−0.330	0.019		85D016

Table 6—**Triplet State ESR Parameters**—Continued

No.	Host	D/cm^{-1}	E/cm^{-1}	$D*/\text{cm}^{-1}$	Ref.
63	**Benzo[*b*]triphenylene**				
	Nonane (2 K)	±0.0984	±0.0129		765139
64	**Biacetyl**				
	Cyclohexane (77 K)	0.208	−0.016		82D180
	EtOH (77 K)	0.214	−0.019		82D180
65	**Biphenyl**				
	3-MP (77 K)	0.110	±0.004	0.110	66D173
	EtOH (77 K)	±0.1094	±0.0036		68D217
66	**Biphenyl, 4,4′-difluoro-**				
	EtOH (77 K)	±0.1140	±0.0011	0.1140	89D149
67	**Biphenyl, 4-hydroxy-**				
	EG/H_2O (77 K)	±0.1058	∓0.0036	0.1060	80D111
68	**2,2′-Bipyridine**				
	Heptane (4.2 K)	0.1125	±0.0128		88D270
	EtOH (77 K)	±0.1104	±0.0121	0.1124	80D098
69	**4,4′-Bipyridine**				
	MeOH/H_2O (77 K)	0.1225	0.0044	0.1209	82E055
70	**2,2′-Biquinoline**				
	EtOH (77 K)	±0.0992	±0.0368	0.1179	80D098
71	**2-Butanone**				
	MCH (77 K)	0.150	0.050		90D074
72	**Butyraldehyde**				
	MCH (77 K)	0.158	0.021		90D074
73	**C_{60}**				
	Toluene (77 K)	±0.0114	±0.00069		91D034
74	**C_{70}**				
	Toluene (77 K)	±0.0052	±0.00069		91D034
75	**Carbazole**				
	Et_2O (77 K)	0.1022	0.0066	0.1043	66D174
	EPA (77 K)			0.1044	625016
76	**2,2′-Carbocyanine, 1,1′-diethyl-**				
	MeOH/H_2O (77 K)	±0.0681	±0.0097	0.063	81D168

Table 6—**Triplet State ESR Parameters**—Continued

No.	Host	D/cm^{-1}	E/cm^{-1}	$D*/cm^{-1}$	Ref.
77	**β-Carotene**				
	Micelle (160 K)	±0.0333	±0.0037		80D131
78	**Chlorophyll** *a*				
	MTHF (4.2 K)	0.0288	0.0042		78D026
	EtOH (100 K)	0.0272	0.0032		78D004
79	**Chlorophyll** *b*				
	MTHF (4.2 K)	0.0294	0.0049		78D026
	Octane (2 K)	±0.0320	±0.0041		775124
80	**Chrysene**				
	2-MTHF (77 K)	0.095	±0.025	0.104	66D175
81	**Cinnoline**				
	EtOH (77 K)	±0.0936	±0.0140		87E535
82	**Codeine**				
	Hexane (1.4 K)	0.0976	±0.0163		88D092
83	**Coronene**				
	EPA (77 K)	.		0.0971	625016
84	**Coumarin**				
	1,4-Dibromobenzene	0.1001	±0.0425		766189
85	**2,2′-Cyanine, 1,1′-diethyl-**				
	MeOH/H_2O (77 K)	±0.0668	±0.0136	0.075	81D168
86	**1,4-Cyclohexanedione**				
	1,4-Cyclohexanedione (1.4 K)	−0.1353	0.0371		87D045
87	**Cyclohexanone**				
	MCH (77 K)	0.143	0.034		90D074
88	**2-Cyclohexen-1-one**				
	CF_3CH_2OH (77 K)	−0.169	±0.00067		88D064
89	**Cyclopentanone**				
	Hexane (4.2 K)	±0.1404	∓0.0271		72D319
90	**2-Cyclopentenone**				
	CF_3CH_2OH (77 K)	−0.185	±0.0015		88D064
91	**2′-Deoxyadenosine**				
	LiCl/H_2O Glass (77 K)			0.133	86D067

Table 6—**Triplet State ESR Parameters**—Continued

No.	Host	D/cm^{-1}	E/cm^{-1}	$D*/\text{cm}^{-1}$	Ref.
92	**2′-Deoxyadenosine 5′-monophosphate**				
	LiCl/H_2O Glass (77 K)			0.134	86D067
93	**Dibenz[*a,h*]anthracene**				
	2-MTHF (77 K)	0.090	±0.025	0.100	66D175
94	**Dibenzofuran**				
	Et_2O (77 K)	0.1071	0.0092	0.1092	66D174
95	**Dibenzothiophene**				
	Et_2O (77 K)	0.1130	0.0021	0.1144	66D174
96	**Dimethyl isophthalate**				
	EtOH/MeOH (4:1) (77 K)	0.136	0.036	0.149	72D317
97	**Dimethyl phthalate**				
	EtOH/MeOH (4:1) (77 K)	0.121	0.034	0.140	72D317
98	**Dimethyl terephthalate**				
	EtOH/MeOH (4:1) (77 K)	0.122	0.005	0.122	72D317
99	**Fluoranthene**				
	2-PrOH (77 K)			0.076	63E021
100	**Fluorene**				
	Et_2O (77 K)	0.1075	0.0033	0.1092	66D174
101	**Fluorescein, 2′,7′-dichloro-**				
	EtOH (77 K)			0.075	755120
102	**Fluorescein dianion**				
	EtOH (77 K)	0.065	0.019	0.075	755120
103	**Fluorescein dianion, 4′,5′-dibromo-2′,7′-dinitro-**				
	EtOH (77 K)			0.082	755120
104	**Fluorescein dianion, 2′,4′,5′,7′-tetrabromo-**				
	EtOH (77 K)			0.081	755120
105	**Formaldehyde**				
	EPA	0.42	0.04		64Y004
106	**Imidazole**				
	Benzoic acid (4.2 K)	0.1077	±0.0293		84D226
107	**1-Indanone**				
	Durene	−0.439	0.011		78E065

Table 6—**Triplet State ESR Parameters**—Continued

No.	Host	D/cm^{-1}	E/cm^{-1}	$D*/cm^{-1}$	Ref.
108	**2-Indanone**				
	Durene (4.2 K)	±0.1303	±0.0359		79D290
109	**Indazole**				
	1,4-Dibromobenzene	0.1003	−0.0382		755396
	Benzoic acid (4.2 K)	0.1077	0.0293		83D218
110	**Indene**				
	1,4-Dibromobenzene	0.1079	−0.0472		755396
111	**Indole**				
	1,4-Dibromobenzene	0.0978	−0.0453		755396
	EtOH (77 K)	0.1011	0.0416	0.1241	705030
112	**Indole, 3-methyl-**				
	EtOH (77 K)	0.0965	0.0436	0.1225	705030
113	**Isoquinoline**				
	EtOH (77 K)	0.0999	−0.0113		88D117
	Durene	±0.1004	∓0.0117		655047
114	**Methylene Blue cation**				
	MeOH/H_2O (98 K)			0.066	757048
115	**Morphine hydrochloride**				
	THF (1.4 K)	0.0994	±0.0163		88D092
116	**2-Naphthaldehyde**				
	EPA	0.094	0.029		695015
117	**Naphthalene**				
	EtOH (77 K)	0.1008	−0.0154		88D117
	Biphenyl (77 K)	±0.0992	∓0.01545		625017
118	**Naphthalene, 2-acetyl-**				
	EPA	0.096	0.027	0.105	695015
119	**Naphthalene, 1-amino-**				
	EtOH (77 K)			±0.0943	82D324
120	**Naphthalene, 2-amino-**				
	EtOH (77 K)			±0.0953	82D324
121	**Naphthalene, 1,4-dimethyl-**				
	EPA	0.0935	−0.0133	0.0959	71D299

Table 6—**Triplet State ESR Parameters**—Continued

No.	Host	D/cm^{-1}	E/cm^{-1}	D^*/cm^{-1}	Ref.
122	**Naphthalene, 1,8-dimethyl-**				
	EPA	0.0948	−0.0137	0.0978	71D299
123	**Naphthalene, 1,4-dinitro-**				
	EPA (77 K)	0.0765	−0.0092		89D004
124	**Naphthalene, 1,8-dinitro-**				
	EPA (77 K)	0.0921	−0.0109		91D249
125	**Naphthalene, 1-hydroxy-**				
	EtOH (77 K)	0.0931	−0.0138		67D239
126	**Naphthalene, 2-hydroxy-**				
	EtOH (77 K)	0.0942	−0.0146		67D239
127	**Naphthalene, 1-methyl-**				
	EPA	0.0963	−0.0143	0.0995	71D299
128	**Naphthalene, 2-methyl-**				
	EPA	0.0958	−0.0137	0.0990	71D299
129	**Naphthalene, 1-nitro-**				
	EPA (77 K)	0.0855	−0.0051		89D004
130	**Naphthalene, 2-nitro-**				
	EPA (77 K)	0.0870	−0.0247		91D249
131	**Naphthalene, 1-phenyl-**				
	EtOH (77 K)	±0.0911	±0.0093	±0.0923	82D324
132	**Naphthalene, 2-phenyl-**				
	EtOH (77 K)	±0.0963	±0.0274	±0.1007	82D324
133	**1,2-Naphthoquinone**				
	EtOH (15 K)	0.114	0.033	0.1275	89D094
134	**1,4-Naphthoquinone**				
	EPA (77 K)	−0.330	0.019		85D016
135	**2,2′-Oxacarbocyanine, 3,3′-diethyl-**				
	MeOH/H$_2$O (77 K)	±0.1022	±0.0165	0.113	81D168
136	**Pentacene**				
	Naphthacene (1.2 K)	0.0463	−0.0014		80D063
137	**3-Pentanone**				
	MCH (77 K)	0.149	0.035		90D074

Table 6—**Triplet State ESR Parameters**—Continued

No.	Host	D/cm^{-1}	E/cm^{-1}	D*/cm^{-1}	Ref.
138	**Phenanthrene**				
	Biphenyl (78 K)	±0.10043	∓0.04658		645023
	EtOH/Et$_2$O/2,2-DMB/Pentane (80 K)	0.1053	−0.0467	0.1325	80D102
139	**9,10-Phenanthrenequinone**				
	EtOH (15 K)	0.0935	0.023	0.1016	89D094
	MTHF (15 K)	0.19	0.035	0.1911	89D094
140	**Phenanthridine**				
	EtOH/Et$_2$O/2,2-DMB/Pentane (80 K)	0.1065	−0.0422	0.1292	80D102
141	**1,10-Phenanthroline**				
	EtOH (77 K)	0.1038	−0.0485		68D217
142	**Phenazine, 1,2,3,4-tetrahydro-**				
	MeOH/H$_2$O (103 K)	0.0942	−0.0180		775232
143	**Phenol**				
	EtOH (77 K)	0.1352	0.0451	0.1561	705030
144	**Phenol, 4-methyl-**				
	EtOH (77 K)	0.1251	0.0590	0.1615	705030
145	**Phenoxazinium, 3,7-bis(dimethylamino)-**				
	MeOH/H$_2$O (98 K)			0.058	757048
146	**Phenoxazinium, 3,7-diamino-**				
	MeOH/H$_2$O (98 K)			0.067	757048
147	**Phenoxide ion**				
	EtOH/0.25M NaOH (77 K)	0.1133	0.0337	0.1275	705030
148	**Phenoxide ion, 4-methyl-**				
	EtOH/0.25M NaOH (77 K)	0.1071	0.0540	0.1422	705030
149	**Phenylalanine**				
	EtOH (77 K)	0.1475	0.0439	0.1517	705030
150	**4-Phenylphenoxide ion**				
	EG/H$_2$O (77 K)	±0.0985	∓0.0062	0.0991	80D111
151	**4-Phenylpyridine**				
	MeOH/H$_2$O (77 K)	0.1125	0.0051	0.1127	82E055

Table 6—**Triplet State ESR Parameters**—Continued

No.	Host	D/cm^{-1}	E/cm^{-1}	$D*/cm^{-1}$	Ref.
152	**Pheophytin** *a*				
	MTHF (100 K)	0.0341	0.0033		78D004
	EtOH (100 K)	0.0344	0.0026		78D004
153	**Pheophytin** *b*				
	Octane (2 K)	±0.0368	±0.0049		775124
154	**Phthalazine**				
	Biphenyl (4.2 K)	−0.012	−0.026	0.038	85D181
	EtOH (77 K)	−0.073	0.022	0.052	85D181
155	**Picene**				
	p-Terphenyl (77 K)	±0.0937	±0.0365		79D033
156	**Piperonal**				
	Durene	−0.1283	0.0310		74E515
157	**Porphine**				
	Octane (1.3 K)	0.0437	0.00664		745445
158	**Porphine, magnesium(II)**				
	Pyridine	±0.0321	±0.0100		79D031
159	**Porphine, tetrakis(4-methylphenyl)-**				
	Octane (80 K)	±0.0369	±0.0076		745458
160	**Porphine, tetrakis(1-methylpyridinium-4-yl)-**				
	Sucrose/H$_2$O/HCl (4.2 K)	0.0418	0.0041		86E546
161	**Porphine, tetrakis(4-sulfonatophenyl)-, zinc(II)**				
	CH$_3$Cl/MeOH (10 K)	0.0298	0.0099		86D238
162	**Porphine, tetrakis(4-sulfonatophenyl)-**				
	H$_2$O/Glycerol (100 K)	0.0391	−0.0075		84D190
163	**Porphine, tetrakis(4-trimethylammoniophenyl)-, zinc(II)**				
	H$_2$O/Glycerol (10 K)	0.0323	0.0095		85R139
164	**Porphine, tetrakis(4-trimethylammoniophenyl)-**				
	H$_2$O/Glycerol (10 K)	0.0400	0.0075		85R139
165	**Porphine, tetraphenyl-**				
	Octane (80 K)	±0.0359	±0.0079		745458
	EtOH/Et$_2$O (100 K)	±0.0369	∓0.0082		745009
166	**Porphine, tetraphenyl-, magnesium(II)**				
	Phase V (120 K)	±0.0304	∓0.0084		80D124

Table 6—**Triplet State ESR Parameters**—Continued

No.	Host	D/cm^{-1}	E/cm^{-1}	D^*/cm^{-1}	Ref.
167	**Porphine, tetraphenyl-, zinc(II)**				
	Phase V (120 K)	±0.0304	∓0.0090		80D124
168	**Porphycene**				
	Liquid crystal (E-7) (108 K)	0.0295	0.0033		87D090
169	**Propionaldehyde**				
	MCH (77 K)	0.163	0.023		90D074
170	**Purine**				
	Benzoic acid (4.2 K)	0.1042	±0.0608		84D226
	1,4-Dibromobenzene	0.1009	−0.0584		755396
171	**Pyrazine, tetrachloro-**				
	Durene (1.8 K)	±0.1398	±0.0197		83E597
172	**Pyrazine, tetramethyl-**				
	Durene	±0.0973	±0.014		84E477
173	**Pyrene**				
	Octane (1.3 K)	0.0858	0.0170		80D122
	EPA (77 K)			0.0929	625016
174	**Pyridazine**				
	EtOH (3.0 K)	−0.138	0.116		85D207
175	**Pyridinium**				
	EG/H$_2$SO$_4$/HCl	0.134	±0.030		776013
176	**Pyridinium, 4-phenyl-**				
	MeOH/H$_2$O (77 K)	0.1032	0.0094	0.1044	82E055
177	**Pyrido[2,3-b]pyrazine**				
	Durene (10 K)	−0.1034			87D046
178	**Quinoline**				
	EtOH (77 K)	0.1014	−0.0164		88D117
	Durene (77 K)	±0.1030	∓0.0162		655047
179	**Quinoline, conjugate acid**				
	EtOH (77 K)	0.0921	−0.0149		88D117
180	**Quinoxaline**				
	Durene (77 K)	±0.1007	∓0.0182		635011
	MeOH/H$_2$O (103 K)	0.0954	−0.0184		775232

Table 6—**Triplet State ESR Parameters**—Continued

No.	Host	D/cm^{-1}	E/cm^{-1}	D^*/cm^{-1}	Ref.
181	**Quinoxaline, conjugate monoacid**				
	EtOH/1-PrOH/Sulfuric acid (103 K)	0.0834	−0.0172		775232
182	*(all-E)*-**Retinal**				
	Polyethylene film (1.2 K)	−0.7272	0.0634		92D032
183	**Rhodamine B, inner salt**				
	EtOH (77 K)	0.058	0.017	0.065	755120
184	**Rhodamine 6G cation**				
	EtOH (77 K)	0.054	0.019	0.063	755120
185	**Rhodamine S cation**				
	EtOH (77 K)	0.057	0.017	0.064	755120
186	**Salicylamide**				
	MCH (77 K)	0.1017	0.012		88D039
	EtOH/Toluene (77 K)	0.0772	0.015		88D039
187	*(E)*-**Stilbene**				
	3-MP (77 K)			0.1122	86D018
	1-Pentanol (77 K)			0.1119	86D018
188	*p*-**Terphenyl**				
	EPA (77 K)			0.0961	625016
189	**Tetracene**				
	Nonane (2 K)	0.0573	0.0043		765139
190	**7,7,8,8-Tetracyanoquinodimethane**				
	Phenazine (300 K)	±0.05975	±0.00760		89D116
191	α-**Tetralone**				
	Durene	0.429	0.017		72D311
192	**2,2′-Thiacarbocyanine, 3,3′-diethyl-**				
	MeOH/H$_2$O (77 K)	±0.1038	±0.0186	0.107	81D168
193	**Thionine cation**				
	MeOH/H$_2$O (98 K)			0.069	757048
194	**Toluene**				
	Cyclohexane/Decalin			0.17	71Z005
	EtOH (77 K)	0.1454	0.0250	0.1517	705030

Table 6—**Triplet State ESR Parameters**—Continued

No.	Host	D/cm^{-1}	E/cm^{-1}	D^*/cm^{-1}	Ref.
195	**1,3,5-Triazine, 2,4-diphenyl-**				
	3-MP (77 K)	0.120	<0.001	0.122	66D173
196	**1,3,5-Triazine, 2-phenyl-**				
	3-MP (77 K)	0.119	±0.003	0.122	66D173
197	**1,3,5-Triazine, 2,4,6-triphenyl-**				
	3-MP (77 K)	0.124	<<0.001	0.125	66D173
198	**Triphenylene**				
	MeOH/H_2O (103 K)	0.1353	0.000		775232
199	**Tryptophan**				
	EtOH (77 K)	0.0984	0.0410	0.1213	705030
200	**Tyrosine**				
	EtOH/0.1M HCl (77 K)	0.1301	0.0558	0.1621	705030
201	**9-Xanthione**				
	Hexane (4.2 K)	−15.9	0.06144		88D018
202	**Xanthone**				
	EPA (77 K)	±0.171	±0.019		88D063
	α-Cyclodextrin (77 K)	−0.150	0.019		89D101

Section 7

Diffusion-Controlled Rate Constants

The time-independent rate constant for reaction between molecules α and β can be computed from Smoluchowski's theory of diffusion-controlled reactions by the expression

$$k_{\text{diff}} = 4\pi N \rho_{\alpha\beta} D \ 10^3 \tag{7-1}$$

in units of L mol^{-1} s^{-1}, where $\rho_{\alpha\beta}$ is reaction radius between α and β, D is the relative diffusion constant between α and β, and N is Avogadro's number. The diffusion constants (D_α and D_β) of molecules α and β can be written as

$$D_\alpha = \frac{k_B T}{\zeta_\alpha} \tag{7-2}$$

from Einstein's relation, where k_B is the Boltzmann constant and ζ_α is the coefficient of friction between the α-molecule and the solvent. The friction coefficient can be computed from the Stokes relation which in the "stick" (non-slippage) limit gives

$$\zeta_\alpha = 6\pi\eta r_\alpha \tag{7-3}$$

where r_α is the radius of the α-molecule and η is the viscosity of the solvent. Substituting Eq. (7-3) into Eq. (7-2) and the resulting equation into Eq. (7-1) gives

$$k_{\text{diff}} = \frac{8RT}{3\eta} \ 10^3 \tag{7-4}$$

where $R = k_B N$ is the gas constant and the T is the absolute temperature. The assumption was made that $\rho_{\alpha\beta} = 2 r_\alpha = 2 r_\beta$ which also implies that $D = 2 D_\alpha$.

Diffusion-controlled rate constants for some common solvents were computed from Eq. (7-4) using viscosities from reference [86Z350]. The results for room temperatures are listed in Table 7.

Table 7

Diffusion-Controlled Rate Constants (Smoluchowski-Stokes-Einstein)

No.	Solvent	η (20 °C) $\times 10^3$ Pa s	k_{diff} (20 °C) L mol^{-1} s^{-1}	η (25 °C) $\times 10^3$ Pa s	k_{diff} (25 °C) L mol^{-1} s^{-1}
1	Isopentane	0.225	2.9×10^{10}	0.215	3.1×10^{10}
2	Diethyl ether	0.242	2.7×10^{10}	0.224	3.0×10^{10}
3	Pentane	0.235	2.8×10^{10}	0.225	2.9×10^{10}
4	Hexane	0.3126	2.1×10^{10}	0.2942	2.2×10^{10}
5	Acetone	0.322	2.0×10^{10}	0.307	2.1×10^{10}
6	Acetonitrile			0.341	1.9×10^{10}
7	Heptane	0.4181	1.6×10^{10}	0.3967	1.7×10^{10}
8	Methylene chloride	0.434	1.5×10^{10}	0.414	1.6×10^{10}
9	Tetrahydrofuran	0.575	1.3×10^{10}	0.460	1.4×10^{10}
10	Isooctane	0.504	1.3×10^{10}		
11	Octane	0.5466	1.2×10^{10}	0.5151	1.3×10^{10}
12	Chloroform	0.564	1.2×10^{10}	0.5357	1.2×10^{10}
13	Methanol	0.5929	1.1×10^{10}	0.5513	1.2×10^{10}
14	Toluene	0.5859	1.1×10^{10}	0.5525	1.2×10^{10}
15	Benzene	0.6487	1.0×10^{10}	0.6028	1.1×10^{10}
16	Methylcyclohexane	0.734	8.8×10^9	0.685	9.6×10^9
17	Decane	0.9284	7.0×10^9	0.8614	7.6×10^9
18	Pyridine	0.952	6.8×10^9	0.884	7.5×10^9
19	Carbon tetrachloride	0.9785	6.6×10^9	0.9004	7.3×10^9
20	Water	1.0019	6.5×10^9	0.89025	7.4×10^9
21	Cyclohexane	0.9751	6.7×10^9	0.898	7.4×10^9
22	Ethanol	1.21	5.4×10^9	1.0826	6.1×10^9
23	Dimethyl sulfoxide	2.2159	2.9×10^9	1.991	3.3×10^9
24	2-Propanol	2.55	2.5×10^9	2.0436	3.2×10^9
25	Glycerol	1412	4.6×10^6	945	7.0×10^6

Section 8

Rate Constants of Singlet-State Quenching

The relatively short lifetimes of excited singlet states of most organic molecules are a natural limit to the amount of data available on the quenching of these excited states. The short lifetimes are due to unimolecular decay processes: fluorescence

$$*S \rightarrow S + h\nu_f \tag{8-I}$$

internal conversion

$$*S \rightarrow S + \text{heat} \tag{8-II}$$

intersystem crossing

$$*S \rightarrow {}^3S + \text{heat} \tag{8-III}$$

and singlet photochemistry

$$*S \rightarrow \text{products} \tag{8-IV}$$

Bimolecular quenching must compete with these unimolecular processes.

The phenomenological approach (see for example [89Z021]) to quenching kinetics

$$*S + Q \underset{k_{-d}}{\overset{k_d}{\rightleftarrows}} *S\text{–}Q \overset{k_{de}}{\rightarrow} \text{products} \tag{8-V}$$

involves the idea of a collision complex *S–Q from which the final products or excited states are formed. The quenching scheme of Process (8-V) leads to a quenching rate constant of

$$k_q = \frac{k_d \, k_{de}}{k_{-d} + k_{de}} \tag{8-1}$$

In the limit that product formation (k_{de}) is much faster than the breakup of the collision complex (k_{-d}) into the original species, k_q in Eq. (8-1) reduces to k_d the diffusion rate constant which can be calculated by Eq. (7-4).

Table 8 contains a *representative* sample of the available quenching rate constants of excited singlet species. The table is ordered by inverted names of the excited state species. Quenchers are listed in alphabetic order under each excited state in the quenching pair. In addition the quenchers are partially grouped by the mechanism of the quenching process. In particular, four quenching mechanisms are distinguished: exciplex formation, energy transfer, oxidation of the excited singlet, and reduction of the excited singlet. There are no sharp lines between the mechanisms since exciplexes often involve partial charge transfer. The collision complex *S–Q itself may be an exciplex, and the existence of redox products signal an oxidative or reductive transfer. Hydrogen abstraction quenching is placed under reductive transfer.

Table 8

Rate Constants for Singlet-State Quenching

No. Quencher	Solvent (pH)	k_q /L mol^{-1} s^{-1}	Ref.
1. Acenaphthene			
O$_2$	Cyclohexane	2.6×10^{10}	88Z003
2. Acetone			
Energy Transfer			
Dibutyldiazene	MeCN	7.0×10^9	756208
1,4-Dioxene	MeCN	1.4×10^9	756080
2-Methoxypropene	MeCN	2.2×10^8	756080
1,3-Pentadiene	Hexane	9×10^7	70E304
3. 2-Adamantanone			
Energy Transfer			
Dibutyldiazene	MeCN	2.9×10^9	756208
Reductive Transfer			
Ethanol	Hexane	1.1×10^6	74A006
2-Propanol	Hexane	1.9×10^6	74A006
Tributylstannane	Hexane	4.8×10^8	74A006
4. Anthracene			
O$_2$	Benzene	3.1×10^{10}	62E014
O$_2$	Cyclohexane	2.5×10^{10}	78A307
O$_2$	EtOH	2.5×10^{10}	62E014
Exciplex Formation			
Triphenylphosphine	Benzene	2.2×10^9	756186
Reductive Transfer			
N,N-Diethylaniline	MeCN	2.1×10^{10}	68E126
5. Anthracene, 9-bromo-			
O$_2$	Cyclohexane	2.9×10^{10}	78A307
Exciplex Formation			
Triethylamine	MeCN	2.3×10^{10}	84B110
Reductive Transfer			
Triphenylphosphine	Benzene	3.7×10^9	756186
6. Anthracene, 9-chloro-			
O$_2$	Cyclohexane	3.1×10^{10}	78A307
Reductive Transfer			
Triphenylphosphine	Benzene	1.9×10^9	756186

Table 8—**Singlet-State Quenching Rates**—Continued

No.	Quencher	Solvent (pH)	k_q /L mol^{-1} s^{-1}	Ref.
7.	**Anthracene, 9-cyano-**			
	O_2	Cyclohexane	6.7×10^9	78A307
	Oxidative Transfer			
	MV^{2+}	MeCN/H$_2$O (9:1)	$\sim 2 \times 10^{10}$	81A070
	Reductive Transfer			
	Triphenylphosphine	Benzene	5.1×10^9	756186
8.	**Anthracene, 9-cyano-10-phenyl-**			
	O_2	Cyclohexane	6.9×10^9	78A307
9.	**Anthracene, 9,10-diacetoxy-**			
	O_2	Cyclohexane	2.5×10^{10}	78A307
10.	**Anthracene, 9,10-dibromo-**			
	O_2	Cyclohexane	2.4×10^{10}	78A307
	Exciplex Formation			
	Biphenyl	MeCN	1.2×10^8	84E393
	Naphthalene	MeCN	1.8×10^9	84E393
	Naphthalene	Toluene	5.2×10^8	84E393
	Pyrene	MeCN	2.4×10^{10}	84E393
	Triethylamine	MeCN	1.8×10^{10}	84B110
	Reductive Transfer			
	Triphenylphosphine	Benzene	3.9×10^9	756186
11.	**Anthracene, 9,10-dichloro-**			
	O_2	Benzene	2.4×10^{10}	62E014
	O_2	EtOH	1.7×10^{10}	62E014
	Exciplex Formation			
	Naphthalene	Toluene	1.5×10^7	84E393
	Pyrene	Toluene	1.6×10^9	84E393
	Reductive Transfer			
	Triphenylphosphine	Benzene	2.5×10^9	756186
12.	**Anthracene, 9,10-dicyano-**			
	O_2	Cyclohexane	4.7×10^9	78A307
	Exciplex Formation			
	Biphenyl	Toluene	2.5×10^7	84E393
	Naphthalene	Toluene	2.1×10^9	84E393
	Pyrene	Toluene	1.7×10^{10}	84E393
	Reductive Transfer			
	Triphenylphosphine	Benzene	1.1×10^{10}	756186

Table 8—**Singlet-State Quenching Rates**—Continued

No.	Quencher	Solvent (pH)	k_q /L mol^{-1} s^{-1}	Ref.
13.	**Anthracene, 9,10-dimethoxy-**			
	O$_2$	Cyclohexane	2.3×10^{10}	78A307
	Exciplex Formation			
	Triphenylphosphine	Benzene	6.0×10^7	756186
14.	**Anthracene, 9,10-dimethyl-**			
	O$_2$	Cyclohexane	2.0×10^{10}	78A307
	Energy Transfer			
	Benzophenone	Benzene	4.5×10^9	81F364
	Oxidative Transfer			
	Benzonitrile	MeCN	4.6×10^7	706216
	1,4-Dicyanobenzene	MeCN	1.6×10^{10}	706216
15.	**Anthracene, 9,10-diphenyl-**			
	O$_2$	Benzene	3.6×10^{10}	68E113
	O$_2$	Cyclohexane	1.7×10^{10}	78A307
	O$_2$	EtOH	2.3×10^{10}	62E014
16.	**Anthracene, 9-methoxy-**			
	O$_2$	Cyclohexane	2.7×10^{10}	78A307
	Exciplex Formation			
	Triphenylphosphine	Benzene	3.6×10^8	756186
17.	**Anthracene, 9-methyl-**			
	O$_2$	Cyclohexane	2.8×10^{10}	78A307
	Exciplex Formation			
	Triphenylphosphine	Benzene	4.7×10^8	756186
18.	**Anthracene, 9-phenyl-**			
	O$_2$	Cyclohexane	1.9×10^{10}	78A307
19.	**1,5-Anthracenedisulfonate ion**			
	Di-*tert*-butylnitroxide	H$_2$O	1.3×10^{10}	85E452
	O$_2$	H$_2$O	1.0×10^{10}	85E452
20.	**1-Anthracenesulfonate ion**			
	Di-*tert*-butylnitroxide	H$_2$O	7.5×10^9	85E452
	O$_2$	H$_2$O	1.2×10^{10}	85E452
21.	**2-Anthracenesulfonate ion**			
	Di-*tert*-butylnitroxide	H$_2$O	1.4×10^{10}	85E452
	O$_2$	H$_2$O	2.0×10^{10}	85E452

Table 8—**Singlet-State Quenching Rates**—Continued

No.	Quencher	Solvent (pH)	k_q /L mol^{-1} s^{-1}	Ref.
22.	**9-Anthroate ion**			
	Oxidative Transfer			
	MV^{2+}	H$_2$O (pH 5.0)	9.0×10^{10}	83A007
23.	**Benz[a]aceanthrylene**			
	O$_2$	Heptane	2.4×10^{10}	82E059
24.	**Benz[c]acridine**			
	Reductive Transfer			
	1,4-Dimethoxybenzene	MeCN	6.7×10^9	706216
	N,N-Dimethylaniline	MeCN	1.7×10^{10}	706216
	Phenol	MeCN	1.1×10^9	706216
	TMPD	MeCN	1.8×10^{10}	706216
25.	**Benz[a]anthracene**			
	O$_2$	Benzene	2.6×10^{10}	88Z003
	O$_2$	Cyclohexane	3.0×10^{10}	706182
	Oxidative Transfer			
	Benzonitrile	MeCN	3×10^6	706216
	Tetracyanoethylene	MeCN	2.4×10^{10}	706216
	Reductive Transfer			
	1,4-Dimethoxybenzene	MeCN	1.2×10^8	706216
	N,N-Dimethylaniline	MeCN	1.4×10^{10}	706216
	TMPD	MeCN	1.5×10^{10}	706216
26.	**Benz[a]anthracene, 9,10-dimethyl-**			
	O$_2$	Benzene	2.8×10^{10}	68E113
	Energy Transfer			
	Azulene	Benzene	8.3×10^9	706079
27.	**Benzene**			
	O$_2$	Cyclohexane	2.3×10^{10}	88Z003
28.	**Benzene, 1,4-dimethoxy-**			
	O$_2$	Cyclohexane	3.1×10^{10}	88Z003
29.	**Benzene, methoxy-**			
	O$_2$	Cyclohexane	3.1×10^{10}	88Z003
30.	**Benzene, 1,2,4-trimethyl-**			
	O$_2$	Cyclohexane	3.1×10^{10}	88Z003

Table 8—**Singlet-State Quenching Rates**—Continued

No.	Quencher	Solvent (pH)	k_q /L mol^{-1} s^{-1}	Ref.
31.	**Benzene, 1,3,5-trimethyl-**			
	O$_2$	Cyclohexane	2.6×10^{10}	88Z003
32.	**Benzidine, *N,N,N′,N′*-tetramethyl-**			
	O$_2$	MeCN	4.0×10^{10}	84B066
	Exciplex Formation			
	Dibenzyl sulfone	MeCN	1.0×10^{10}	84F074
	Oxidative Transfer			
	1,4-Dicyanonaphthalene	MeCN	2.2×10^{10}	84B066
33.	**Benzo[*b*]fluoranthene**			
	O$_2$	Heptane	5.1×10^{9}	82E059
34.	**Benzo[*j*]fluoranthene**			
	O$_2$	Heptane	1.8×10^{10}	82E059
35.	**Benzo[*k*]fluoranthene**			
	O$_2$	Heptane	2.3×10^{10}	82E059
36.	**Benzo[*rst*]pentaphene**			
	O$_2$	Cyclohexane	2.7×10^{10}	706182
37.	**Benzo[*ghi*]perylene**			
	Oxidative Transfer			
	1,4-Dicyanobenzene	MeCN	8.7×10^{9}	706216
	Tetracyanoethylene	MeCN	2.1×10^{10}	706216
38.	**Benzo[*g*]pteridine-2,4-dione, 3,7,8,10-tetramethyl-**			
	Exciplex Formation			
	Triethylamine	MeCN	1.1×10^{10}	82F433
	Trimethylamine	Dioxane	3.7×10^{9}	82F433
39.	**Benzo[*g*]pteridine-2,4-dione, 7,8,10-trimethyl-**			
	Reductive Transfer			
	1,3-Dimethoxybenzene	MeOH	1.1×10^{10}	81E041
	EDTA	H$_2$O (pH 7)	5.8×10^{9}	85A374
	Imidazole	H$_2$O (pH 7)	5.8×10^{9}	85A374
	Naphthalene	MeOH	8.9×10^{9}	81E041
40.	**Benzo[*a*]pyrene**			
	O$_2$	Cyclohexane	2.9×10^{10}	706182

Table 8—**Singlet-State Quenching Rates**—Continued

No.	Quencher	Solvent (pH)	k_q /L mol^{-1} s^{-1}	Ref.
41.	**Benzo[*h*]quinoline**			
	Reductive Transfer			
	Phenol	Cyclohexane	1.2×10^{10}	86A205
	Phenol	MeCN	2.4×10^9	706216
42.	**Benzo[*b*]triphenylene**			
	O_2	Cyclohexane	2.6×10^{10}	706182
43.	**Biacetyl**			
	Exciplex Formation			
	1,3-Dioxole	MeCN	1.0×10^9	87A031
	SCN$^-$	H_2O	4.2×10^9	756251
	Reductive Transfer			
	N,N-Diethylaniline	Benzene	1.1×10^{10}	696078
	N,N-Diethylaniline	EtOH	1.6×10^{10}	696078
	N,N-Diethylaniline	MeCN	1.6×10^{10}	696078
	Phenol	Benzene	2.1×10^9	696077
	Phenol	EtOH	1×10^8	696078
	Phenol	MeCN	1×10^8	696078
	Triphenylamine	Benzene	4.9×10^9	69A001
	Triphenylamine	EtOH	1.3×10^{10}	696078
	Triphenylamine	MeCN	1.3×10^{10}	69A001
44.	**Bicyclo[2.2.1]heptan-2-one, 7,7-dimethyl-**			
	Energy Transfer			
	Dibutyldiazene	MeCN	5.2×10^9	756208
45.	**1,1′-Binaphthyl**			
	O_2	Cyclohexane	3.3×10^{10}	88Z003
46.	**2,2′-Binaphthyl**			
	O_2	Cyclohexane	2.6×10^{10}	88Z003
47.	**Biphenyl**			
	O_2	Cyclohexane	2.8×10^{10}	88Z003
	Reductive Transfer			
	SCN$^-$	MeCN	8.6×10^9	736067
48.	**Biphenyl, 4-benzyl-**			
	O_2	Cyclohexane	3.0×10^{10}	88Z003

Table 8—**Singlet-State Quenching Rates**—Continued

No.	Quencher	Solvent (pH)	k_q /L mol^{-1} s^{-1}	Ref.
49.	**Biphenyl, 4-methoxy-**			
	O$_2$	Cyclohexane	3.3×10^{10}	88Z003
50.	**Biphenyl, 4-phenoxy-**			
	O$_2$	Cyclohexane	3.2×10^{10}	88Z003
51.	**1,3-Butadiene, 1,4-diphenyl-**			
	EtI	Cyclohexane	8.3×10^{9}	82E365
	O$_2$	Cyclohexane	6.0×10^{10}	82E365
52.	**Carbazole**			
	Reductive Transfer			
	Pyridine	Cyclohexane	1.0×10^{10}	84F039
53.	**Chrysene**			
	O$_2$	Cyclohexane	2.9×10^{10}	706182
	Energy Transfer			
	Benzophenone	Benzene	1.1×10^{10}	81F364
54.	**Coronene**			
	Exciplex Formation			
	Pb^{2+}	DMF	3.3×10^{8}	84E533
	Oxidative Transfer			
	Eu^{3+}	MeCN	4.0×10^{10}	78A163
	Reductive Transfer			
	N,N-Diethylaniline	MeCN	1.9×10^{10}	68E126
55.	**Decacyclene**			
	O$_2$	Benzene	1.7×10^{10}	68E113
56.	**Dibenz[*a,h*]anthracene**			
	O$_2$	Cyclohexane	2.9×10^{10}	706182
57.	**Dibenzo[*def,mno*]chrysene**			
	O$_2$	Benzene	3.5×10^{10}	68E113
58.	**Dibenzofuran**			
	O$_2$	Cyclohexane	2.9×10^{10}	88Z003
59.	**Diphenylmethane**			
	O$_2$	Cyclohexane	2.9×10^{10}	88Z003
60.	**Fluoranthene**			
	O$_2$	Cyclohexane	4.4×10^{9}	88Z003
	Energy Transfer			
	Diphenyl disulfide	Benzene	3×10^{8}	766276

Table 8—**Singlet-State Quenching Rates**—Continued

No.	Quencher	Solvent (pH)	k_q /L mol^{-1} s^{-1}	Ref.
60.	**Fluoranthene**—Continued			
	Reductive Transfer			
	Br$^-$	MeCN	1.1×10^9	746283
	NO$_3^-$	MeCN	4×10^6	746283
61.	**Fluorene**			
	Reductive Transfer			
	Br$^-$	MeCN	1.1×10^9	746283
	SCN$^-$	MeCN	2.2×10^9	746283
62.	**9-Fluorenone**			
	Reductive Transfer			
	Triethylamine	Benzene	7×10^9	69F399
63.	**Fluorescein dianion, 2′,4′,5′,7′-tetrabromo-**			
	Reductive Transfer			
	Phenoxide ion	H$_2$O	1.5×10^{10}	61A005
64.	**1,3,5-Hexatriene, 1,6-diphenyl-**			
	EtI	Cyclohexane	1.7×10^8	82E365
	O$_2$	Cyclohexane	2.0×10^{10}	82E365
65.	**Indole**			
	O$_2$	H$_2$O	6.5×10^9	733187
	Succinimide	H$_2$O	4.8×10^9	84E335
	Energy Transfer			
	Anthracene	Cyclohexane	7×10^{10}	81E082
66.	**Indole, 5-methoxy-**			
	Succinimide	H$_2$O	4.6×10^9	84E335
67.	**Indole, 1-methyl-**			
	Succinimide	H$_2$O	5.2×10^9	84E335
68.	**Indole, 3-methyl-**			
	Succinimide	H$_2$O	4.6×10^9	84E335
69.	**Methylene Blue cation, conjugate monoacid**			
	Reductive Transfer			
	Fe^{II}	MeCN/H$_2$O (1:1) (pH 2)	3.2×10^9	80A038

Table 8—**Singlet-State Quenching Rates**—Continued

No.	Quencher	Solvent (pH)	k_q /L mol^{-1} s^{-1}	Ref.
70.	**Naphthalene**			
	O_2	Cyclohexane	2.7×10^{10}	88Z003
	Succinimide	H_2O	4.0×10^8	84E335
	Energy Transfer			
	Biacetyl	Cyclohexane	1.0×10^{10}	70E320
	Oxidative Transfer			
	Benzonitrile	MeCN	4×10^6	706216
	1,4-Dicyanobenzene	MeCN	1.8×10^{10}	706216
	Reductive Transfer			
	Br$^-$	MeCN	1.0×10^9	746283
	ClO$_4^-$	MeCN	5×10^6	746283
71.	**Naphthalene, 1-cyano-**			
	Oxidative Transfer			
	MV^{2+}	MeCN/H$_2$O (9:1)	$\sim2 \times 10^{10}$	81A070
72.	**Naphthalene, 1,4-dicyano-**			
	O_2	MeCN	1.3×10^{10}	84B066
	Exciplex Formation			
	1,4-Dimethoxybenzene	Heptane	4.3×10^{10}	84E236
	2,5-Dimethyl-2,4-hexadiene	MeCN	1.9×10^{10}	84B066
	Mesitylene	Heptane	1.8×10^{10}	84E236
	Reductive Transfer			
	Triethylamine	MeCN	1.6×10^{10}	84B066
73.	**Naphthalene, 2,3-dimethyl-**			
	O_2	Cyclohexane	2.9×10^{10}	88Z003
74.	**Naphthalene, 2,6-dimethyl-**			
	O_2	Cyclohexane	2.7×10^{10}	88Z003
75.	**Naphthalene, 1,4-diphenyl-**			
	O_2	Cyclohexane	2.7×10^{10}	88Z003
76.	**Naphthalene, 1,5-diphenyl-**			
	O_2	Cyclohexane	2.9×10^{10}	88Z003
77.	**Naphthalene, 2-hydroxy-**			
	Oxidative Transfer			
	Pyridine	MeCN	2.6×10^9	706216

Table 8—**Singlet-State Quenching Rates**—Continued

No.	Quencher	Solvent (pH)	k_q /L mol^{-1} s^{-1}	Ref.
78.	**Naphthalene, 1-methyl-**			
	O$_2$	Cyclohexane	3.2×10^{10}	88Z003
79.	**Naphthalene, 2-methyl-**			
	O$_2$	Cyclohexane	2.5×10^{10}	88Z003
80.	**Naphthalene, 1-phenyl-**			
	O$_2$	Cyclohexane	2.4×10^{10}	88Z003
81.	**Naphthalene, 1-styryl-, (*E*)-**			
	Exciplex Formation			
	Diethylamine	Hexane	1.8×10^{10}	776378
82.	**Naphthalene, 2-styryl-, (*E*)-**			
	Exciplex Formation			
	Diethylamine	Hexane	9.0×10^{9}	776378
83.	**Norcamphor**			
	Exciplex Formation			
	1,4-Dioxene	MeCN	5.5×10^{8}	756080
	1,3-Pentadiene	Benzene	5×10^{7}	70E304
84.	**1,3,5,7-Octatetraene, 1,8-diphenyl-**			
	EtI	Cyclohexane	6.0×10^{7}	82E365
	O$_2$	Cyclohexane	1.9×10^{10}	82E365
	O$_2$	MeOH	2.6×10^{10}	82E365
85.	**Oxazole, 2,5-diphenyl-**			
	CCl$_4$	Cyclohexane	2.0×10^{9}	80E439
86.	**Pentahelicene**			
	O$_2$	MCH	1.6×10^{10}	79A237
87.	**2-Pentanone**			
	Energy Transfer			
	1,3-Pentadiene	Hexane	4×10^{7}	70E304
88.	**Perylene**			
	I$^-$	MeCN	1.1×10^{10}	746283
	O$_2$	Benzene	2.7×10^{10}	68E113
	O$_2$	Cyclohexane	2.2×10^{10}	88Z003
	O$_2$	EtOH	2.7×10^{10}	62E014
	Reductive Transfer			
	N,*N*-Diethylaniline	MeCN	2.0×10^{10}	68E126

Table 8—**Singlet-State Quenching Rates**—Continued

No.	Quencher	Solvent (pH)	k_q /L mol^{-1} s^{-1}	Ref.
89.	**Phenanthrene**			
	I$^-$	MeCN	4.7×10^9	746283
	O$_2$	Cyclohexane	2.3×10^{10}	88Z003
	Energy Transfer			
	Di-*tert*-butyl disulfide	MeCN	2.2×10^8	766276
90.	**Phenol**			
	Succinimide	H$_2$O	5.5×10^9	84E335
91.	**Phenoxazinium, 3,7-diamino-**			
	Reductive Transfer			
	EDTA	H$_2$O (pH 8.0)	6.2×10^9	777063
92.	**Phenyl ether**			
	O$_2$	Cyclohexane	3.1×10^{10}	88Z003
93.	**Phthalocyanine, magnesium(II)**			
	Oxidative Transfer			
	Nitrobenzene	Dioxane	1.0×10^9	65A001
94.	**Pinacolone**			
	Energy Transfer			
	1,3-Pentadiene	Benzene	2×10^7	70E304
95.	**Porphine, tetrakis(4-sulfonatophenyl)-**			
	Oxidative Transfer			
	1,4-Benzoquinone	H$_2$O (pH 7)	2.7×10^{10}	81E084
	Nitrobenzene	H$_2$O (pH 7)	1.9×10^{10}	81E084
96.	**Porphine-2,18-dipropanoic acid, 7,12-diethenyl-3,8,13,17-tetramethyl-, dimethyl ester**			
	O$_2$	Benzene	1.8×10^{10}	771078
97.	**Pyrene**			
	I$^-$	MeCN	4.5×10^9	746283
	O$_2$	Cyclohexane	2.5×10^{10}	706182
	Energy Transfer			
	Benzophenone	Benzene	8×10^9	81F364
	DABCO	Cyclohexane	3.2×10^9	79E109
	N,N-Diethylaniline	Toluene	1.0×10^{10}	736162
	Oxidative Transfer			
	Benzonitrile	MeCN	4×10^6	706216
	1,4-Dicyanobenzene	MeCN	2.1×10^{10}	706216
	Reductive Transfer			
	SCN$^-$	MeCN	4.4×10^7	746283

Table 8—**Singlet-State Quenching Rates**—Continued

No.	Quencher	Solvent (pH)	k_q /L mol^{-1} s^{-1}	Ref.
98.	**Pyrene, 1-hydroxy-**			
	Oxidative Transfer			
	Pyridine	MeCN	2.0×10^9	706216
99.	**1-Pyrenecarboxylic acid**			
	Reductive Transfer			
	1,4-Dimethoxybenzene	MeCN	1.1×10^{10}	706216
	N,N-Dimethylaniline	MeCN	1.8×10^{10}	706216
	TMPD	MeCN	1.6×10^{10}	706216
100.	**2-Pyrenecarboxylic acid**			
	Reductive Transfer			
	1,4-Dimethoxybenzene	MeCN	7.0×10^7	706216
	N,N-Dimethylaniline	MeCN	1.7×10^{10}	706216
	TMPD	MeCN	1.8×10^{10}	706216
101.	**Pyrimido[4,5-*b*]quinoline-2,4-dione, 3,7,8,10-tetramethyl-**			
	Reductive Transfer			
	1,2,3-Trimethoxybenzene	MeOH	6.3×10^9	81E041
102.	**Rubrene**			
	O_2	Benzene	1.2×10^{10}	68E113
103.	**_p_-Terphenyl**			
	I$^-$	MeCN	2.5×10^{10}	746283
	Reductive Transfer			
	Br$^-$	MeCN	1.8×10^{10}	746283
	ClO$_4^-$	MeCN	1.1×10^9	746283
104.	**Tetracene**			
	O_2	Benzene	2.4×10^{10}	68E113
	Oxidative Transfer			
	Eu^{3+}	MeCN	4.5×10^{10}	78A163
	Reductive Transfer			
	N,N-Diethylaniline	MeCN	1.2×10^{10}	68E126
105.	**Thionine cation, conjugate monoacid**			
	Reductive Transfer			
	Fe^{2+}	Sulfuric acid (pH 2.5)	3.5×10^9	776118
106.	**Thioxanthen-9-one**			
	Exciplex Formation			
	Br$^-$	MeCN/H$_2$O (3:2)	2.5×10^9	85E351

Table 8—**Singlet-State Quenching Rates**—Continued

No.	Quencher	Solvent (pH)	k_q /L mol^{-1} s^{-1}	Ref.
107.	**Toluene**			
	O_2	Cyclohexane	2.8×10^{10}	88Z003
108.	**Triphenylene**			
	O_2	Cyclohexane	2.0×10^{10}	88Z003
	Energy Transfer			
	Dibutyldiazene	Benzene	1.1×10^{10}	756208
	Dibutyl disulfide	Benzene	5.7×10^8	766276
109.	**Tryptophan**			
	I^-	H_2O	1.9×10^9	733187
	O_2	H_2O	5.9×10^9	733187
110.	**L-Tryptophan**			
	Succinimide	H_2O	3.5×10^9	84E335
111.	**L-Tryptophan, *N*-acetyl-**			
	Succinimide	H_2O	4.4×10^9	84E335
112.	**L-Tryptophanamide, *N*-acetyl-**			
	Succinimide	H_2O	3.9×10^9	84E335
113.	***m*-Xylene**			
	O_2	Cyclohexane	2.6×10^{10}	88Z003
114.	***o*-Xylene**			
	O_2	Cyclohexane	2.6×10^{10}	88Z003
115.	***p*-Xylene**			
	O_2	Cyclohexane	2.9×10^{10}	88Z003

Section 9

Rate Constants of Triplet-State Quenching

The literature on the quenching of triplet states of organic molecules is much more extensive than that of excited singlet states. The unimolecular lifetimes of the triplet states are usually long enough so that diffusion of excited states and quenchers toward each other can compete favorably with intersystem crossing and phosphorescence. The long lifetimes of triplet states are a consequence of the spin-forbidden nature of the transitions from the lowest triplet state.

In Table 9, the triplet species are listed primarily by inverted name with the quenchers collected under each excited state. The quenchers are in turn collected by quenching mechanism. The mechanisms are charge transfer, energy transfer, oxidation of the triplet state, and reduction of the triplet state. Since the triplet exciplex has not achieved the status of the singlet exciplex, quenching whereby the collision complex *S–Q in Process (8-V) is strongly suspected of being a partially ionized pair is labeled as a *charge-transfer* mechanism. This mechanism may be interpreted by some workers as an exciplex. In the classification scheme of Table 9, the *charge-transfer* mechanism is to be distinguished from the *oxidative* or *reductive-transfer* mechanisms in that the latter two mechanisms have produced, or are very likely to produce, separated redox products. (For redox potentials of common quenchers and equations for estimating quenching ΔG's see Section 10b.)

Hydrogen abstraction has been subsumed under *reductive transfer*. Alcohols and amines both reduce triplet ketones but with rate constants three orders of magnitude different. Triplet ketones are thought to abstract hydrogens directly from alcohols ROH to form ketyl radicals,

$$>C=O^* + HO-R \;\rightarrow\; >\dot{C}-OH + \dot{O}-R \qquad (9\text{-I})$$

but amines are thought to react by a sequential electron transfer/proton transfer mechanism

$$>C=O^* + H-N< \;\rightarrow\; >C-O^- + H-N^+< \;\rightarrow\; >\dot{C}-OH + \dot{N}< \qquad (9\text{-II})$$

to give an effective hydrogen abstraction. Both mechanisms of the types in processes (9-I) and (9-II) are grouped together as *reductive transfer* in Table 9. Where groups of quenchers of both types have been measured for a given triplet state, one or more of each type has been selected from the database to be displayed in Table 9.

Most of the oxidative quenching of triplet states has been done with methyl viologen MV^{2+}, but some work with nitrile and quinone quenchers has also been done. Although oxygen quenching of triplets is usually via energy transfer to form singlet oxygen and the ground state of the molecule originally excited, oxygen can also react efficiently with the triplet species. In this work no attempt was made assess the mechanism of oxygen quenching, and oxygen is listed in Table 9 along with quenchers of unspecified mechanisms.

Table 9

Rate Constants for Triplet-State Quenching

No.	Quencher	Solvent (pH)	k_q /L mol^{-1} s^{-1}	Ref.
1.	**Acenaphthylene**			
	Oxidative Transfer			
	MV^{2+}	MeCN/H$_2$O (9:1)	2.0×10^9	83A113
2.	**Acetone**			
	O_2	Cyclohexane	8.4×10^9	80E563
	O_2	MeCN	5.4×10^{10}	84B051
	Energy Transfer			
	Dibutyldiazene	MeCN	3.1×10^9	756208
	I$^-$	H$_2$O	7.1×10^9	767189
	1,3-Pentadiene	MeCN	4.4×10^9	80F439
	Oxidative Transfer			
	Tetrachloroethylene	MeCN	6.9×10^8	776369
	Reductive Transfer			
	Benzene	MeCN	1.7×10^6	737292
	Ethanol	MeCN	5.4×10^5	737292
	Triethylamine	H$_2$O	2.7×10^8	727296
3.	**Acetophenone**			
	O_2	Benzene	4.5×10^9	85A268
	O_2	MeCN	3.7×10^9	85A268
	Energy Transfer			
	Naphthalene	Benzene	7.7×10^9	696034
	Oxidative Transfer			
	MV^{2+}	MeCN/H$_2$O (9:1)	5.2×10^9	84B033
	Reductive Transfer			
	I$^-$	H$_2$O/MeCN (4:1)	2.3×10^9	82A082
	2-Propanol	MeCN	2.1×10^6	78A183
	Triethylamine	Benzene	1.9×10^9	78A343
4.	**Acetophenone, 4′-amino-**			
	O_2	Benzene	9.4×10^9	85A268
5.	**Acetophenone, 4′-cyano-**			
	O_2	Benzene	5.2×10^8	85A268
	Energy Transfer			
	2,5-Dimethyl-2,4-hexadiene	Benzene	5.8×10^9	85B078
	Oxidative Transfer			
	MV^{2+}	MeCN/H$_2$O (9:1)	1.5×10^9	81A078

Table 9—**Triplet-State Quenching Rates**—Continued

No.	Quencher	Solvent (pH)	k_q /L mol^{-1} s^{-1}	Ref.
5.	**Acetophenone, 4′-cyano-**—Continued			
	Reductive Transfer			
	Phenoxide ion	MeCN/H$_2$O (1:1)	9.5×10^9	81A114
6.	**Acetophenone, 4′-fluoro-**			
	O$_2$	Benzene	3.8×10^9	85A268
	Oxidative Transfer			
	MV^{2+}	MeCN/H$_2$O (9:1)	3.1×10^9	84B033
7.	**Acetophenone, 4′-methoxy-**			
	Di-*tert*-butylnitroxide	Benzene	1.8×10^9	85B078
	O$_2$	Benzene	6.0×10^9	84F051
	Energy Transfer			
	Azulene	Benzene	7.5×10^9	85B078
	Oxidative Transfer			
	MV^{2+}	MeCN/H$_2$O (9:1)	8.0×10^9	84B033
	Reductive Transfer			
	Phenoxide ion	MeCN/H$_2$O (1:1)	6.8×10^9	81A114
	Triethylamine	Benzene	1.0×10^8	84F051
8.	**Acetophenone, 4′-methyl-**			
	O$_2$	Benzene	4.0×10^9	85A268
	Energy Transfer			
	β-Ionone	Toluene	3.2×10^9	85E293
	Oxidative Transfer			
	MV^{2+}	MeCN/H$_2$O (9:1)	4.1×10^9	84B033
	Reductive Transfer			
	2-Propanol	Benzene	1.3×10^5	737198
9.	**Acetophenone, 2,2,2-trifluoro-**			
	Charge Transfer			
	Anisole	MeCN	1.9×10^8	80F190
	Oxidative Transfer			
	MV^{2+}	MeCN/H$_2$O (9:1)	2.6×10^9	81A078
	Reductive Transfer			
	1,4-Dimethoxybenzene	Benzene	4.1×10^9	81F215
10.	**Acetophenone, 4′-(trifluoromethyl)-**			
	O$_2$	Benzene	2.5×10^9	85A268
	Charge Transfer			
	p-Xylene	MeCN	4.7×10^6	86A400

Table 9—**Triplet-State Quenching Rates**—Continued

No.	Quencher	Solvent (pH)	k_q /L mol^{-1} s^{-1}	Ref.
10.	**Acetophenone, 4′-(trifluoromethyl)-**—Continued			
	Reductive Transfer			
	Cyclohexane	MeCN	2.0×10^6	78A183
	2-Propanol	MeCN	8.8×10^6	78A183
11.	**Acridine**			
	O$_2$	Benzene	1.8×10^9	78E263
	Energy Transfer			
	(E)-Azobenzene	Benzene	5.1×10^9	81E012
	Cr(acac)$_3$	Benzene	1.3×10^9	81E374
	Reductive Transfer			
	2-Naphthol	Cyclohexane	2.9×10^9	737113
12.	**Acridine Orange, conjugate monoacid**			
	Reductive Transfer			
	Triphenylamine	MeOH	2.5×10^8	79E219
13.	**Acridine Orange, free base**			
	Reductive Transfer			
	TMPD	MeOH	5.0×10^9	79E219
14.	**9-Acridinethione, 10-methyl-**			
	O$_2$	Cyclohexane	4.8×10^9	84E342
15.	**Acridinium, 3,6-bis(dimethylamino)-10-methyl-**			
	Reductive Transfer			
	N,N-Dimethylaniline	MeOH	1.3×10^9	79E219
16.	**Aniline, N,N-dimethyl-**			
	O$_2$	Cyclohexane	$\sim 1 \times 10^{10}$	761069
17.	**Aniline, N,N-diphenyl-**			
	O$_2$	MCH	1.5×10^{10}	72B002
18.	**Anthracene**			
	Di-*tert*-butylnitroxide	Benzene	8.8×10^6	84E319
	O$_2$	Cyclohexane	3.9×10^9	82B121
	Energy Transfer			
	Azulene	Benzene	6.7×10^9	84E319
	β-Carotene	Hexane	1.1×10^{10}	66E089
	Cr(acac)$_3$	Benzene	1.3×10^9	81E374
	Ferrocene	Benzene	4.2×10^9	84E319
	(all-E)-Retinal	Benzene	3.0×10^{10}	777434

Table 9—**Triplet-State Quenching Rates**—Continued

No.	Quencher	Solvent (pH)	k_q /L mol^{-1} s^{-1}	Ref.
18.	**Anthracene**—Continued			
	Oxidative Transfer			
	1,1'-Dibenzyl-4,4'-bipyridinium	MeOH	4.1×10^9	83A113
	MV^{2+}	MeOH	5.1×10^9	83A113
19.	**Anthracene, 9-bromo-**			
	Energy Transfer			
	Azulene	Benzene	4.5×10^9	81F275
20.	**Anthracene, 9,10-dibromo-**			
	Energy Transfer			
	(*E*)-Azobenzene	Benzene	2.4×10^8	81E012
21.	**Anthracene, 9,10-dimethyl-**			
	O$_2$	Benzene	4.0×10^9	706079
22.	**Anthracene, 9,10-diphenyl-**			
	Di-*tert*-butylnitroxide	Benzene	5.1×10^6	84E319
	O$_2$	Benzene	1.9×10^9	84E319
	Energy Transfer			
	Anthracene	Toluene	3.0×10^8	83E281
	Oxidative Transfer			
	MV^{2+}	MeCN/H$_2$O (9:1)	4.3×10^9	83A113
23.	**1,8-Anthracenedisulfonate ion**			
	I$^-$	H$_2$O	1.1×10^5	85E452
	O$_2$	MeOH	1.5×10^9	85E452
24.	**1-Anthracenesulfonate ion**			
	I$^-$	H$_2$O	4.4×10^4	85E452
	O$_2$	MeOH	2.4×10^9	85E452
25.	**2-Anthracenesulfonate ion**			
	I$^-$	H$_2$O	4.0×10^4	85E452
	O$_2$	MeOH	2.4×10^9	85E452
26.	**9,10-Anthraquinone**			
	O$_2$	Benzene	1.4×10^9	771014
	Energy Transfer			
	Anthracene	Benzene	2.0×10^{10}	720392
	Biacetyl	Benzene	1.2×10^9	720392
	Reductive Transfer			
	2-Propanol	Benzene	2.1×10^7	720392

Table 9—**Triplet-State Quenching Rates**—Continued

No.	Quencher	Solvent (pH)	k_q /L mol^{-1} s^{-1}	Ref.
27.	**9,10-Anthraquinone-2-sulfonate ion**			
	Charge Transfer			
	I$^-$	H$_2$O	4.2×10^9	83B054
28.	**9-Anthroate ion**			
	Oxidative Transfer			
	MV^{2+}	H$_2$O (pH 5.0)	2.2×10^9	83A007
29.	**Anthrone**			
	O$_2$	Benzene	2×10^9	82B102
	Energy Transfer			
	Cu(acac)$_2$	Benzene	5.0×10^9	82B102
	1,3-Octadiene	Benzene	6.8×10^9	82B102
30.	**Azulene**			
	O$_2$	Benzene	6×10^9	81F275
	O$_2$	MeCN	8×10^9	81F275
	Energy Transfer			
	Anthracene	Benzene	7.5×10^7	84E491
	Perylene	Benzene	9.5×10^9	84E491
31.	**Benzaldehyde**			
	Reductive Transfer			
	Benzene	MeCN	2.2×10^7	746020
	SCN$^-$	H$_2$O/MeCN (4:1)	1.7×10^{10}	82A082
32.	**Benz[*a*]anthracene**			
	Di-*tert*-butylnitroxide	Benzene	2.5×10^7	84E237
	O$_2$	Cyclohexane	1.9×10^9	706182
	Energy Transfer			
	Azulene	Benzene	8.3×10^9	84E237
	(*all-E*)-Retinol	Hexane	5×10^9	69E217
	Oxidative Transfer			
	MV^{2+}	MeCN/H$_2$O (9:1)	5.2×10^9	83A113
33.	**Benz[*a*]anthracene, 9,10-dimethyl-**			
	O$_2$	Benzene	3.0×10^9	706079
34.	**Benzanthrone**			
	Energy Transfer			
	Cu(acac)$_2$	Benzene	2.3×10^9	66E093
	Ferrocene	Benzene	5.7×10^9	66E093

Table 9—**Triplet-State Quenching Rates**—Continued

No.	Quencher	Solvent (pH)	k_q /L mol^{-1} s^{-1}	Ref.
35.	**Benzene**			
	Energy Transfer			
	Benzophenone	Benzene	1.3×10^{11}	751124
36.	**Benzene, methoxy-**			
	O_2	H_2O (pH 8.5)	6.3×10^9	757161
37.	**Benzene, nitro-**			
	Energy Transfer			
	(*E*)-Piperylene	THF	1×10^9	84E530
	Reductive Transfer			
	Triethylamine	Benzene	1.2×10^9	84F625
38.	**Benzene, 1,3,5-trimethyl-**			
	O_2	Hexane	7×10^8	82E258
39.	**Benzene, 1,3,5-triphenyl-**			
	O_2	Benzene	1.1×10^9	736097
40.	**Benzidine, *N*,*N*,*N*′,*N*′-tetramethyl-**			
	O_2	MeCN	2.2×10^{10}	84B066
	Energy Transfer			
	Naphthalene	Cyclohexane	2×10^9	767177
	Oxidative Transfer			
	Duroquinone	MeOH	2.3×10^{10}	767177
	MV^{2+}	MeCN	8.4×10^9	84B066
41.	**Benzil**			
	Charge Transfer			
	$Cu(acac)_2$	MeOH	9.5×10^8	83F123
	1,3-Cyclohexadiene	Benzene	3.0×10^8	66E093
	β-Ionone	Toluene	3.5×10^8	85E293
	Reductive Transfer			
	Phenoxide ion	MeCN/H_2O (1:1)	3.7×10^9	81A114
	Triethylamine	Cyclohexane	8.8×10^8	78A343
	1,3,5-Trimethoxybenzene	MeCN	8.0×10^5	81F215
42.	**Benzo[*rst*]pentaphene**			
	O_2	Cyclohexane	2.9×10^9	706182
	Energy Transfer			
	Rubrene	Benzene	1.7×10^9	81E346

Table 9—**Triplet-State Quenching Rates**—Continued

No.	Quencher	Solvent (pH)	k_q /L mol^{-1} s^{-1}	Ref.
43.	**Benzo[*ghi*]perylene**			
	TEMPO	MeCN	1.2×10^7	80E116
	Charge Transfer			
	1,4-Benzoquinone	Benzene	8.2×10^9	79A093
44.	**Benzo[*c*]phenanthrene**			
	O_2	Cyclohexane	1.0×10^9	706182
45.	**Benzophenone**			
	O_2	Benzene	2.3×10^9	84F005
	O_2	MeCN	2.3×10^9	85A268
	Energy Transfer			
	Azulene	Benzene	1.2×10^{10}	84E491
	Cu(acac)$_2$	MeOH	3.6×10^9	83F123
	Ferrocene	MeCN	1.8×10^{10}	81F275
	Naphthalene	Benzene	4.7×10^9	70E288
	(*E*)-Stilbene	Benzene	6×10^9	81E214
	Oxidative Transfer			
	Fumaronitrile	MeCN	1.2×10^9	776369
	MV^{2+}	H_2O	3.7×10^9	85A361
	Reductive Transfer			
	Cyclohexene	MeCN	9.6×10^7	81A275
	DABCO	Benzene	3.1×10^9	84A201
	Phenol	Benzene	1.3×10^9	81A174
	2-Propanol	Benzene	1.1×10^6	727064
	Triethylamine	Benzene	4.1×10^9	81A016
46.	**Benzophenone, 4,4′-bis(dimethylamino)-**			
	O_2	Benzene	1.3×10^{10}	85A268
	O_2	MeCN	1.1×10^{10}	85A268
	Energy Transfer			
	β-Ionone	Toluene	4.6×10^9	85E293
	Naphthalene	Cyclohexane	9.9×10^9	777603
47.	**Benzophenone, 4-(carboxymethyl)-, ion(1-)**			
	Oxidative Transfer			
	Fumaronitrile	CCl$_4$	6.2×10^9	82F225

Table 9—**Triplet-State Quenching Rates**—Continued

No.	Quencher	Solvent (pH)	k_q /L mol^{-1} s^{-1}	Ref.
48.	**Benzophenone, 4-chloro-**			
	O$_2$	MeCN	1.9×10^9	85A268
	Oxidative Transfer			
	MV^{2+}	MeCN/H$_2$O (9:1)	2.8×10^9	84B033
	Reductive Transfer			
	Triethylamine	MeCN/H$_2$O (9:1)	2.6×10^9	86A248
49.	**Benzophenone, 4-cyano-**			
	Reductive Transfer			
	Quadricyclane	MeCN	6.5×10^9	87F368
50.	**Benzophenone, 4,4′-dichloro-**			
	O$_2$	MeCN	1.4×10^9	85A268
	Oxidative Transfer			
	MV^{2+}	MeCN/H$_2$O (9:1)	2.1×10^9	84B033
	Reductive Transfer			
	Triethylamine	MeCN/H$_2$O (9:1)	3.3×10^9	86A248
51.	**Benzophenone, 4,4′-dimethoxy-**			
	O$_2$	MeCN	5.6×10^9	85A268
	Energy Transfer			
	β-Ionone	Toluene	4.2×10^9	85E293
	Reductive Transfer			
	Triethylamine	MeCN/H$_2$O (9:1)	5.5×10^8	86A248
52.	**Benzophenone, 4-fluoro-**			
	O$_2$	MeCN	2.3×10^9	85A268
	Oxidative Transfer			
	MV^{2+}	MeCN/H$_2$O (9:1)	1.3×10^9	84B033
	Reductive Transfer			
	Triethylamine	MeCN/H$_2$O (9:1)	2.7×10^9	86A248
53.	**Benzophenone, 4-methoxy-**			
	O$_2$	Benzene	3.7×10^9	85A268
	Energy Transfer			
	β-Ionone	Toluene	3.0×10^9	85E293
	Oxidative Transfer			
	MV^{2+}	MeCN/H$_2$O (9:1)	5.6×10^9	84B033
	Reductive Transfer			
	Triethylamine	MeCN/H$_2$O (9:1)	1.2×10^9	86A248

Table 9—**Triplet-State Quenching Rates**—Continued

No.	Quencher	Solvent (pH)	k_q /L mol^{-1} s^{-1}	Ref.
54.	**Benzophenone, 4-methyl-**			
	Di-*tert*-butyl peroxide	Benzene	4.1×10^9	84F051
	O_2	Benzene	2.3×10^9	85A268
	Energy Transfer			
	2,5-Dimethyl-2,4-hexadiene	Benzene	6.2×10^9	84F051
	Oxidative Transfer			
	MV^{2+}	MeCN/H$_2$O (9:1)	3.3×10^9	84B033
	Reductive Transfer			
	Triethylamine	MeCN/H$_2$O (9:1)	1.5×10^9	86A248
55.	**Benzophenone, 4-phenyl-**			
	Energy Transfer			
	2-Nitrothiophene	MeCN	2.3×10^9	82A153
	Reductive Transfer			
	2,3-Dimethyl-2-butene	MeCN	5.2×10^6	776369
56.	**Benzophenone, 4-(trifluoromethyl)-**			
	O_2	MeCN	1.6×10^9	85A268
	Oxidative Transfer			
	MV^{2+}	MeCN/H$_2$O (9:1)	1.7×10^9	84B033
	Reductive Transfer			
	Triethylamine	MeCN/H$_2$O (9:1)	4.2×10^9	86A248
57.	**Benzo[*g*]pteridine-2,4-dione**			
	O_2	H$_2$O	1.7×10^9	777617
58.	**Benzo[*g*]pteridine-2,4-dione, 3,7,8,10-tetramethyl-**			
	Reductive Transfer			
	Allylthiourea	H$_2$O (pH 7)	2.8×10^9	84F284
	Triethylamine	MeCN	5.8×10^9	82F433
59.	**Benzo[*g*]pteridine-2,4-dione, 7,8,10-trimethyl-**			
	O_2	H$_2$O (pH 7.2)	1.2×10^9	88A385
	Reductive Transfer			
	Adenine	H$_2$O (pH 7.2)	1.6×10^9	85A186
	EDTA	H$_2$O (pH 7)	3.1×10^8	85A374
	Indole	H$_2$O (pH 7.2)	3.7×10^9	88A385
	Tryptophan	H$_2$O (pH 7.2)	3.7×10^9	85A186

Table 9—**Triplet-State Quenching Rates**—Continued

No.	Quencher	Solvent (pH)	k_q /L mol^{-1} s^{-1}	Ref.
60.	**Benzo[*g*]pteridine-2,4-dione, 7,8,10-trimethyl-, conjugate monoacid**			
	Reductive Transfer			
	EDTA	H$_2$O (pH 3)	1.4×10^8	85A374
	Naphthalene	MeOH	7.9×10^7	81E041
61.	**Benzo[*g*]pteridine-2,4-dione, 7,8,10-trimethyl-, ion(1-)**			
	Reductive Transfer			
	EDTA	H$_2$O (pH 14)	1.5×10^8	85A374
62.	**Benzo[*a*]pyrene**			
	Di-*tert*-butylnitroxide	Benzene	8.0×10^6	84E319
	O$_2$	Benzene	3.2×10^9	84E319
	O$_2$	Cyclohexane	2.6×10^9	706182
	Energy Transfer			
	Azulene	Benzene	5.6×10^9	84E319
	Biacetyl	Benzene	4.9×10^7	64E011
	Ferrocene	Benzene	4.5×10^9	84E319
	Oxidative Transfer			
	MV^{2+}	MeCN/H$_2$O (9:1)	5.4×10^9	83A113
63.	**Benzo[*e*]pyrene**			
	Energy Transfer			
	Methyl cinnamate	Benzene	7.3×10^7	86A322
64.	**1,4-Benzoquinone**			
	O$_2$	Heptane	3×10^8	82A370
	O$_2$	MeCN	3×10^8	82A370
65.	**1,4-Benzoquinone, 2,5-bis(4-chlorophenyl)-**			
	O$_2$	Heptane	3.6×10^9	82A370
	O$_2$	MeCN	1.1×10^9	82A370
66.	**1,4-Benzoquinone, 4-chlorophenyl-**			
	O$_2$	MeCN	9×10^8	82A370
67.	**1,4-Benzoquinone, 2,6-diphenyl-**			
	Charge Transfer			
	Triphenylamine	Toluene	5.0×10^9	79B044
68.	**1,4-Benzoquinone, tetrachloro-**			
	Charge Transfer			
	Naphthalene	Benzene	1.4×10^8	88E648
	Reductive Transfer			
	Indole	MeCN	1.3×10^{10}	85A336

Table 9—**Triplet-State Quenching Rates**—Continued

No.	Quencher	Solvent (pH)	k_q /L mol^{-1} s^{-1}	Ref.
69.	**1,4-Benzoquinone, tetramethyl-**			
	O_2	Cyclohexane	1.3×10^9	717520
	O_2	MeCN	7×10^8	82A370
	Energy Transfer			
	Anthracene	Benzene	6.8×10^9	717520
	$Cu(acac)_2$	MeOH	3.8×10^9	83F123
	I$^-$	H_2O (pH 7)	9×10^9	80C005
	Reductive Transfer			
	I$^-$	MeCN	9.9×10^9	82A154
	Triphenylamine	MeCN	1.0×10^{10}	777602
70.	**Benzo[*b*]triphenylene**			
	O_2	Benzene	1.8×10^9	736097
	O_2	Cyclohexane	1.5×10^9	706182
	Energy Transfer			
	$Cr(acac)_3$	Benzene	1.2×10^9	81E374
	β-Ionone	Toluene	5.0×10^8	85E293
	Methyl cinnamate	Benzene	2.4×10^6	86A322
	Oxidative Transfer			
	MV^{2+}	MeCN/H_2O (9:1)	6.1×10^9	83A113
71.	**Biacetyl**			
	Energy Transfer			
	Acridine	Benzene	9×10^9	69A001
	Benzil	Cyclohexane	1.5×10^9	73E373
	1,8-Dinitronaphthalene	Benzene	2.6×10^9	64E011
	I$^-$	H_2O	5.5×10^9	756251
	Piperylene	CCl_4	2.5×10^8	741013
	Reductive Transfer			
	Aniline	Benzene	2.9×10^8	704011
	2-Propanol	MeCN	1.3×10^4	696078
	Triethylamine	Benzene	5.0×10^7	69A001
72.	**9,9′-Bicarbazole**			
	Energy Transfer			
	Naphthalene	Cyclohexane	6.2×10^9	78A368

Table 9—**Triplet-State Quenching Rates**—Continued

No.	Quencher	Solvent (pH)	k_q /L mol^{-1} s^{-1}	Ref.
73.	**Bilirubin**			
	O$_2$	Benzene	8.2×10^8	761168
74.	**2,2'-Binaphthyl**			
	Energy Transfer			
	Biacetyl	Benzene	1.3×10^9	64E011
75.	**Biphenyl**			
	I$^-$	MeCN	6.2×10^3	746194
	O$_2$	Benzene	1.4×10^9	78E263
	Energy Transfer			
	β-Carotene	Cyclohexane	8.4×10^9	83B121
	(*all-E*)-Retinal	Hexane	2.0×10^{10}	776412
	Oxidative Transfer			
	MV^{2+}	MeCN/H$_2$O (9:1)	9.3×10^9	83A113
	Reductive Transfer			
	Diethyl sulfide	Cyclohexane	4.8×10^6	84A363
	Diethyl sulfide	MeCN	1.0×10^7	84A363
76.	**2,2'-Bipyridine**			
	O$_2$	Cyclohexane	5.2×10^8	88E253
77.	**1,3-Butadiene, 1,4-diphenyl-**			
	O$_2$	Benzene	5.0×10^9	84E319
	O$_2$	Cyclohexane	5.3×10^9	82E365
	O$_2$	MeCN	4.0×10^9	82E429
	Energy Transfer			
	Azulene	Benzene	2.9×10^9	84E319
	Ferrocene	Benzene	1.5×10^9	82E429
78.	**1,3-Butadiene, 1,1,4,4-tetraphenyl-**			
	Energy Transfer			
	β-Carotene	Toluene	4.3×10^9	84E144
79.	**2-Butanone**			
	O$_2$	Hexane	3.4×10^{10}	84B051
	O$_2$	MeCN	4.3×10^{10}	84B051
80.	**Butyraldehyde**			
	Energy Transfer			
	2,5-Dimethyl-2,4-hexadiene	Pentane	1.5×10^{10}	81E442

Table 9—**Triplet-State Quenching Rates**—Continued

No.	Quencher	Solvent (pH)	k_q /L mol^{-1} s^{-1}	Ref.
81.	**Carbazole**			
	O$_2$	Benzene	5.7×10^9	771014
	Energy Transfer			
	Ferrocene	Cyclohexane	6×10^9	78A368
	Reductive Transfer			
	Pyridine	Cyclohexane	4.9×10^7	84F039
82.	**β-*apo*-14′-Carotenal**			
	Di-*tert*-butylnitroxide	MCH	1.0×10^9	84E180
	O$_2$	Cyclohexane	4.7×10^9	84F005
83.	**β-Carotene**			
	Di-*tert*-butylnitroxide	MCH	6.8×10^8	84E180
	O$_2$	Benzene	3.6×10^9	73E347
	Energy Transfer			
	I$_2$	Benzene	2.5×10^9	84A286
	I$_2$	MeCN	1.5×10^9	84A286
84.	**Chlorophyll *a***			
	O$_2$	Benzene	1.1×10^9	573003
	Energy Transfer			
	β-Carotene	Benzene	1.2×10^9	73E347
	Tetracene	Toluene	6×10^8	697266
	Oxidative Transfer			
	1,4-Benzoquinone	Benzene	2.4×10^9	573003
	α-Tocopherylquinone	EtOH	7×10^8	70F737
	Reductive Transfer			
	TMPD	EtOH	2.7×10^6	78E589
85.	**Chlorophyll *b***			
	Energy Transfer			
	1,6-Diphenyl-1,3,5-hexatriene	Hexane	2×10^5	761088
	Oxidative Transfer			
	1,4-Benzoquinone	Benzene	1.5×10^9	573003
	Reductive Transfer			
	Phenylhydrazine	Toluene	5.0×10^4	697266

Table 9—**Triplet-State Quenching Rates**—Continued

No.	Quencher	Solvent (pH)	k_q /L mol^{-1} s^{-1}	Ref.
86.	**Chrysene**			
	Di-*tert*-butylnitroxide	Hexane	9.6×10^8	736099
	O$_2$	Benzene	1.4×10^9	736097
	O$_2$	Cyclohexane	1.0×10^9	706182
	Energy Transfer			
	Azulene	Benzene	8×10^9	81E214
	Azulene	MeCN	1.3×10^{10}	81E214
	Cr(acac)$_3$	Benzene	1.5×10^9	81E374
	Oxidative Transfer			
	MV^{2+}	MeCN/H$_2$O (9:1)	7.4×10^9	83A113
87.	**Coronene**			
	Di-*tert*-butylnitroxide	Hexane	3.3×10^8	736099
	O$_2$	Benzene	4.2×10^8	736097
	Energy Transfer			
	Cr(acac)$_3$	Benzene	1.6×10^9	81E374
	Methyl cinnamate	Benzene	3.1×10^8	86A322
	Oxidative Transfer			
	MV^{2+}	MeCN/H$_2$O (9:1)	5.4×10^9	83A113
88.	**Coumarin, 3-benzoyl-**			
	Energy Transfer			
	Methyl cinnamate	Benzene	3.7×10^9	86A322
89.	**Coumarin, 3,3′-carbonylbis-**			
	Energy Transfer			
	Methyl cinnamate	Benzene	2.8×10^9	86A322
90.	**Coumarin, 3,3′-carbonylbis(5,7-diethoxy-**			
	Energy Transfer			
	Methyl cinnamate	Benzene	1.5×10^9	86A322
91.	**Coumarin, 3,3′-carbonylbis(7-diethylamino-**			
	Energy Transfer			
	Methyl cinnamate	Benzene	7.7×10^6	86A322
92.	**Coumarin, 3,3′-carbonylbis(5,7-dimethoxy-**			
	Energy Transfer			
	Methyl cinnamate	Benzene	1.6×10^9	86A322

Table 9—**Triplet-State Quenching Rates**—Continued

No.	Quencher	Solvent (pH)	k_q /L mol^{-1} s^{-1}	Ref.
93.	**Coumarin, 7-(diethylamino)-4-methyl-**			
	Energy Transfer			
	Anthracene	EtOH	6.9×10^9	747049
94.	**Crocetin**			
	O_2	H_2O (pH 8)	1.4×10^9	83B068
95.	**Cyclobutene, 1,2-diphenyl-**			
	O_2	Benzene	3.4×10^9	81E214
96.	**1,3-Cycloheptadiene**			
	Energy Transfer			
	β-Carotene	Toluene	9.0×10^9	84E144
97.	**1,3-Cyclohexadiene**			
	Energy Transfer			
	Azulene	Benzene	1.0×10^{10}	84E491
	Oxidative Transfer			
	MV^{2+}	MeOH	2.6×10^9	82E484
98.	**Cyclohexanone**			
	Energy Transfer			
	9,10-Dibromoanthracene	Cyclohexane	4.5×10^9	84E208
99.	**Cyclohexene, 1-phenyl-**			
	Oxidative Transfer			
	MV^{2+}	MeOH	1.4×10^9	82E526
100.	**2-Cyclohexenethione, 3,5,5-trimethyl-**			
	O_2	Benzene	3.7×10^9	86A240
101.	**2-Cyclohexen-1-one**			
	O_2	Cyclohexane	5×10^9	80B055
	Reductive Transfer			
	Triethylamine	Cyclohexane	9.2×10^7	88E675
	Triethylamine	MeCN	9.0×10^7	88E675
102.	**2-Cyclohexen-1-one, 4,4-dimethyl-**			
	Reductive Transfer			
	Triethylamine	MeCN	3.7×10^7	88E675
103.	**2-Cyclohexen-1-one, 4,4,6,6-tetramethyl-**			
	Reductive Transfer			
	Triethylamine	Cyclohexane	1.3×10^8	88E675

Table 9—**Triplet-State Quenching Rates**—Continued

No.	Quencher	Solvent (pH)	k_q /L mol^{-1} s^{-1}	Ref.
104.	**Cyclopentadiene**			
	Energy Transfer			
	Anthracene	Benzene	9.8×10^9	81E270
	Naphthalene	Benzene	2.0×10^7	81E270
105.	**Cyclopentene, 1-phenyl-**			
	Oxidative Transfer			
	MV^{2+}	MeOH	4.5×10^9	82E526
106.	**2-Cyclopentenone**			
	O_2	Cyclohexane	5×10^9	80B055
107.	**Dibenz[*a,h*]anthracene**			
	O_2	Benzene	1.6×10^9	736097
	O_2	Cyclohexane	1.3×10^9	706182
	Energy Transfer			
	Ferrocene	EtOH	5.0×10^9	747190
	Rubrene	Benzene	2.2×10^9	81E346
	Oxidative Transfer			
	MV^{2+}	MeCN/H_2O (9:1)	7.5×10^9	83A113
108.	**Dibenzo[*b,def*]chrysene**			
	Di-*tert*-butylnitroxide	Hexane	1.9×10^7	736099
	O_2	Benzene	2.8×10^9	736097
	Energy Transfer			
	(*E*)-Azobenzene	Benzene	1.0×10^7	81E012
109.	**Dibenzo[*def,mno*]chrysene**			
	Energy Transfer			
	Ferrocene	Benzene	5.8×10^8	756237
	Rubrene	Benzene	1.6×10^9	81E346
110.	**Dibenzo[*g,p*]chrysene**			
	Energy Transfer			
	Methyl cinnamate	Benzene	3.4×10^5	86A322
111.	**2,6-Dimethylquinoline**			
	Energy Transfer			
	Methyl cinnamate	Benzene	4.0×10^9	86A322
112.	**1,4-Dioxin, 2,3,5,6-tetraphenyl-**			
	O_2	Benzene	1.7×10^9	79A241

Table 9—**Triplet-State Quenching Rates**—Continued

No.	Quencher	Solvent (pH)	k_q /L mol^{-1} s^{-1}	Ref.
113.	**Dodecapreno-β-carotene**			
	O_2	Benzene	5.6×10^9	73E347
114.	**Ethene, tetraphenyl-**			
	O_2	Benzene	2.4×10^9	82E204
	O_2	MeCN	2.5×10^9	82E204
115.	**Flavine mononucleotide**			
	Reductive Transfer			
	Glycyl-L-tyrosine	H_2O (pH 7.0)	1.0×10^9	79A260
	L-Methionine	H_2O (pH 7.0)	1.7×10^7	79A260
116.	**Fluoranthene**			
	TEMPO	MeCN	4.3×10^8	80E116
	Energy Transfer			
	Biacetyl	Benzene	2.1×10^7	64E011
	Oxidative Transfer			
	MV^{2+}	MeCN/H_2O (9:1)	1.0×10^{10}	83A113
117.	**Fluorene**			
	O_2	Benzene	4.9×10^9	771014
	Energy Transfer			
	Dibutyldiazene	Benzene	1.8×10^8	756208
	Ruthenocene	Benzene	6.4×10^8	777439
	Oxidative Transfer			
	MV^{2+}	MeCN/H_2O (9:1)	8.2×10^9	83A113
118.	**9-Fluorenone**			
	Norbornadiene	Benzene	4.6×10^5	78E131
	Energy Transfer			
	Azulene	Benzene	8×10^9	81F275
	Cu(acac)$_2$	Benzene	1.0×10^9	66E093
	Ferrocene	Benzene	5.1×10^9	66E093
	Reductive Transfer			
	Triethylamine	Cyclohexane	1.0×10^7	78A343
	Triethylamine	MeCN	4.7×10^8	78A343
119.	**Fluorescein, conjugate monoacid**			
	O_2	H_2O (pH −4)	2×10^8	647009
	Oxidative Transfer			
	Fe^{3+}	H_2O (pH 1.6)	1×10^9	60A001

Table 9—**Triplet-State Quenching Rates**—Continued

No.	Quencher	Solvent (pH)	k_q /L mol^{-1} s^{-1}	Ref.
120.	**Fluorescein dianion**			
	O_2	H_2O (pH 12)	1.7×10^9	647009
	Reductive Transfer			
	Allylthiourea	H_2O (pH 12)	2×10^5	60A001
	p-Phenylenediamine	H_2O (pH 12)	5×10^9	60A001
121.	**Fluorescein dianion, 2′,4′,5′,7′-tetrabromo-**			
	O_2	H_2O	$\sim 3 \times 10^8$	677361
	Energy Transfer			
	Ferrocene	EtOH	3.5×10^9	747190
	Oxidative Transfer			
	MV^{2+}	EtOH/H_2O (1:1)	3.2×10^9	87F258
	Reductive Transfer			
	Allylthiourea	H_2O (pH 7.1)	4.9×10^5	747304
	Aniline	H_2O (pH 10.4)	1.4×10^9	687222
	2-Naphthol	H_2O (pH 10.5)	1.5×10^9	687222
	L-Tryptophan	H_2O (pH 7.1)	8.4×10^8	747304
122.	**Fluorescein dianion, 3,4,5,6-tetrachloro-2′,4′,5′,7′-tetraiodo-**			
	O_2	H_2O (pH 7.2)	9.8×10^8	88A385
	Reductive Transfer			
	EDTA	H_2O (pH 5.0)	1.5×10^6	85A166
123.	**Fluorescein monoanion**			
	O_2	H_2O (pH 4.5)	1.2×10^9	647009
	Reductive Transfer			
	EDTA	H_2O (pH 5.0)	3.9×10^6	85A166
124.	**Fluorescein monoanion, 2′,4′,5′,7′-tetrabromo-**			
	Oxidative Transfer			
	I^-	H_2O (pH 6.5)	1.5×10^7	81A170
	Reductive Transfer			
	$Fe(CN)_6^{4-}$	H_2O (pH 6.5)	5.7×10^8	81A170
125.	**2-Furoic acid, 5-nitro-**			
	O_2	Acetone	1.7×10^9	81A140
	Energy Transfer			
	Azulene	Acetone	1.2×10^{10}	81A140
	Reductive Transfer			
	I^-	MeCN	6.6×10^9	81A140
	TMPD	Me_2CO	6.0×10^9	81A140

Table 9—**Triplet-State Quenching Rates**—Continued

No.	Quencher	Solvent (pH)	k_q /L mol^{-1} s^{-1}	Ref.
126.	**D-Glucose phenylosazone**			
	O_2	EtOH	1×10^9	83E018
127.	**2-Heptanone**			
	Styrene	Benzene	4.0×10^9	79P084
128.	**(*E,E,E*)-2,4,6-Heptatrienal, 5-methyl-7-(2,6,6-trimethyl-1-cyclohexen-1-yl)-**			
	O_2	Cyclohexane	3.3×10^9	84F005
	O_2	MeOH	4.0×10^9	84F005
	Energy Transfer			
	Azulene	Toluene	3.1×10^7	84E180
	Tetracene	Toluene	3.3×10^9	84E180
129.	**(*E,E*)-2,4-Hexadiene**			
	Oxidative Transfer			
	MV^{2+}	MeOH	1.0×10^9	82E484
130.	**2,4-Hexadiene, 2,5-dimethyl-**			
	Energy Transfer			
	Anthracene	Benzene	2.8×10^9	81E270
	Oxidative Transfer			
	MV^{2+}	MeOH	2.0×10^9	82E484
131.	**1,3,5-Hexatriene, 1,6-diphenyl-**			
	EtI	Cyclohexane	7.0×10^4	82E365
	EtI	MeOH	5.2×10^4	82E365
	O_2	Benzene	5.6×10^9	82E429
	O_2	Cyclohexane	4.4×10^9	82E365
	O_2	MeCN	5.8×10^9	82E429
132.	**Indene, 2-phenyl-**			
	O_2	Benzene	3.8×10^9	81E214
133.	**Indeno[2,1-*a*]indene**			
	O_2	Benzene	3.4×10^9	81E214
	Energy Transfer			
	Azulene	MeCN	1.5×10^{10}	80A274
134.	**Indole**			
	O_2	Cyclohexane	1.6×10^{10}	88R179
	O_2	EtOH (abs)	1.1×10^{10}	88R179
	Energy Transfer			
	Anthracene	Cyclohexane	9×10^9	81E082

Table 9—**Triplet-State Quenching Rates**—Continued

No.	Quencher	Solvent (pH)	k_q /L mol^{-1} s^{-1}	Ref.
135.	**Indole, 5-methoxy-**			
	O$_2$	Cyclohexane	2.1×10^{10}	88R179
136.	**Indole, 1-methyl-**			
	O$_2$	Benzene	1.4×10^{10}	771014
	Charge Transfer			
	9,10-Anthraquinone	Benzene	7.8×10^9	771021
137.	**β-Ionone**			
	Di-*tert*-butylnitroxide	MCH	4.2×10^9	84E180
	O$_2$	Toluene	5.1×10^9	85E293
	Energy Transfer			
	β-Carotene	Toluene	3.1×10^9	85E293
138.	**Isobenzofuran, 1,3-diphenyl-**			
	Energy Transfer			
	Azulene	Benzene	6.5×10^7	81E346
139.	**(*all-E*)-Lycopene**			
	O$_2$	Hexane	5.4×10^9	733001
140.	**Mesoporphyrin, dimethyl ester**			
	O$_2$	Benzene	1.4×10^9	80E200
141.	**Methylene Blue cation**			
	O$_2$	MeCN	1.7×10^9	84E216
	Energy Transfer			
	Fe(acac)$_3$	Benzene/EtOH (9:1)	5.4×10^8	767094
	Oxidative Transfer			
	Benzenediazonium cation	MeCN	6.7×10^6	86A507
	Reductive Transfer			
	Allylthiourea	MeCN	8.3×10^6	86A507
	DABCO	MeCN	1.9×10^8	86A507
	Triethylamine	MeOH	4.7×10^7	767574
	Triphenylamine	MeCN	3.0×10^9	77A203
142.	**Methylene Blue cation, conjugate monoacid**			
	Reductive Transfer			
	EDTA	H$_2$O (pH 4.5)	5×10^8	747039
	Fe^{II}(CN)$_6^{4-}$	H$_2$O (pH 4.4)	1.4×10^{10}	80A171
	Ferrocene	EtOH/H$_2$O (19:1)	4.9×10^9	80A171
	Methylene Blue cation	H$_2$O	1.1×10^8	81A087

Table 9—**Triplet-State Quenching Rates**—Continued

No.	Quencher	Solvent (pH)	k_q /L mol^{-1} s^{-1}	Ref.
142.	**Methylene Blue cation, conjugate monoacid**—Continued			
	Thionine cation	MeCN/H$_2$O (1:1)	9.0×10^7	82E232
143.	**2-Naphthaldehyde**			
	Oxidative Transfer			
	MV^{2+}	MeCN/H$_2$O (9:1)	3.8×10^9	81A078
	Reductive Transfer			
	Triethylamine	Benzene	1.3×10^7	78A343
144.	**Naphthalene**			
	O$_2$	Benzene	1.5×10^9	78E263
	O$_2$	MeCN	2×10^9	88R179
	TEMPO	MeCN	9.6×10^8	80E116
	Energy Transfer			
	Anthracene	EtOH	1.3×10^{10}	747049
	Cr(hfac)$_3$	Benzene	8.2×10^9	83E054
	Ferrocene	EtOH	7.0×10^9	747190
	(*all-E*)-Retinal	Benzene	5.5×10^9	81B008
	Oxidative Transfer			
	MV^{2+}	MeCN/H$_2$O (9:1)	8.5×10^9	83A113
	Reductive Transfer			
	Diethyl sulfide	MeCN	1.9×10^6	84A363
145.	**Naphthalene, 2-acetyl-**			
	Di-*tert*-butylnitroxide	Benzene	7.6×10^8	84A344
	O$_2$	Benzene	1.7×10^9	84A344
	Energy Transfer			
	Azulene	Benzene	9.7×10^9	84A344
	Azulene	MeCN	1.4×10^{10}	81E214
	Cu(acac)$_2$	Benzene	1.9×10^9	66E093
	Ferrocene	Benzene	7.6×10^9	84A344
	β-Ionone	Benzene	2.2×10^9	84A344
	Quadricyclane	Benzene	1.0×10^6	78E131
	Oxidative Transfer			
	MV^{2+}	MeCN/H$_2$O (9:1)	5.4×10^9	81A078
	Reductive Transfer			
	Phenoxide ion	MeCN/H$_2$O (1:1)	2.6×10^9	81A114
	Triethylamine	Cyclohexane	1.2×10^5	78A343

Table 9—**Triplet-State Quenching Rates**—Continued

No.	Quencher	Solvent (pH)	k_q /L mol^{-1} s^{-1}	Ref.
146.	**Naphthalene, 1-bromo-**			
	Energy Transfer			
	Biacetyl	Benzene	3.4×10^9	64E011
147.	**Naphthalene, 1-chloro-**			
	Energy Transfer			
	Biacetyl	Cyclohexane	4.7×10^9	73E373
148.	**Naphthalene, 1,4-dicyano-**			
	O_2	MeCN	2.1×10^9	84B066
	Energy Transfer			
	2,5-Dimethyl-2,4-hexadiene	MeCN	1.7×10^9	84B066
149.	**Naphthalene, 1,4-dinitro-**			
	O_2	EtOH	2.0×10^9	81B064
	O_2	Hexane	1.7×10^9	81B064
	Energy Transfer			
	Tetracene	Benzene	1.0×10^{10}	767270
150.	**Naphthalene, 1,8-dinitro-**			
	O_2	EtOH	2.1×10^9	776194
	Energy Transfer			
	Tetracene	Benzene	6.7×10^9	776194
	Reductive Transfer			
	Triethylamine	MeCN	4.6×10^9	776194
151.	**Naphthalene, 2-hydroxy-**			
	O_2	Cyclohexane	2.6×10^9	737113
	Oxidative Transfer			
	Pyridine	Cyclohexane	1.5×10^9	737113
152.	**Naphthalene, 1-methyl-**			
	Energy Transfer			
	β-Ionone	Toluene	2.7×10^9	85E293
	Oxidative Transfer			
	MV^{2+}	MeCN/H_2O (9:1)	7.8×10^9	83A113
153.	**Naphthalene, 1-[2-(1-naphthyl)ethenyl]-, (*E*)-**			
	O_2	Benzene	4.4×10^9	84E237
	Energy Transfer			
	Azulene	Benzene	2.9×10^9	84E237

Table 9—**Triplet-State Quenching Rates**—Continued

No.	Quencher	Solvent (pH)	k_q /L mol^{-1} s^{-1}	Ref.
154.	**Naphthalene, 1-[2-(2-naphthyl)ethenyl]-, (*E*)-**			
	O$_2$	Benzene	3.8×10^9	84E237
	Energy Transfer			
	Azulene	Benzene	3.7×10^9	84E237
155.	**Naphthalene, 2-[2-(2-naphthyl)ethenyl]-, (*E*)-**			
	O$_2$	Benzene	6.0×10^9	84E237
	Energy Transfer			
	Azulene	Benzene	4.4×10^9	84E237
156.	**Naphthalene, 1-[1-(1-naphthyl)ethenyl]-**			
	O$_2$	Benzene	3.7×10^9	84B007
	Energy Transfer			
	Ferrocene	Benzene	4.4×10^9	84B007
157.	**Naphthalene, 1-nitro-**			
	O$_2$	EtOH	3.3×10^9	81B064
	O$_2$	Hexane	1.3×10^9	81B064
	Energy Transfer			
	Triphenylethylene	Benzene	1.8×10^9	82E204
158.	**Naphthalene, 2-nitro-**			
	O$_2$	EtOH	1.6×10^9	81B064
	O$_2$	Hexane	1.7×10^9	81B064
	Energy Transfer			
	Azulene	Benzene	7×10^9	81E214
	Azulene	MeCN	1.4×10^{10}	81E214
	Tetracene	Benzene	7.4×10^9	767269
159.	**2-Naphthalenesulfonate ion**			
	Charge Transfer			
	Fe(CN)$_6^{4-}$	H$_2$O	1.4×10^7	767189
160.	**1,4-Naphthoquinone, 2-methyl-**			
	O$_2$	H$_2$O (pH 7.0)	1.2×10^9	83E311
	Reductive Transfer			
	Thymine	H$_2$O (pH 7.0)	2.7×10^9	83E311
161.	**1,3,5,7-Octatetraene, 1,8-diphenyl-**			
	O$_2$	Benzene	6.0×10^9	82E429
	O$_2$	Cyclohexane	4.8×10^9	82E365
	O$_2$	MeCN	6.4×10^9	82E429

Table 9—**Triplet-State Quenching Rates**—Continued

No.	Quencher	Solvent (pH)	k_q /L mol^{-1} s^{-1}	Ref.
162.	**Orotate ion**			
	O_2	H_2O (pH 6.2)	3.0×10^9	71E367
163.	**Orotic acid**			
	O_2	H_2O (pH 1.1)	2.2×10^9	71E367
164.	**1,3,4-Oxadiazole, 2,5-diphenyl-**			
	O_2	Benzene	1.6×10^9	777265
165.	**Oxazole, 2,5-bis(4-biphenylyl)-**			
	O_2	Benzene	1.7×10^9	777265
166.	**Oxazole, 2,5-diphenyl-**			
	O_2	Benzene	2.5×10^9	777265
	Energy Transfer			
	Anthracene	Cyclohexane	4.7×10^9	80E439
167.	**Oxazole, 2-(1-naphthyl)-5-phenyl-**			
	O_2	Benzene	2.3×10^9	777265
168.	**Oxirane, 2,3-di-(2-naphthyl)-, (Z)-**			
	O_2	Benzene	1.5×10^9	84A344
	Energy Transfer			
	Azulene	Benzene	9.7×10^9	84A344
169.	**Pentacene**			
	O_2	Benzene	1.7×10^9	736097
	Charge Transfer			
	Cr(hfac)$_3$	Benzene	6.0×10^8	83E054
170.	**1,3-Pentadiene, 2-methyl-, (*E*)-**			
	Energy Transfer			
	β-Carotene	Toluene	5.0×10^9	84E144
	Oxidative Transfer			
	MV^{2+}	MeOH	4×10^8	82E484
171.	**Pentahelicene**			
	O_2	MCH	5.7×10^8	79A237
172.	**2-Pentanone**			
	O_2	Hexane	3.4×10^{10}	84B051
	O_2	MeCN	3.2×10^{10}	84B051
	O_2	MeOH	2.9×10^{10}	84B051

Table 9—**Triplet-State Quenching Rates**—Continued

No.	Quencher	Solvent (pH)	k_q /L mol^{-1} s^{-1}	Ref.
173.	**Perylene**			
	TEMPO	MeCN	2.1×10^6	80E116
	Energy Transfer			
	Cr(acac)$_3$	Benzene	4.7×10^7	81E374
	Ferrocene	EtOH	1.3×10^9	747190
	Oxidative Transfer			
	MV^{2+}	MeCN/H$_2$O (9:1)	1.9×10^9	83A113
174.	**Phenalene, 2,3-dihydro-**			
	O$_2$	EPA	2×10^9	81F390
175.	**Phenanthrene**			
	O$_2$	Benzene	2.0×10^9	736097
	Energy Transfer			
	Azulene	Benzene	6.6×10^9	81E214
	Azulene	MeCN	1.0×10^{10}	81E214
	Cu(acac)$_2$	MeOH	2.2×10^9	83F123
	Oxidative Transfer			
	Cu^{2+}	MeOH/H$_2$O (9:1)	1.0×10^8	78A333
	Reductive Transfer			
	tert-Butyl hydroperoxide	Benzene	2.3×10^7	83A110
176.	**Phenazine**			
	Energy Transfer			
	Ferrocene	EtOH	4.6×10^9	747190
177.	**Phenol**			
	O$_2$	H$_2$O (pH 7.1)	6.1×10^9	757161
178.	**Phenol, 4-methyl-**			
	O$_2$	H$_2$O (pH 7.5)	5.3×10^9	757161
179.	**Phenothiazine**			
	O$_2$	MeOH	2.4×10^{10}	757353
	Oxidative Transfer			
	Cu^{2+}	MeOH	6.0×10^9	757353
180.	**Phenothiazine, 10-methyl-**			
	Oxidative Transfer			
	Cu^{2+}	H$_2$O/EtOH (2:1)	1.0×10^9	79N005

Table 9—**Triplet-State Quenching Rates**—Continued

No.	Quencher	Solvent (pH)	k_q /L mol^{-1} s^{-1}	Ref.
181.	**Phenothiazinium, 3-(dimethylamino)-7-(methylamino)-**			
	Reductive Transfer			
	EDTA	H$_2$O (pH 8.2)	5.5×10^7	747039
182.	**Phenothiazinium, 3-(dimethylamino)-7-(methylamino)-, conjugate monoacid**			
	Reductive Transfer			
	EDTA	H$_2$O (pH 4.5)	9×10^8	747039
183.	**Phenoxazinium, 3,7-bis(diethylamino)-**			
	O$_2$	EtOH	8.3×10^8	82E456
	Energy Transfer			
	1,3,5,7-Cyclooctatetraene	EtOH	1.7×10^7	82E456
184.	**Phenoxazinium, 3,7-bis(diethylamino)-, conjugate monoacid**			
	O$_2$	EtOH	3.0×10^8	82E456
	Energy Transfer			
	1,3,5,7-Cyclooctatetraene	EtOH	8.7×10^6	82E456
185.	**Phenoxazinium, 3,7-diamino-**			
	Reductive Transfer			
	Allylthiourea	MeOH	1.8×10^6	767246
	EDTA	H$_2$O (pH 8.0)	1.5×10^7	777063
186.	**Phenylalanine**			
	O$_2$	H$_2$O (pH 7.5)	3.3×10^9	757162
187.	**Phenylalanine, *N*-acetyl-**			
	O$_2$	H$_2$O (pH 8.1)	3.9×10^9	757162
188.	**Phthalazine**			
	O$_2$	H$_2$O (pH 7.1)	1.4×10^9	757309
189.	**Picene**			
	O$_2$	Benzene	1.4×10^9	736097
190.	**Pivalophenone, 4′-methoxy-**			
	Energy Transfer			
	2,5-Dimethyl-2,4-hexadiene	Benzene	5.0×10^9	85E449
191.	**Pivalothiophenone, 4-chloro-**			
	Di-*tert*-butylnitroxide	Benzene	2.4×10^9	87A340
	Energy Transfer			
	Ferrocene	Benzene	6.6×10^9	87A340
	Reductive Transfer			
	Triethylamine	Benzene	7.8×10^6	87A340

Table 9—**Triplet-State Quenching Rates**—Continued

No.	Quencher	Solvent (pH)	k_q /L mol^{-1} s^{-1}	Ref.
192.	**Pivalothiophenone, 4-methoxy-**			
	Di-*tert*-butylnitroxide	Benzene	2.0×10^9	87A340
	Energy Transfer			
	Ferrocene	Benzene	5.5×10^9	87A340
	Reductive Transfer			
	Triethylamine	Benzene	4.1×10^6	87A340
193.	**Porphine, tetrakis(4-sulfonatophenyl)-**			
	O_2	H_2O (pH 7)	1.9×10^9	81E084
	Oxidative Transfer			
	1,4-Benzoquinone	H_2O (pH 7)	3.4×10^9	81E084
	Nitrobenzene	H_2O (pH 7)	3.2×10^6	81E084
194.	**Porphine-2,18-dipropanoic acid, 7,12-diethenyl-3,8,13,17-tetramethyl-, dimethyl ester**			
	O_2	Benzene	2.7×10^9	771078
	Energy Transfer			
	β-Carotene	Benzene	1.8×10^9	771078
	Tetracene	Benzene	5×10^8	771078
195.	**Porphine-2,18-dipropanoic acid, 3,7,12,17-tetramethyl-, dimethyl ester**			
	O_2	Benzene	2.3×10^9	80E200
196.	**Propiophenone**			
	Di-*tert*-butylnitroxide	Benzene	2.3×10^9	79A284
	Charge Transfer			
	Cu(acac)$_2$	MeOH	7.6×10^9	83F123
	1-Methylnaphthalene	Benzene	1.4×10^{10}	84E018
	Oxidative Transfer			
	MV^{2+}	MeCN/H$_2$O (9:1)	5.5×10^9	84E018
197.	**Propiophenone, 2,3-epoxy-4′-methoxy-3-phenyl-**			
	O_2	Benzene	3.9×10^9	85B078
	Energy Transfer			
	Azulene	Benzene	8.3×10^9	85B078
	(*E*)-Stilbene	Benzene	6.0×10^9	85B078
	Reductive Transfer			
	4-Methoxyphenol	Benzene	4.9×10^9	85B078
198.	**Psoralen**			
	O_2	H_2O (pH 8)	3.3×10^9	83B068
	Energy Transfer			
	(*all-E*)-Retinol	EtOH	1.1×10^9	79E678

Table 9—**Triplet-State Quenching Rates**—Continued

No.	Quencher	Solvent (pH)	k_q /L mol^{-1} s^{-1}	Ref.
199.	**Psoralen, 3-carbethoxy-**			
	O$_2$	H$_2$O	3.3×10^9	83E324
	Energy Transfer			
	(*all-E*)-Retinol	EtOH	9.6×10^8	82E133
	Reductive Transfer			
	Tryptophan	H$_2$O	3.6×10^9	82E133
200.	**Psoralen, 3-carbethoxy-4′,5′-dihydro-**			
	O$_2$	H$_2$O	3×10^9	82E133
	Reductive Transfer			
	Tryptophan	H$_2$O	2.6×10^9	82E133
201.	**Psoralen, 8-methoxy-**			
	O$_2$	H$_2$O	3.3×10^9	79A114
	Tryptophan	MeOH	3.5×10^8	79B042
	Energy Transfer			
	(*all-E*)-Retinol	EtOH	5.3×10^9	79E678
202.	**Psoralen, 4,5′,8-trimethyl-**			
	Tryptophan	MeOH	6.9×10^8	79B042
203.	**4-Pteridinone, 2-amino-**			
	O$_2$	H$_2$O (pH 9.2)	1.3×10^9	81E151
	Reductive Transfer			
	Tryptophan	H$_2$O (pH 9.2)	4.9×10^9	81E151
204.	**Pyranthrene**			
	O$_2$	Toluene	4.2×10^9	83F075
	Energy Transfer			
	Rubrene	Toluene	2.3×10^9	81E716
205.	**Pyrazine**			
	O$_2$	H$_2$O (pH 7.1)	3.2×10^9	747233
	Reductive Transfer			
	I$^-$	H$_2$O (pH 7.1)	1.0×10^{10}	757309
206.	**Pyrene**			
	O$_2$	Benzene	2.2×10^9	736097
	O$_2$	Cyclohexane	1.6×10^9	706182
	TEMPO	MeCN	4.0×10^7	80E116
	Charge Transfer			
	1,4-Benzoquinone	Benzene	9.9×10^9	79A093
	Cu(acac)$_2$	MeOH	1.9×10^9	83F123

Table 9—**Triplet-State Quenching Rates**—Continued

No.	Quencher	Solvent (pH)	k_q /L mol^{-1} s^{-1}	Ref.
206.	**Pyrene**—Continued			
	2,5-Dimethyl-2,4-hexadiene	Benzene	4.0×10^5	81E270
	Ferrocene	EtOH	6.0×10^9	747190
	Oxidative Transfer			
	MV^{2+}	MeCN/H$_2$O (9:1)	7.0×10^9	83A113
207.	**1-Pyrenecarboxaldehyde**			
	O$_2$	Benzene	1.9×10^9	84F005
	O$_2$	Cyclohexane	1.8×10^9	84F005
	Energy Transfer			
	Azulene	Benzene	8.6×10^9	84E237
208.	**Pyrimidine**			
	O$_2$	H$_2$O (pH 7.0)	4.6×10^9	757309
	Reductive Transfer			
	2-Propanol	H$_2$O (pH 7.0)	8.6×10^7	757309
209.	**Pyruvic acid**			
	Reductive Transfer			
	Ethanol	Benzene	2.3×10^6	727392
210.	**Pyruvic acid, ethyl ester**			
	Energy Transfer			
	1-Methylnaphthalene	MeCN	8.0×10^9	86A164
211.	**Quinoxaline**			
	O$_2$	H$_2$O (pH 7.0)	4.6×10^9	757309
212.	**(*all-E*)-Retinal**			
	O$_2$	Cyclohexane	3.7×10^9	84F005
	O$_2$	MeOH	4.6×10^9	84F005
	Energy Transfer			
	β-Carotene	Toluene	6.2×10^9	84E180
	Ferrocene	Toluene	3.6×10^7	84E180
	Tetracene	Toluene	2.6×10^9	84E180
213.	**(*all-E*)-Retinoic acid**			
	O$_2$	MeOH	1.4×10^9	82A205
214.	**(*all-E*)-Retinol**			
	Di-*tert*-butylnitroxide	MCH	8.5×10^8	84E180
	O$_2$	Hexane	4.7×10^9	733001

Table 9—**Triplet-State Quenching Rates**—Continued

No.	Quencher	Solvent (pH)	k_q /L mol^{-1} s^{-1}	Ref.
214.	**(*all-E*)-Retinol**—Continued			
	Energy Transfer			
	Azulene	Toluene/EtI	1×10^7	84E180
	β-Carotene	Toluene/EtI	8.0×10^9	84E180
	Ferrocene	Toluene/EtI	6.5×10^7	84E180
	Tetracene	Toluene/EtI	2.4×10^9	84E180
215.	**Retinyl acetate**			
	O_2	MeOH	1.0×10^9	82A205
216.	**Rhodamine 6G cation**			
	O_2	EtOH	1.6×10^9	747050
	Oxidative Transfer			
	1,4-Benzoquinone	H_2O (pH 6)	2.5×10^9	78A304
	Reductive Transfer			
	Ascorbic acid	H_2O (pH 6)	8.0×10^8	78A304
	p-Phenylenediamine	H_2O (pH 6)	1.0×10^9	78A304
217.	**Rubrene**			
	O_2	Toluene	3.4×10^9	82E072
218.	**Stilbene**			
	O_2	Benzene	9.0×10^9	81E214
	O_2	MeCN	9.5×10^9	81E214
	O_2	MeOH	8.6×10^9	81E214
219.	**(*E*)-Stilbene**			
	Energy Transfer			
	Anthracene	Benzene	2.6×10^9	720447
	β-Carotene	Toluene	3.8×10^9	84E144
220.	**(*Z*)-Stilbene**			
	O_2	Benzene	3.8×10^9	771014
221.	**(*E*)-Stilbene, 4,4′-dinitro-**			
	Energy Transfer			
	Ferrocene	Benzene	9.1×10^9	79Z027
	Reductive Transfer			
	DABCO	MeCN	3.5×10^9	86A043

Table 9—**Triplet-State Quenching Rates**—Continued

No.	Quencher	Solvent (pH)	k_q /L mol^{-1} s^{-1}	Ref.
222.	**(*E*)-Stilbene, 4-nitro-**			
	Energy Transfer			
	Ferrocene	Benzene	4.8×10^9	79Z027
	Reductive Transfer			
	DABCO	MeCN	3.6×10^9	86A043
223.	**Styrene, β-methyl-, (*E*)-**			
	Oxidative Transfer			
	MV^{2+}	MeOH	6.5×10^8	82E526
224.	**Styrene, α-phenyl-**			
	O$_2$	Benzene	9.0×10^9	82E204
	O$_2$	MeCN	6.9×10^9	82E204
225.	**p-Terphenyl**			
	O$_2$	Benzene	1.2×10^9	78E263
	Energy Transfer			
	2-Nitrothiophene	Acetone	1.0×10^9	82A153
	Oxidative Transfer			
	MV^{2+}	MeCN/H$_2$O (9:1)	9.2×10^9	83A113
226.	**α-Terthienyl**			
	Energy Transfer			
	Anthracene	Cyclohexane	1.7×10^9	86A357
	Oxidative Transfer			
	MV^{2+}	MeOH	6.8×10^9	86A357
227.	**Testosterone**			
	O$_2$	EtOH	2.2×10^9	80B055
228.	**Testosterone acetate**			
	Reductive Transfer			
	Diphenylamine	MeCN	3.9×10^9	88E675
229.	**Tetracene**			
	Di-*tert*-butylnitroxide	Hexane	3.5×10^7	736099
	O$_2$	Benzene	3.1×10^9	736097
	Energy Transfer			
	Ferrocene	Benzene	3.6×10^7	756237
	Rubrene	Toluene	2.0×10^9	81E716
	Oxidative Transfer			
	MV^{2+}	MeCN/H$_2$O (9:1)	1.5×10^8	83A113

Table 9—**Triplet-State Quenching Rates**—Continued

No.	Quencher	Solvent (pH)	k_q /L mol^{-1} s^{-1}	Ref.
230.	**Thiobenzophenone**			
	2,5-Dimethyl-2,4-hexadiene	Benzene	4.2×10^8	84A221
	O$_2$	Benzene	2.9×10^9	84A221
	Energy Transfer			
	Azulene	Benzene	4.2×10^9	84A221
	β-Carotene	Benzene	1.1×10^{10}	84A221
231.	**Thiobenzophenone, 4,4′-bis(dimethylamino)-**			
	2,5-Dimethyl-2,4-hexadiene	Benzene	1.7×10^8	84A221
	Energy Transfer			
	Azulene	Benzene	9.3×10^9	84A221
232.	**Thiobenzophenone, 4,4′-dimethoxy-**			
	2,5-Dimethyl-2,4-hexadiene	Benzene	2.7×10^8	84A221
	O$_2$	Benzene	3.7×10^9	84A221
	Energy Transfer			
	Azulene	Benzene	4.6×10^9	84A221
	β-Carotene	Benzene	1.0×10^{10}	84A221
233.	**Thiocoumarin**			
	1,4-Cyclohexadiene	Benzene	1.1×10^9	86A240
	O$_2$	Benzene	3.0×10^9	86A240
	Reductive Transfer			
	N,N-Dimethylaniline	MeCN	1.0×10^9	86A240
	Tributylstannane	Benzene	3.6×10^9	86A240
234.	**Thioindigo**			
	O$_2$	Benzene	3.2×10^9	78F030
235.	**Thionine cation**			
	O$_2$	H$_2$O (pH 8)	4.5×10^8	707174
	Energy Transfer			
	Crystal Violet	MeOH	1.9×10^8	697141
	Reductive Transfer			
	Allylthiourea	MeOH	8.0×10^6	777242
	Azulene	MeOH	2.0×10^9	777242
	DABCO	MeOH	8.0×10^8	777242
	Diphenylamine	MeCN	5.4×10^9	78A447
	I$^-$	H$_2$O (pH 7.1)	4.7×10^9	747304
	L-Phenylalanine	H$_2$O (pH 7.1)	2.2×10^7	747304

Table 9—**Triplet-State Quenching Rates**—Continued

No.	Quencher	Solvent (pH)	k_q /L mol^{-1} s^{-1}	Ref.
235.	**Thionine cation**—Continued			
	Triphenylamine	MeOH	6.4×10^9	767584
236.	**Thionine cation, conjugate monoacid**			
	O_2	H_2O (pH 4.62)	2.6×10^8	707174
	Reductive Transfer			
	Allylthiourea	MeOH	7.0×10^8	777242
	Azulene	MeOH	4.0×10^9	777242
237.	**Thiophene, 2-nitro-**			
	O_2	Acetone	1.0×10^9	82A153
	Energy Transfer			
	Azulene	MeCN	1.4×10^{10}	82A153
	Oxidative Transfer			
	CCl_4	MeCN	4.4×10^6	82A153
	Reductive Transfer			
	I$^-$	H_2O (pH 7)	8.6×10^9	82A154
	1,3,5-Trimethoxybenzene	MeCN	5.1×10^9	82A153
238.	**4-Thiouridine**			
	Cysteine	H_2O	4×10^8	83E625
	I$^-$	H_2O	6×10^9	83E625
	Methionine	H_2O	1.5×10^9	83E625
	Tryptophan	H_2O	1.5×10^9	83E625
239.	**Thioxanthen-9-one**			
	Styrene	Benzene	3×10^9	81A294
	Energy Transfer			
	1,1-Diphenylethylene	Benzene	5.5×10^9	82E204
	I$^-$	MeCN/H_2O (3:2)	6.6×10^9	85E351
	Reductive Transfer			
	Ethyl 4-(dimethylamino)benzoate	Benzene	6×10^9	81A294
240.	**Thioxanthione**			
	O_2	Benzene	2.8×10^9	84A221
	Energy Transfer			
	Azulene	Benzene	5.8×10^9	84A221
	β-Carotene	Benzene	1.0×10^{10}	84A221

Table 9—**Triplet-State Quenching Rates**—Continued

No.	Quencher	Solvent (pH)	k_q /L mol^{-1} s^{-1}	Ref.
241.	**Thioxanthylium, 3,6-bis(dimethylamino)-**			
	Oxidative Transfer			
	1,4-Benzoquinone	H_2O (pH 7.2)	2.8×10^9	82E680
	Reductive Transfer			
	Allylthiourea	MeOH	5.4×10^5	80A369
	N,N-Dimethylaniline	MeCN	8.2×10^9	80A369
	EDTA	H_2O (pH 7.2)	1.4×10^6	82E680
242.	**Thymidine**			
	Energy Transfer			
	(*all-E*)-Retinol	MeCN	6×10^9	79B087
	Oxidative Transfer			
	Metronidazole	MeCN	7.3×10^9	87A090
243.	**Thymidine 5′-monophosphate**			
	Energy Transfer			
	(*all-E*)-Retinol	EtOH	2×10^9	79B087
244.	**Thymine**			
	Energy Transfer			
	(*all-E*)-Retinol	MeCN	$\sim 6 \times 10^9$	757510
	Oxidative Transfer			
	Metronidazole	MeCN	4.3×10^9	87A090
245.	**1,3,5-Triazine**			
	O_2	MeCN	5.0×10^9	757066
	Reductive Transfer			
	2-Propanol	MeCN	1.4×10^8	757066
246.	**Triphenylene**			
	Di-*tert*-butylnitroxide	Hexane	2.2×10^9	736099
	O_2	Benzene	1.1×10^9	736097
	Energy Transfer			
	Azulene	Benzene	7×10^9	81E214
	Cu(acac)$_2$	Benzene	1.3×10^9	66E093
	1,3-Cyclohexadiene	Benzene	1.3×10^9	66E093
	Ferrocene	EtOH	6.5×10^9	747190
	Naphthalene	Hexane	1.3×10^9	61E005
	Oxidative Transfer			
	MV^{2+}	MeCN/H_2O (9:1)	7.1×10^9	83A113

Table 9—**Triplet-State Quenching Rates**—Continued

No. Quencher	Solvent (pH)	k_q /L mol^{-1} s^{-1}	Ref.
247. Triphenylethylene			
O$_2$	Benzene	4.8×10^9	82E204
248. Tris(2,2′-bipyridine)ruthenium(II) ion			
O$_2$	H$_2$O	3.3×10^9	767180
Energy Transfer			
(*all-E*)-Retinol	MeCN	2.5×10^9	767180
249. Tryptamine			
O$_2$	H$_2$O (pH 7.5)	5.7×10^9	757163
250. Tryptophan			
O$_2$	H$_2$O (pH 7.5)	5.0×10^9	757163
Energy Transfer			
Anthracene	EtOH	4.0×10^9	757163
251. Tryptophan, *N*-methyl-			
O$_2$	H$_2$O	5×10^9	88R179
252. L-Tryptophanamide, *N*-acetyl-			
O$_2$	H$_2$O	5×10^9	88R179
253. Tyrosine			
O$_2$	H$_2$O (pH 7.5)	4.8×10^9	757161
Energy Transfer			
Tryptophan	H$_2$O (pH 7.3)	6.0×10^9	757161
Oxidative Transfer			
Cysteine	H$_2$O (pH 7.0)	5.2×10^8	757161
Reductive Transfer			
Hydroxide ion	H$_2$O (pH 7-10.3)	2.0×10^{10}	757161
254. Uracil			
Energy Transfer			
(*all-E*)-Retinol	MeCN	$\sim 8 \times 10^9$	757510
Oxidative Transfer			
Fumaronitrile	MeCN	5.3×10^9	87A090
Metronidazole	MeCN	1.3×10^{10}	87A090
255. Uridine			
Energy Transfer			
(*all-E*)-Retinol	MeCN	6×10^9	79B087

Table 9—**Triplet-State Quenching Rates**—Continued

No.	Quencher	Solvent (pH)	k_q /L mol^{-1} s^{-1}	Ref.
256.	**Uridine 5′-monophosphate**			
	O$_2$	H$_2$O	3×10^9	79B087
	Energy Transfer			
	(*all-E*)-Retinol	EtOH	5×10^9	79B087
257.	**9-Xanthione**			
	O$_2$	Benzene	2.8×10^9	84A221
	Energy Transfer			
	Azulene	Benzene	8.2×10^9	84A221
	Reductive Transfer			
	Ethanol	EtOH	1.1×10^4	79F100
	Triethylamine	Benzene	2.1×10^8	87A340
258.	**Xanthone**			
	O$_2$	Benzene	5.6×10^9	767171
	Energy Transfer			
	Azulene	Benzene	8.5×10^9	81F275
	Biphenyl	MeOH	8.8×10^9	85E449
	I$^-$	MeCN/H$_2$O (3:2)	7.1×10^9	85E351
	Naphthalene	Benzene	9.5×10^9	767171
	(*E*)-Stilbene	Benzene	7×10^9	81E214
	Reductive Transfer			
	Cyclohexane	CCl$_4$	7.7×10^6	80A338
	Indole	Benzene	1.1×10^{10}	78A170
	2-Propanol	CCl$_4$	1.1×10^8	80A338
	Tributylstannane	CCl$_4$	1.5×10^9	80A338

Section 10

Ionization Energies, Electron Affinities, and Redox Potentials of Organic Compounds

The tables in this section consist of critical compilations of parameters related to the study of exciplexes and electron-transfer processes. Subsection 10a contains tables of data derived from gas-phase measurements. The data in this subsection are ionization energies and electron affinities. Subsection 10b contains the electrochemistry data on organic solutions. The data in this section consist mainly of oxidation and reduction halfwave potentials.

Section 10a

Ionization Energies and Electron Affinities

In the theory of charge-transfer complexes, the energy of the charge-transfer state is computed by the following set of processes involving the electron donor, D, and the electron acceptor, A:

$$D \rightarrow D^+ + e^- \tag{10-I}$$

$$A + e^- \rightarrow A^- \tag{10-II}$$

$$D^+ + A^- \rightarrow [D^+ \cdots A^-] \tag{10-III}$$

The energy required to remove an electron to infinity in process (10-I) is the ionization energy (ionization potential), I_p. The energy to form the negative ion, starting with A and e^- at infinity, is *minus* the electron affinity, E_a, of A. The energy of process (10-III) is the Coulomb energy of the ion-pair (or complex) at the distance of closest approach, $R_{D^+A^-}$. The total energy of forming the complex in a medium of dielectric constant, ε, starting with D and A is

$$E_{D^+A^-} = I_p - E_a - \frac{e_0^2}{\varepsilon R_{D^+A^-}} \tag{10-1}$$

This could represent a first approximation

$$E_{[DA]^*} \approx E_{D^+A^-} \tag{10-2}$$

to the energy of an exciplex (an excited DA complex with no stable ground state) or excited DA complex with a higher local excitation energy.

The ionization energies of some common electron donors are collected in Table 10a-1. Electron affinities of electron acceptors are emphasized, along with their ionization energies, in Table 10a-4. The two intermediate tables contain unsaturated organic compounds. Many of the aromatics in Table 10a-3 can act as either donors or acceptors, depending on the nature of the excited state-quencher pair.

Table 10a-1

Ionization Energies[a] of Some Electron Donors

No.	Compound	Adiabatic /eV	Vertical /eV
1	Ammonia	10.16	10.85
2	*tert*-Amylamine	8.5	9.20
3	Aniline	7.72	8.05
4	Aniline, *N,N*-dimethyl-	7.12	7.37
5	Aniline, *N,N*-diphenyl-	6.80	7.00
6	Aniline, *N*-methyl-	7.33	-
7	Aniline, *N*-phenyl-	7.16	7.44
8	Benzene, methoxy-	8.21	8.42
9	Benzidine, *N,N,N′,N′*-tetramethyl-	6.40	-
10	Benzylamine	8.64	-
11	Butylamine	8.71	9.40
12	*sec*-Butylamine	8.7	9.30
13	*tert*-Butylamine	8.6	9.25
14	Carbazole	7.57	-
15	Cyclohexylamine	8.6	9.16
16	Diethylmethylamine	7.5	8.22
17	Dimethylamine	8.23	8.93
18	Ethanamine, *N*-ethyl-	8.01	8.63
19	Ethylamine	8.86	9.47
20	Isobutylamine	8.7	9.30
21	Methylamine	8.97	9.66
22	Naphthalene, 1-methoxy-	7.70	7.72
23	Naphthalene, 2-methoxy-	7.4	7.87
24	Neopentylamine	8.5	9.25
25	*p*-Phenylenediamine, *N,N,N′,N′*-tetramethyl-	6.20	6.75
26	Piperidine	8.05	8.66
27	Tributylamine	7.4	7.90
28	Triethylamine	7.50	8.08
29	Trimethylamine	7.82	8.53

[a]From references [76M467, 82Z365, 88Z502, 90Z548].

Table 10a-2

Ionization Energies[a] and Electron Affinities[b] of Alkenes, Dienes, and Alkynes

No.	Compound	Adiabatic I_p /eV	Vertical I_p /eV	Electron affinity /eV
1	Acetylene	11.40	11.43	−1.8
2	Bicyclo[2.1.0]pentane	8.7	-	-
3	Bicyclo[2.1.0]pent-2-ene	8.0	8.6	-
4	Butadiene	9.07	9.03	−0.65
5	1,3-Butadiene, 2,3-dimethyl-	8.71	8.72	-
6	(*E*)-2-Butene	9.10	9.11	-
7	(*Z*)-2-Butene	9.11	9.11	-
8	1-Butyne	10.18	-	-
9	2-Butyne	9.56	9.79	-
10	1,3-Cyclohexadiene	8.25	8.25	−0.73
11	1,4-Cyclohexadiene	8.82	8.82	−1.75
12	Cyclohexene	8.95	9.12	−2.70
13	1,3-Cyclooctadiene	8.4	-	-
14	Cyclopentadiene	8.56	-	−1.05
15	Cyclopentene	9.01	9.01	-
16	Cyclopropane	9.86	-	-
17	Ethylene	10.51	10.50	−1.55
18	(*E,E*)-2,4-Hexadiene	8.18	-	-
19	(*E,Z*)-2,4-Hexadiene	8.24	-	-
20	(*Z,Z*)-2,4-Hexadiene	8.3	-	-
21	Isoprene	8.84	8.87	-
22	Norbornadiene	8.35	8.70	-
23	Norbornylene	8.31	8.95	−1.70
24	(*E*)-Piperylene	8.59	-	-
25	(*Z*)-Piperylene	8.63	8.60	-
26	Propylene	9.73	9.91	-
27	Propyne	10.36	10.36	-

[a]From references [82Z365, 88Z502, 90Z548].   [b]From reference [84Z133].

Table 10a-3

Ionization Energies[a] and Electron Affinities[b] of Aromatics

No.	Compound	Adiabatic I_p /eV	Vertical I_p /eV	Electron affinity /eV
1	Acenaphthene	7.7	7.76	-
2	Acenaphthylene	8.22	-	0.45
3	Acetylene, phenyl-	8.81	8.82	-
4	Anthracene	7.45	7.41	0.52
5	Anthracene, 9-methyl-	7.24	7.24	-
6	Anthracene, 9-phenyl-	-	7.25	-
7	Azulene	7.41	7.44	0.65
8	Benz[a]anthracene	7.43	7.41	0.63
9	Benzene	9.24	9.24	−1.15
10	Benzene, isopropyl-	8.73	8.75	−1.08
11	Benzene, 1,2,4,5-tetramethyl-	8.04	8.05	0.07
12	Benzene, 1,3,5-trimethyl-	8.41	8.45	−1.03
13	Benzo[b]chrysene	7.14	7.20	-
14	Benzo[a]coronene	7.1	7.08	-
15	Benzo[rst]pentaphene	7.0	-	-
16	Benzo[ghi]perylene	7.15	7.15	-
17	Benzo[c]phenanthrene	7.60	7.60	0.54
18	Benzo[a]pyrene	7.12	7.41	0.68
19	Benzo[e]pyrene	7.41	-	0.60
20	Benzo[b]triphenylene	7.39	7.39	-
21	Biphenyl	7.95	8.34	0.15
22	Biphenylene	7.56	7.60	-
23	Carbazole	7.57	-	-
24	Chrysene	7.59	7.59	0.40
25	Coronene	7.29	7.29	-
26	Dibenz[a,h]anthracene	7.38	7.38	0.60
27	Dibenz[a,j]anthracene	7.40	7.40	0.59
28	Fluoranthene	7.95	-	0.63
29	Fluorene	7.89	7.93	0.22
30	Furan	8.88	-	−1.76
31	Indan	8.3	8.45	-
32	Indene	8.14	8.15	0.16

Table 10a-3—**Ionization Energies and Electron Affinities of Aromatics**—Continued

No.	Compound	Adiabatic I_p /eV	Vertical I_p /eV	Electron affinity /eV
33	Isoquinoline	8.53	8.54	−0.42
34	Naphthalene	8.14	8.15	0.15
35	Naphthalene, 1-methyl-	7.85	8.01	0.13
36	Naphthalene, 2-methyl-	7.8	8.01	0.16
37	Ovalene	6.71	6.71	-
38	Pentacene	6.61	6.61	-
39	Pentaphene	7.27	7.27	-
40	Perylene	6.90	6.97	-
41	Phenanthrene	7.86	7.86	0.31
42	Pyrazine	9.29	9.63	−0.80
43	Pyrene	7.41	7.41	0.58
44	Pyridazine	8.64	9.31	−0.49
45	Pyridine	9.25	-	−0.59
46	Pyrimidine	9.23	9.73	−0.33
47	Quinoline	8.62	8.62	−0.60
48	(*E*)-Stilbene	7.70	7.90	+0.38
49	Styrene	8.43	8.50	−0.75
50	*m*-Terphenyl	8.01	-	-
51	*o*-Terphenyl	8.00	-	-
52	*p*-Terphenyl	7.78	-	-
53	Tetrabenz[*a*,*c*,*h*,*j*]anthracene	7.43	7.43	-
54	Tetracene	6.97	6.97	0.88
55	Toluene	8.82	8.85	−0.4
56	Triphenylene	7.84	7.88	0.28
57	*m*-Xylene	8.56	8.55	−1.06
58	*o*-Xylene	8.56	8.57	−1.12
59	*p*-Xylene	8.44	8.43	−1.07

[a]From references [82Z365, 88Z502, 90Z548].   [b]From reference [84Z133].

Table 10a-4

Ionization Energies[a] and Electron Affinities[b] of Some Electron Acceptors

No.	Compound	Adiabatic I_p /eV	Vertical I_p /eV	Electron affinity /eV
1	Anthracene-9-carboxaldehyde	7.69	7.67	+1.02
2	9,10-Anthraquinone	9.25	–	+1.55
3	Benzene, 1-cyano-4-nitro-	10.2	-	+1.82
4	Benzene, 1,2-dicyano-	9.90	10.27	+0.95
5	Benzene, 1-methoxy-4-nitro-	8.8	9.08	+0.81
6	Benzene, nitro-	9.86	9.88	+2.1
7	Benzene, 1,2,4,5-tetracyano-	-	-	+1.6
8	Benzonitrile	9.62	9.71	+0.25
9	Benzophenone	9.05	-	+0.63
10	1,4-Benzoquinone, tetrachloro-	9.74	-	+1.37
11	1,4-Benzoquinone	10.0	9.99	+1.83
12	Biacetyl	9.24	9.55	+0.75
13	Ethene, 1,2-dicyano-, (*E*)-	11.16	-	+0.96
14	Ethene, 1,1-dicyano-	-	11.38	+1.54
15	Ethene, tetracyano-	11.77	11.79	+2.9
16	9-Fluorenone	8.36	-	+1.19
17	Maleic anhydride	10.8	11.45	+1.41
18	1-Naphthaldehyde	8.3	-	+0.68
19	Naphthalene, 2-acetyl-	-	8.23	+0.6
20	1,4-Naphthoquinone	9.56	-	+1.71
21	4-Nitrotoluene	9.4	9.54	+0.89
22	9,10-Phenanthrenequinone	8.64	-	+1.83
23	Tetracyano-1,4-benzoquinone	-	-	+1.8
24	7,7,8,8-Tetracyanoquinodimethane	-	-	+2.84

[a]From references [82Z365, 88Z502, 90Z548].

[b]From references [84Z133, 85C011].

Section 10b

Redox Potentials

The sign conventions used in these tables are as such that the standard potentials, E^0, are for the half-cell reactions written as reductions:

$$O + e^- \rightarrow R, \qquad (10\text{-}IV)$$

where O is the oxidized form and R is the reduced form of some species in a half-cell. With this convention, the standard potentials are approximately equal to either the halfwave potentials for reduction,

$$E^0 \approx E_{1/2}^{red} \qquad (10\text{-}3)$$

or *minus* the halfwave potentials for oxidation,

$$E^0 \approx -E_{1/2}^{ox} \qquad (10\text{-}4)$$

The standard free energy, ΔG^0, of forming a pair of separated ions from a neutral donor-acceptor pair (processes 10-I and 10-II) is then approximately

$$\Delta G^0 \approx -23.06[\, -E_{1/2}^{ox}\,(D) + E_{1/2}^{red}\,(A)\,], \qquad (10\text{-}5)$$

where the ΔG^0 will be in kcal/mole if the halfwave potentials are in volts.

In the theory of electron-transfer quenching of excited states, the free energies for the processes,

$$D* + A \rightarrow [D^+ \cdots A^-] \qquad (10\text{-}V)$$

$$A* + D \rightarrow [D^+ \cdots A^-] \qquad (10\text{-}VI)$$

are central for the quenching of excited states of donors and acceptors, respectively. The free energy change in either of these electron-transfer quenching processes has been related to the standard potentials of D and A in a theory due to Rehm and Weller. [706216] Making use of the approximation in Eq. (10-5), their equation can be written as

$$\Delta G = 23.06 \left[E_{1/2}^{ox}\,(D) - E_{1/2}^{red}\,(A) - \frac{e_0^2}{\varepsilon R_{D^+ A^-}} \right] - \Delta E_{0,0} \qquad (10\text{-}6)$$

where $\Delta E_{0,0}$ is the excited state energy of the donor (process 10-V) or acceptor (process 10-VI).

The electron-transfer rate constant, k_{elt}, can be written classically as an activated rate constant, [84Z150]

$$k_{elt} = k_0 \exp\left\{\frac{-\Delta G^{\ddagger}}{RT}\right\},$$ (10-7)

where k_0 is the reciprocal of the dielectric relaxation time. Rehm and Weller [706216] proposed an empirical relation,

$$\Delta G^{\ddagger} = \left\{\left[\frac{\Delta G}{2}\right]^2 + [\Delta G^{\ddagger}(0)]^2\right\}^{\frac{1}{2}} + \frac{\Delta G}{2}$$ (10-8)

for the free energy of activation of the process of electron-transfer quenching. In Eq. (10-8), $\Delta G^{\ddagger}(0)$ is the free energy of activation when the free energy change, ΔG, for the overall quenching process is zero. ΔG is given by Eq. (10-6). Equations such as these (along with various modifications) have proved fruitful in elucidating electron-transfer processes in excited states.

Four points can be noted with regard to the selection and use of the tables:

(1)   In the tables we have stayed exclusively with halfwave potentials. Thus we have left out important compounds having only E_p's such as N,N-diethylaniline, $E_p = +0.7$ V, and carbazole, $E_p = +1.16$ V.

(2)   Also we have uniformly tried to list $E_{\frac{1}{2}}$ with respect to standard calomel electrodes (SCE). There are large differences between SCE and the various Ag electrodes that are popular to use in organic solvents. Only in the case of the nitriles (Table 10a-6) did we quote $E_{\frac{1}{2}}$ vs. Ag electrodes. Unfortunately these $E_{\frac{1}{2}}$'s are not compatible for use in Eq. (10-6) when the other $E_{\frac{1}{2}}$ has been measured relative to SCE. These $E_{\frac{1}{2}}$ vs. Ag electrodes can still be used to make qualitative correlations between the various nitriles as quenchers.

(3)   We tried consistently to choose $E_{\frac{1}{2}}^{red}$'s in N,N-dimethylformamide (DMF) solutions and $E_{\frac{1}{2}}^{ox}$'s in CH_3CN. These choices were made since they gave the largest selection of $E_{\frac{1}{2}}$'s in nonaqueous solvents. Given the uncertainties in the measurements, mixing DMF and CH_3CN results are not nearly as serious as mixing $E_{\frac{1}{2}}$'s from different reference electrodes. For example, in CH_3CN, $E_{\frac{1}{2}}$'s measured relative to Ag electrodes can vary anywhere from ± 0.3 V compared to SCE. [70C003]

(4)   The ΔG's and the $E_{\frac{1}{2}}$'s in the theory are for reversible processes. It is not clear whether the all nitrile results included are reversible.

Table 10b-1

Halfwave Oxidation Potentials of Some Electron Donors

No.	Compound	$E_{1/2}^{ox}$ /V CH$_3$CN SCE	Ref
1	Aniline	+0.98	75C006
2	Aniline, 2,4-dimethoxy-*N,N*-dimethyl-	+0.27	64C002
3	Aniline, 3,4-dimethoxy-*N,N*-dimethyl-	+0.20	64C002
4	Aniline, 3,5-dimethoxy-*N,N*-dimethyl-	+0.50	64C002
5	Aniline, *N,N*-dimethyl-	+0.53	64C002
6	Aniline, *N,N*-diphenyl-	+0.92	75C006
7	Aniline, 2-methoxy-*N,N*-dimethyl-	+0.48	64C002
8	Aniline, 3-methoxy-*N,N*-dimethyl-	+0.49	64C002
9	Aniline, 4-methoxy-*N,N*-dimethyl-	+0.33	64C002
10	Aniline, 4-methyl-	+0.78	75C006
11	Aniline, *N*-phenyl-	+0.86	75C006
12	Anthracene, 2-amino-	+0.44	75C006
13	Anthracene, 9-amino-	+0.15	75C006
14	Benzene, 1,2-dimethoxy-	+1.45	64C001
15	Benzene, 1,4-dimethoxy-	+1.34	64C001
16	Benzene, hexamethoxy-	+1.24	64C001
17	Benzene, methoxy-	+1.76	64C001
18	Benzene, pentamethoxy-	+1.07	64C001
19	Benzene, 1,2,3,4-tetramethoxy-	+1.25	64C001
20	Benzene, 1,2,3,5-tetramethoxy-	+1.09	64C001
21	Benzene, 1,2,4,5-tetramethoxy-	+0.81	64C001
22	Benzene, 1,2,3-trimethoxy-	+1.42	64C001
23	Benzene, 1,2,4-trimethoxy-	+1.12	64C001
24	Benzidine, *N,N,N′,N′*-tetramethyl-	+0.43	75C006
25	*N,N,N′,N′*-Tetramethyl-*m*-phenylenediamine	+0.32	64C002
26	*N,N,N′,N′*-Tetramethyl-*o*-phenylenediamine	+0.28	64C002
27	*N,N,N′,N′*-Tetramethyl-*p*-phenylenediamine	+0.32	64C002
28	Triethylamine	+1.15	77C008

Table 10b-2

Halfwave Potentials of Aromatic Hydrocarbons

No.	Compound	$E_{\frac{1}{2}}^{ox}$ /V CH$_3$CN[a] SCE	$E_{\frac{1}{2}}^{red}$ /V DMF[b] SCE
1	Acenaphthene	+1.21	−2.67
2	Acetylene, diphenyl-	-	−2.11*
3	Anthracene	+1.09	−1.95*
4	Anthracene, 9,10-diphenyl-	+1.22*	−1.94*
5	Anthracene, 9-methyl-	+0.96	−1.97[c]
6	Anthracene, 9-phenyl-	-	−1.86
7	Azulene	+0.71	−1.65[c]
8	Benz[a]anthracene, 10-methyl-	+1.14	−2.03
9	Benz[a]anthracene, 11-methyl-	+1.14	−2.00
10	Benz[a]anthracene, 12-methyl-	+1.07	−1.98
11	Benz[a]anthracene, 1-methyl-	+1.14	−2.03
12	Benz[a]anthracene, 2-methyl-	+1.14	−2.00
13	Benz[a]anthracene, 3-methyl-	+1.14	−2.00
14	Benz[a]anthracene, 5-methyl-	+1.15	−2.02
15	Benz[a]anthracene, 6-methyl-	+1.15	−2.01
16	Benz[a]anthracene, 7-methyl-	+1.08	−1.99
17	Benz[a]anthracene, 8-methyl-	+1.13	−2.02
18	Benz[a]anthracene, 9-methyl-	+1.15	−2.02
19	Benz[a]anthracene	+1.18	-
20	Benzene	+2.30*	-
21	Benzo[c]phenanthrene, 1-methyl-	-	−2.17
22	Benzo[c]phenanthrene, 3-methyl-	-	−2.22
23	Benzo[c]phenanthrene, 4-methyl-	-	−2.20
24	Benzo[c]phenanthrene, 5-methyl-	-	−2.22
25	Benzo[c]phenanthrene, 6-methyl-	-	−2.20
26	Benzo[c]phenanthrene	-	−2.20
27	Benzo[a]pyrene	+0.94	−2.10
28	Benzo[e]pyrene	+1.27	−2.13[c]
29	Benzo[b]triphenylene	+1.25	−2.08[c]
30	Biphenyl	-	−2.55*
31	Biphenylene	-	−2.28[c]

Table 10b-2—**Halfwave Potentials of Aromatic Hydrocarbons**—Continued

No.	Compound	$E_{1/2}^{ox}$ /V CH$_3$CN[a] SCE	$E_{1/2}^{red}$ /V DMF[b] SCE
32	Chrysene	+1.35	−2.25
33	Coronene	+1.23	−2.07
34	Dibenz[*a,h*]anthracene	+1.19	−2.10[c]
35	Fluoranthene	+1.45	−1.74
36	Naphthalene	+1.54	−2.49*
37	Naphthalene, 1,2-dimethyl-	-	−2.60[c]
38	Naphthalene, 1,4-dimethyl-	-	−2.57[c]
39	Naphthalene, 1,6-dimethyl-	-	−2.60[c]
40	Naphthalene, 1,7-dimethyl-	-	−2.61[c]
41	Naphthalene, 2,3-dimethyl-	+1.35	−2.64[c]
42	Naphthalene, 2,6-dimethyl-	+1.36	−2.62[c]
43	Naphthalene, 1-methyl-	+1.43	−2.58*
44	Naphthalene, 2-methyl-	+1.45	−2.58*
45	Naphtho[1,2,3,4-*def*]chrysene	+1.01	−1.91[c]
46	Perylene	+0.85	−1.67
47	Phenanthrene	+1.50	−2.44
48	Pyrene	+1.16	−2.09
49	Rubrene	-	−1.41
50	(*E*)-Stilbene	+1.43*	−2.08*
51	(*Z*)-Stilbene	-	−2.07*
52	Styrene	-	−2.60*
53	Tetracene	+0.77	−1.58
54	Triphenylene	+1.55	−2.46[c]

*Selected from reference [77C008].

[a]From [63C001].

[b]From [70C003] in *N,N*-dimethylformamide.

[c]Values (reference [629025]) were measured relative to mercury-pool reference electrodes and adjusted to SCE by subtracting 0.55 V.

Table 10b-3

Halfwave Potentials and Singlet Energies of Naphthalenes*

No.	Compound	$E_{\frac{1}{2}}^{ox}$ /V CH$_3$CN[a] SCE	$E_{\frac{1}{2}}^{red}$ /V 75% DO[b] SCE	$E_{\frac{1}{2}}^{red}$ /V DMF[c] SCE	E_s /eV[d]
1	Naphthalene	+1.70	-2.437	-	3.99
2	Naphthalene, 1-allyl-	-	-2.447	-	-
3	Naphthalene, 2-allyl-	-	-2.444	-	-
4	Naphthalene, 1-amino-	+0.54	-	-	-
5	Naphthalene, 1-*tert*-butyl-	-	-2.493	-	-
6	Naphthalene, 1-cyano-	vs. Ag/Ag$^+$	see	Table 10b-6	
7	Naphthalene, 1,4-dicyano-	vs. Ag/Ag$^+$	see	Table 10b-6	
8	Naphthalene, 1,3-dimethoxy-	+1.265	-	-2.61	-
9	Naphthalene, 1,4-dimethoxy-	+1.10	-	-2.69	-
10	Naphthalene, 1,5-dimethoxy-	+1.28	-	-2.76	-
11	Naphthalene, 1,6-dimethoxy-	+1.28	-	-2.68	-
12	Naphthalene, 1,7-dimethoxy-	+1.28	-	-2.67	-
13	Naphthalene, 1,8-dimethoxy-	+1.17	-	-2.72	-
14	Naphthalene, 2,3-dimethoxy-	+1.39	-	-2.73	-
15	Naphthalene, 2,6-dimethoxy-	+1.33	-	-2.60	-
16	Naphthalene, 2,7-dimethoxy-	+1.47	-	-2.68	-
17	Naphthalene, 1,2-dimethyl-	-	-2.479	-	3.84
18	Naphthalene, 1,3-dimethyl-	-	-2.483	-	3.85
19	Naphthalene, 1,4-dimethyl-	-	-2.471	-	3.85
20	Naphthalene, 1,5-dimethyl-	-	-2.475	-	3.86
21	Naphthalene, 1,6-dimethyl-	-	-2.476	-	3.85
22	Naphthalene, 1,7-dimethyl-	-	-2.469	-	3.85
23	Naphthalene, 1,8-dimethyl-	-	-2.521	-	3.85
24	Naphthalene, 2,3-dimethyl-	+1.38	-2.501	-	3.87
25	Naphthalene, 2,6-dimethyl-	+1.36	-2.476	-	3.83
26	Naphthalene, 2,7-dimethyl-	-	-2.485	-	3.86
27	Naphthalene, 1-(dimethylamino)-	+0.75	-	-2.58	-
28	Naphthalene, 2-(dimethylamino)-	+0.67	-	-2.63	-
29	Naphthalene, 2-hydroxy-	-	-	-2.52[e]	-
30	Naphthalene, 1-methoxy-	+1.38	-	-2.65	-
31	Naphthalene, 2-methoxy-	+1.52	-	-2.60	-

Table 10b-3—**Halfwave Potentials and Singlet Energies of Naphthalenes**—Continued

No.	Compound	$E_{1/2}^{ox}$ /V CH$_3$CN[a] SCE	$E_{1/2}^{red}$ /V 75% DO[b] SCE	$E_{1/2}^{red}$ /V DMF[c] SCE	E_s /eV[d]
32	Naphthalene, 1-methyl-	+1.43	−2.458	-	3.91
33	Naphthalene, 2-methyl-	+1.45	−2.460	-	3.87
34	Naphthalene, 1-(methylthio)-	+1.32	-	−2.25	-
35	Naphthalene, 2-(methylthio)-	+1.365	-	−2.28	-
36	Naphthalene, 1-nitro-	-	-	−0.97[f]	-
37	Naphthalene, 2-nitro-	-	-	−0.98[f]	-
38	Naphthalene, 1,3,7-trimethyl-	-	−2.496	-	-
39	Naphthalene, 1,4,5-trimethyl-	-	−2.529	-	-
40	Naphthalene, 1,6,7-trimethyl-	-	−2.515	-	-
41	Naphthalene, 2,3,6-trimethyl-	-	−2.523	-	-
42	Naphthalene, 1-vinyl-	-	−2.09	-	-
43	Naphthalene, 2-vinyl-	-	−2.15	-	3.66

*The values for the reduction potentials have been extracted mainly from reference [70C003]. The values for the energies of the lowest excited singlet states have been extracted from references [64Z006, 706229, 71Z014].

[a]Selected from reference [77C008] in acetonitrile.

[b]From references [60C003, 63C003]. Solvent: 75% dioxane-water; Suporting electrolyte: 0.1 mol/L tetrabutylammonium iodide.

[c]From reference [67C005]. Solvent: *N,N*-dimethylformamide; Electrolyte: 0.1 mol/L tetrabutylammonium perchlorate.

[d]Energy of lowest excited singlet state.

[e]Reference [77C008], measured relative to Hg-pool reference electrode but subtracted 0.55 V (reference [629025]).

[f]Reference [77C008].

Table 10b-4

Halfwave Reduction Potentials of Azaaromatics

No.	Compound	$E_{1/2}^{red}$ /V CH$_3$CN SCE[a]
1	Acridine	−1.62
2	Benzo[c]cinnoline	−1.55
3	Benzo[f]quinoline	−2.14
4	Benzo[h]quinoline	−2.21
5	Benzo[f]quinoxaline	−1.74
6	2,2'-Bipyridinium, 1,1'-dimethyl-	−0.73[b]
7	3,3'-Bipyridinium, 1,1'-dimethyl-	−0.84[b]
8	4,4'-Bipyridinium, 1,1'-dimethyl-	−0.46[b]
9	Cinnoline	−1.69
10	Isoquinoline	−2.22
11	Phenanthridine	−2.12
12	1,10-Phenanthroline	−2.05
13	1,7-Phenanthroline	−2.09
14	4,7-Phenanthroline	−2.04
15	Phenazine	−1.23
16	Phthalazine	−1.98
17	Pyrazine	−2.08
18	Pyridazine	−2.12
19	Pyridine	−2.62
20	Pyrimidine	−2.08
21	Quinazoline	−1.80
22	Quinoline	−2.11
23	Quinoxaline	−1.70

[a]From reference [75C005].

[b]From reference [79C010].

Table 10b-5

Halfwave Reduction Potentials of Carbonyls

No.	Compound	$E_{1/2}^{red}$ /V DMF (or CH$_3$CN)* SCE[a]	$E_{1/2}^{red}$ /V EtOH/H$_2$O 1/1[b] SCE
1	Acetophenone	−2.14*	−1.66
2	Acetophenone, 4′-chloro-	−2.10*	-
3	Acetophenone, 4′-cyano-	−1.58*	-
4	Acetophenone, 4′-methoxy-	−2.23*	-
5	Acetophenone, 4′-methyl-	−2.19*	-
6	9,10-Anthraquinone	−0.86	-
7	Anthrone	−1.51	-
8	Benzaldehyde	−1.93	-
9	Benzil	-	−0.71
10	Benzoic acid	−2.24	-
11	Benzoic acid, methyl ester	−2.32	-
12	Benzophenone	−1.83*	−1.55
13	Benzophenone, 4-chloro-	−1.75*	-
14	Benzophenone, 4-cyano-	−1.42*	-
15	Benzophenone, 4,4′-dimethoxy-	−2.02*	-
16	Benzophenone, 4,4′-dimethyl-	−1.90*	-
17	1,4-Benzoquinone	−0.45	-
18	Biacetyl	-	−1.03
19	2-Cyclopentenone	−2.16	-
20	9-Fluorenone	−1.29	−1.21
21	Maleic anhydride	−0.85	-
22	2-Naphthaldehyde	-	−1.34
23	Naphthalene, 2-acetyl-	-	−1.72
24	1,4-Naphthoquinone	−0.63	-
25	9,10-Phenanthrenequinone	−0.67[c]	-
26	Phthalic anhydride	−1.27	-
27	Xanthone	−1.77[c]	-

*Reference [86A400]. [a]Selected from reference [77C008]. [b]Reference [78E131], pH 12.65 buffer. [c]Measurements w/ Hg-pool, adjusted to SCE by subtracting 0.55V (reference [629025]).

Table 10b-6

Halfwave Reduction Potentials[a] of Nitriles

No.	Compound	$E_{1/2}^{red}$ /V DMF[b] Ag electrode
1	Anthracene, 9-cyano-	vs. SCE*
2	Anthracene, 9,10-dicyano-	vs. SCE*
3	Benzene, 1-cyano-3,5-dinitro-	−0.96
4	Benzene, 1-cyano-4-nitro-	−1.25
5	Benzene, 1,2-dicyano-	−2.12
6	Benzene, 1,3-dicyano-	−2.17
7	Benzene, 1,4-dicyano-	−1.97
8	Benzene, 1,2,4,5-tetracyano-	−1.02
9	Benzonitrile	−2.74
10	Benzonitrile, 4-chloro-	−2.4
11	Benzonitrile, 4-methoxy-	−3
12	Benzonitrile, 4-methyl-	−2.75
13	Benzoylacetonitrile	−2
14	4-Cyanobenzoic acid	−1.91
15	Ethene, tetracyano-	−0.2[d]
16	Naphthalene, 1-cyano-	−2.33[c]
17	Naphthalene, 1,4-dicyano-	−1.67[c]
18	Pyridine, 4-cyano-	−2.03

[a]Most values have been selected from pp. 340-341 of reference [70C003].

[b]In *N,N*-dimethylformamide vs. Ag/AgClO$_4$ electrode; supporting electrolyte: 0.1 mol/L tetrapropylammonium perchlorate.

[c]From reference [767370]. In acetonitrile vs. Ag/AgNO$_3$ electrode; supporting electrolyte: 0.1 mol/L tetraethylammonium perchlorate.

[d]In acetonitrile.

*See Table 10b-8 for value in acetonitrile vs. SCE.

Table 10b-7

Halfwave Reduction Potentials[a] of Some Halogenated Benzenes

Substituent	Iodobenzenes	Bromobenzenes	Chlorobenzenes
H	−1.21	−1.81	−2.13
m-Br	−0.96	−1.45	-
p-Br	−1.08	−1.54	-
m-CF$_3$	−1.00	−1.52	-
p-CF$_3$	−1.01	−1.53	-
p-CHO	−0.96	-	−1.20
m-C$_6$H$_5$	−1.15	−1.58	-
p-C$_6$H$_5$	−1.16	−1.56	-
m-CN	-	−1.29	-
p-CN	-	−1.26	−1.36
p-COC$_6$H$_5$	−0.96	−1.06	-
m-COMe	-	−1.19	-
p-COMe	−1.04	−1.15	-
m-Cl	−0.98	−1.53	-
p-Cl	−1.06	−1.61	−1.85
m-I	−0.92	-	-
p-I	−1.01	-	-
m-Me	−1.22	−1.85	−2.16
p-Me	−1.23	−1.84	−2.16
p-NHCOMe	-	−1.88	−2.09
p-NH$_2$	-	−1.96	-
m-NMe$_2$	-	-	−2.23
p-NMe$_2$	−1.35	−1.97	-
m-NMe$_3^+$	-	-	−1.48
p-NMe$_3^+$	−0.91	−1.34	-
p-OC$_6$H$_5$	-	−1.73	-
p-OEt	-	−1.82	-
m-OMe	−1.19	−1.76	-
p-OMe	−1.25	−1.84	−2.15

[a]From reference [68C005] ($E_{1/2}^{red}$ /V). DMF solutions with values relative to a Ag/AgBr electrode. Suporting electrolyte: 0.02 mol/L tetraethylammonium bromide.

Table 10b-8

Miscellaneous Halfwave Reduction Potentials

No.	Compound	$E_{\frac{1}{2}}^{\text{red}}$ /V SCE	Solvent	Ref.
1	Anthracene, 9-cyano-	−1.58	CH_3CN	78E649
2	Anthracene, 9,10-dicyano-	−0.98	CH_3CN	78E649
3	Benzaldehyde, 4-nitro-	−0.86	CH_3CN	61C002
4	Benzene, 1-chloro-4-nitro-	−1.06	CH_3CN	61C002
5	Benzene, 1,4-dinitro-	−0.69	CH_3CN	61C002
6	Benzene, 1-methyl-2-nitro-	−1.28	DMF	77C008
7	Benzene, nitro-	−1.08	DMF	77C008
8	Benzene, 1,1'-sulfonylbis-	−2.05	DMF	77C008
9	1,3-Butadiene, 1,4-diphenyl-	−1.98	Dioxane 75%	75C006
10	β-Carotene	−1.63	THF	78C018
11	Cyclopentadiene	−2.91	DMF	77C008
12	Diazene, diphenyl-	−1.36	DMF	77C008
13	1-Fluoro-4-nitrobenzene	−1.13	CH_3CN	61C002
14	1,3,5-Hexatriene, 1,6-diphenyl-	−1.76	Dioxane 96%	75C006
15	Isoprene	−2.70	DMF	77C008
16	Methyl 4-nitrobenzoate	−0.95	CH_3CN	61C002
17	4-Nitrotoluene	−1.14	DMF	77C008
18	1,3,5,7-Octatetraene, 1,8-diphenyl-	−1.62	Dioxane 96%	75C006
19	Phthalimide	−1.47	DMF	70C003
20	Piperylene	−2.76	DMF	77C008
21	Porphine, 2,7,12,17-tetraethyl-3,8,13,18-tetramethyl-	−1.37	DMF	75C006
22	Porphine, tetraphenyl-	−1.08	DMF	75C006
23	Pyridine-N-oxide	−2.30	DMF	77C008
24	Retinal	−1.42	THF	78C018
25	Thiofluorenone	−1.07	CH_3CN	85Z114
26	Thioxanthione	−1.05	CH_3CN	85Z114
27	Tris(2,2'-bipyridine)ruthenium(II) ion	−1.33	CH_3CN	79C010

Section 11

Bond Dissociation Energies

Bond dissociation energies, D_e, are often useful for the estimation of the enthalpy of a reaction. Three tables are included here that cover many of the bonds commonly encountered in photochemical reactions. All of the energies in these tables are in units of kJ/mol.

Table 11-1

Bond Dissociation Energies (kJ/mol) of Single Bonds[a]

	H–	CH_3–	C_2H_5–	iso-Pr–	tert-Bu–	C_6H_5–	$C_6H_5CH_2$–	CH_3CO–	CF_3–
H–	436	435	410	398	385	461	368	360	444
CH_3–	435	368	356	351	343	418	301	339	423
C_2H_5–	410	356	343	335	322	410	289	326	-
iso-Pr–	398	351	335	324	305	404	282	314	-
tert-Bu–	385	343	322	305	282	389	282	297	-
C_6H_5–	461	418	410	404	389	494	328	406	-
$C_6H_5CH_2$–	368	301	289	282	268	328	-	264	-
CH_3CO–	360	339	326	314	297	406	264	282	-
CF_3–	444	423	-	-	-	-	-	-	405
$CH_2=CHCH_2$–	362	-	-	-	-	-	-	-	-
cyclo-C_3H_5–	445	-	-	-	-	-	-	-	-
cyclo-C_4H_7–	404	-	-	-	-	-	-	-	-
cyclo-C_5H_9–	395	-	-	-	-	-	-	-	-
cyclo-C_6H_{11}–	400	-	-	-	-	-	-	-	-
CH_3COCH_2–	411	-	-	-	-	-	-	-	-
CH_3OCH_2–	389	-	-	-	-	-	-	-	-
$HOCH_2$–	393	-	-	-	-	-	-	-	-
N≡C–	502	518	-	-	-	-	-	-	-
$(CH_3)_2COH$–	381	-	-	-	-	-	-	-	-
N≡CCH_2–	389	304	322	-	-	-	275	-	-
Cl_3C–	401	-	-	-	-	-	-	-	-

$CH_2=CH-$	452	406	393	385	372	473	-	-	-
$CH\equiv C-$	523	-	-	-	-	-	-	-	-
NH_2CH_2-	396	-	-	-	-	-	286	-	-
$Br-$	366	293	285	285	280	313	230	280	295
$Cl-$	431	351	343	339	331	398	289	339	360
$F-$	570	452	444	444	-	523	-	498	544
$I-$	298	236	224	222	207	268	188	209	223
H_2N-	460	364	351	356	351	435	301	414	-
$ON-$	207	176	176	172	172	216	-	-	130
$HO-$	498	383	383	385	379	460	322	452	-
CH_3O-	437	339	339	339	339	410	-	406	-
C_2H_5O-	436	339	339	343	339	423	-	-	-
$tert\text{-}BuO-$	439	-	-	-	-	-	-	-	-
C_6H_5O-	368	280	268	-	-	368	-	-	-
CH_3COO-	443	301	301	-	-	389	-	-	-
$HOO-$	365	-	-	-	-	-	-	-	-

[a]Values selected from references [76Z049, 85Z456].

Table 11-2

Bond Dissociation Energies (kJ/mol) of Small Molecules[a]

	H	F	Cl	Br	I	OH
H	436	570	431	366	298	498
F	570	158	256	250	-	-
Cl	431	256	243	219	211	251
Br	366	250	219	194	179	237
I	298	-	211	179	153	234
OH	498	-	251	237	234	213

[a]Values selected from reference [85Z456].

Table 11-3

Bond Dissociation Energies of Peroxides and Multiple Bonds[a]

Type of Bond	D_e /kJ mol^{-1}
N≡N	945
O=O	498
CO	1077
NO	631
C≡C	962
C=C	720
O=C=O	532
HO–OH	213
$C_2H_5O-OC_2H_5$	159
$(CH_3)_2CHO-OCH(CH_3)_2$	158
$(CH_3)_3CO-OC(CH_3)_3$	159
$CH_3C(=O)O-O(=O)CH_3$	127
CH_3CH_2O-OH	184
$(CH_3)_3CO-OH$	151

[a]Values selected from references [76Z049, 85Z456].

Section 12

Solvent Properties

This section contains four tables useful for the choice of an appropriate solvent and for an understanding of photochemical and spectroscopic behavior in that solvent.

The first table includes molecular mass, boiling and melting points, density (d), refractive index (n_D), viscosity (η), dielectric constant (ε), and Kosower's Z and Dimroth's E solvent parameters. The second table contains data regarding end absorption of many solvents. Since many studies are done in the presence of oxygen the concentration of oxygen at 0.2095 atmospheres and 1.00 atmospheres are included to facilitate calculation of quenching constants. The final table contains a listing of transparent organic glasses, along with some data on these glasses at low temperature.

Table 12-1

Physical Properties[a] of Solvents

No.	Solvent	Mol. wt. g/mol	bp °C	mp °C	d^b g/cc	$n_D{}^c$	η^d 10^{-3} Pa·s	ε^e	Kosower Z^f	Dimroth E^g
1	Acetic acid	60.05	117.9	16.66	1.0492	1.3719	1.13[h]	6.15	79.2	51.1
2	Acetic acid, ethyl ester	88.11	77.11	-83.55	0.9006	1.37239	0.4508	6.053	64.0	38.1
3	Acetone	58.08	56.2	-94.7	0.7900	1.35868	0.303[h]	20.7[h]	65.7	42.2
4	Acetonitrile	41.05	81.60	-43.84	0.7822	1.34411	0.345[h]	35.94[h]	71.3	46.0
5	Benzene	78.12	80.1	5.53	0.8790	1.50112	0.649	2.284	54	34.5
6	Benzene, bromo-	157.02	155.91	-30.82	1.4959	1.55680	1.196[i]	5.40[h]	-	37.5
7	Benzene, chloro-	112.56	131.69	-45.58	1.1063	1.5248	0.799	5.621[h]	58.0	37.5
8	Benzene, 1,2-dimethyl-	106.17	144.429	-25.182	0.88014	1.50545	0.809	2.568	-	-
9	Benzene, 1,3-dimethyl-	106.17	139.120	-47.872	0.86436	1.49722	0.617	2.3742	-	-
10	Benzene, 1,4-dimethyl-	106.17	138.359	13.263	0.86098	1.49582	0.644	2.2699	-	-
11	Benzene, isopropyl-	120.194	152.41	-96.03	0.8618	1.49145	0.791	2.383	-	-
12	Benzene, methoxy-	108.14	153.60	-37.5	0.9940	1.51700	1.32	4.33[h]	-	37.2
13	Benzene, 1,3,5-trimethyl-	120.20	164.74	-44.72	0.8652	1.49937	1.154	2.279	-	-
14	Benzonitrile	103.12	191.1	-12.75	1.0006[h]	1.52823	1.24[h]	25.20[h]	65.0	42.0
15	1-Butanol	74.12	117.73	-88.62	0.8096	1.3993	2.948	17.51[h]	77.7	50.2
16	2-Butanol	74.12	99.51	-114.7	0.8065	1.3971	3.632	16.56[h]	75.4	-
17	2-Butanone	72.11	79.58	-86.69	0.8049	1.3788	0.399	18.51	64.0	41.3
18	Carbon disulfide	76.14	46.22	-111.6	1.2632	1.62746	0.363	2.64	-	32.6
19	Carbon tetrachloride	153.82	76.64	-22.82	1.5940	1.4601	0.969	2.238	-	32.5
20	Chloroform	119.38	61.18	-63.52	1.4832	1.4459	0.58	4.806	63.2	39.1

#	Solvent								
21	Cyclohexane	84.16	80.73	6.72	0.7786	1.42623	0.975	2.023	60.1	31.2
22	Cyclohexane, methyl-	98.19	100.93	−126.59	0.7694	1.42312	0.734	2.020	-	-
23	Cyclohexanol	100.16	161.10	25.15	0.9624	1.4641	68.	15.0[h]	75.0	-
24	Cyclopentane	70.14	49.26	−93.88	0.7454	1.40645	0.439	1.969	-	-
25	Decahydronaphthalene	138.25	191.7	−124	0.8865	1.4758	2.415[h]	2.1542[h]	-	-
26	Decane	142.29	174.15	−29.64	0.7300	1.41189	0.9284	1.991	-	-
27	Dichloromethane	84.93	39.64	−94.92	1.3266	1.42416	0.449[i]	8.93[h]	64.2	41.1
28	Diethyl ether	74.12	34.43	−116.3	0.7136	1.35243	0.242	4.335	-	34.6
29	1,2-Dimethoxyethane	90.12	84.5	−69	0.8691	1.37963	0.455[h]	7.20[h]	61.2	38.2
30	N,N-Dimethylformamide	73.09	153	−60.43	0.9439	1.43047	0.924	36.71[h]	68.5	43.8
31	Dimethyl sulfoxide	78.13	189.0	18.54	1.1014	1.4793	1.991[h]	46.45[h]	70.2	45.0
32	1,4-Dioxane	88.11	101.3	11.80	1.0336	1.42241	1.439[i]	2.209[h]	64.6	36.0
33	Dodecane	170.34	216.32	−9.58	0.7487	1.42167	1.508	2.015	-	-
34	Ethanamine, N-ethyl-	73.14	55.55	−49.8	0.707	1.3845	0.306	3.894	-	-
35	Ethane, 1,2-dichloro	98.96	83.48	−35.66	1.2521	1.4448	0.800[k]	10.37[h]	63.4	41.9
36	Ethane, 1,1,2,2-tetrachloro-1,2-difluoro-	203.83	92.8	26.55	1.64470[h]	1.41297[h]	1.21[h]	2.52[h]	-	-
37	Ethane, 1,1,2-trichloro-1,2,2-trifluoro-	187.38	47.633	−36.4	1.56354[h]	1.35572[h]	0.711	2.41[h]	-	-
38	Ethanol	46.07	78.29	−114.5	0.7892	1.36143	1.200	24.55[h]	79.6	51.9
39	Ethylene glycol	62.07	197.54	−12.6	1.1135	1.4318	19.9	37.7[h]	85.1	56.6

Table 12-1—**Physical Properties of Solvents**—Continued

No.	Solvent	Mol. wt. g/mol	bp °C	mp °C	d^b g/cc	n_D^c	η^d 10^{-3} Pa·s	ε^e	Kosower Z^f	Dimroth E^g
40	Formamide	45.04	210.5	2.55	1.1334	1.44754	3.764	111.0	83.3	56.6
41	Glycerol	92.095	290.0	18.18	1.2613	1.4746	1412.	42.5[h]	82.7	-
42	Heptane	100.21	98.42	−90.58	0.6838	1.38764	0.4181	1.9246	-	-
43	Hexamethylphosphoramide	179.20	233	7.20	1.027	1.4588	3.47	29.3	62.8	40.9
44	Hexane	86.18	68.74	−95.32	0.6594	1.37486	0.3126	1.8863	-	30.9
45	Methanol	32.04	64.55	−97.68	0.7910	1.32840	0.5929	32.66[h]	83.6	55.5
46	2-Methylbutane	72.15	27.88	−159.9	0.6193	1.35373	0.225	1.843	-	-
47	2-Methyl-2-butanol	88.15	102.0	−8.8	0.8096	1.4050	3.548[h]	5.78[h]	70.7	-
48	3-Methyl-1-butanol	88.15	130.5	−117.2	0.8104	1.4072	3.738[h]	15.19[h]	77.6	47.0
49	3-Methylpentane	86.18	62.28	glass	0.6643	1.37652	0.307[h]	1.895	-	-
50	2-Methyl-1-propanol	74.12	107.89	−108	0.8016	1.39591	3.333[h]	17.93[h]	77.7	-
51	2-Methyl-2-propanol	74.12	82.35	25.62	0.7887	1.3878	5.942	12.47[h]	71.3	43.9[i]
52	Naphthalene, decahydro-, (E)-	138.25	187.27	−30.40	0.8697	1.46949	2.128	2.172	-	-
53	Naphthalene, decahydro-, (Z)-	138.25	195.77	−43.01	0.8967	1.48098	3.381	2.197	-	-
54	Octane	114.23	125.67	−56.76	0.7025	1.39743	0.5466	1.948	60.1	-
55	Pentane	72.15	36.07	−129.73	0.6262	1.35748	0.235	1.844	-	-
56	1-Pentanol	88.15	137.983	−78.2	0.8145	1.4100	3.5128[h]	13.9[h]	-	-
57	2-Pentanol	88.15	119.0	glass	0.8094	1.4064	5.307[i]	13.71[h]	-	-
58	1-Propanol	60.10	97.151	−126.2	0.80361	1.38556	1.9430[h]	20.45[h]	78.3	50.7
59	2-Propanol	60.10	82.242	−88.0	0.78545	1.3772	2.0436[h]	19.92[h]	76.3	48.6
60	Pyridine	79.10	115.254	−41.55	0.98319	1.51016	0.952	12.92[h]	64.0	40.2

#	Solvent									
61	Tetrahydrofuran	72.11	65.965	−108.39	0.8892	1.40716	0.575	7.58[h]	58.8	37.4
62	Tetramethylene sulfone	120.17	287.3	28.45	1.2604[j]	1.4833[j]	10.286[j]	43.26[j]	77.5	44.0
63	Toluene	92.14	110.630	−94.991	0.86683	1.49693	0.5859	2.379[h]	-	33.9
64	Trifluoroacetic acid	114.02	71.78	−15.216	1.4890	1.2850	0.926	8.55	-	-
65	2,2,2-Trifluoroethanol	100.04	74.05	−43.5	1.3826[l]	-	1.995	26.67[l]	-	-
66	2,2,4-Trimethylpentane	114.23	99.238	−107.388	0.69193	1.39145	0.504	1.940	-	-
67	Water	18.015	100.	0.	0.9982	1.332988	1.0019	80.16	94.6	63.1

[a]Values for molecular weight, boiling point, melting point, density, refractive index, viscosity and dielectric constant are at 20 °C unless otherwise noted, and have been extracted from references [85Z456, 86Z350]. Values for Kosower's Z and Dimroth's E solvent parameters are at 25 °C and have been extracted from references [68Z006] and [71M447], respectively. For a discussion of Z, E, and other solvent parameters see references [79M363, 81Z335].

[b]Density at 20 °C unless otherwise noted.

[c]Refractive index at 20 °C at the average sodium D line unless otherwise noted.

[d]Viscosity at 20 °C unless otherwise noted. For values at 25 °C see Table 8-1.

[e]Dielectric constant at 20 °C unless otherwise noted. Limiting values at low frequencies.

[f]Kosower's solvent parameter derived from the wavelength of the charge-transfer band in the visible spectrum of 1-ethyl-4-methoxycarbonylpyridinium iodide ($Z = 2.859 \times 10^4/\lambda$ where λ is the position of the absorption maximum in nanometers). See references [79M363, 81Z335] for a discussion of this parameter.

[g]Dimroth's solvent parameter derived from the visible spectrum of two pyridinium betaines. See references [79M363, 81Z335] for a discussion of this parameter.

[h]Measured at 25 °C. [i]Measured at 15 °C. [j]Measured at 30 °C. [k]Measured at 19 °C. [l]From reference [85Z457].

Table 12-2

Ultraviolet Transmission[a] of Solvents

No.	Solvent	%T[b] 254 nm	%T 313 nm	%T 366 nm	T = 10%[c] at λ(nm) =
1	Acetic acid, ethyl ester	<10	99	100	255
2	Acetone	0	0	100	329
3	Acetonitrile	98	100	100	190
4	Benzene	0	94	100	280
5	Benzonitrile	0	85	100	299
6	Carbon disulfide	0	0	0	380
7	Carbon tetrachloride	0	100	100	265
8	Chloroform	80	100	100	245
9	Cyclohexane	100	100	100	205
10	Cyclohexane, methyl-	100	100	100	207
11	Dichloromethane	98	100	100	232
12	Diethyl ether	84	100	100	215
13	*N,N*-Dimethylformamide	0	93	100	270
14	Dimethyl sulfoxide	0	96	100	262
15	1,4-Dioxane	64	100	100	215
16	Ethane, 1,2-dichloro	97	98	100	226
17	Ethanol	98	100	100	205
18	Heptane	100	100	100	197
19	Hexane	100	100	100	195
20	Methanol	100	100	100	205
21	2-Methylbutane	100	100	100	192
22	2-Propanol	98	100	100	205
23	Pyridine	0	88	100	305
24	Tetrahydrofuran	57	99	100	233
25	Toluene	0	90	100	285
26	2,2,4-Trimethylpentane	100	100	100	197

[a]Values taken from reference [71Z014], numbers M1-M19 and 97, 98.

[b]% transmission through 1 cm of neat solvent at indicated wavelength.

[c]Wavelength (in nm) at which the percent transmission has dropped to 10% or where the absorption is equal to 1 (for 1 cm pathlength of neat solvent).

Table 12-3

O$_2$ Concentration in Solvents

No.	Solvent	Mol. mass g/mol	T °C	d g/cc[a]	[O$_2$] /10^{-3} mol/L (1 atm O$_2$)[b]	[O$_2$] /10^{-3} mol/L (0.21 atm O$_2$)[c]
1	Acetic acid					
		60.052	20	1.0495	8.11	1.7
2	Acetic acid, ethyl ester					
		88.106	20	0.9006	8.89	1.9
3	Acetone					
		58.08	20	0.7900	11.4	2.4
			25	0.7844	11.3	2.4
4	Acetonitrile					
		41.052	24	0.7765[d]	9.1[e]	1.9[e]
5	Aniline					
		93.128	25	1.0175	2.47	0.52
6	Aniline, *N,N*-dimethyl-					
		121.182	25	0.9523	5.64	1.2
7	Aniline, *N*-methyl-					
		107.155	25	0.9822	2.67	0.56
8	Benzene					
		78.113	20	0.8790	9.02	1.9
			25	0.8736	9.06	1.9
9	Benzene, bromo-					
		157.01	25	1.4882	7.09	1.5
10	Benzene, chloro-					
		112.559	25	1.1009	8.78	1.8
11	Benzene, 1,2-dimethyl-					
		106.167	25	0.8759	9.22	1.9
12	Benzene, 1,3-dimethyl-					
		106.167	25	0.8601	9.69	2.0
13	Benzene, 1,4-dimethyl-					
		106.167	25	0.8566	10.0	2.1
14	Benzene, ethyl-					
		106.167	25	0.8625	9.91	2.1
15	Benzene, fluoro-					
		96.104	25	1.022[g]	16.0	3.4
16	Benzene, hexafluoro-					
		186.056	20	1.6187	21.1	4.4
			25	1.6073	20.8	4.4

Table 12-3—**O$_2$ Concentration in Solvents**—Continued

No.	Solvent	Mol. mass g/mol	T °C	d g/cc[a]	$[O_2]/10^{-3}$ mol/L (1 atm O$_2$)[b]	$[O_2]/10^{-3}$ mol/L (0.21 atm O$_2$)[c]
17	Benzene, isopropyl-	120.194	25	0.8574	9.90	2.1
18	Benzene, nitro-	123.111	25	1.1983	4.82	1.0
19	1-Butanol	74.122	20	0.8096	8.77	1.8
			25	0.8057	8.65	1.8
20	2-Butanone	72.107	25	0.7997	11.2	2.4
21	Butylamine	73.138	20	0.7392	11.3	2.4
22	Carbon disulfide	76.131	25	1.2555	7.24	1.5
23	Carbon tetrachloride	153.823	20	1.5940	12.4	2.6
			25	1.5844	12.4	2.6
24	Chloroform	119.378	20	1.4891	11.6	2.4
25	Cyclohexane	84.161	20	0.7786	11.5	2.4
			25	0.7739	11.5	2.4
26	Cyclohexane, methyl-	98.188	20	0.7694	12.3	2.6
			25	0.7651	12.5	2.6
27	Cyclohexanol	100.16	26	0.9684[d]	8.27	1.7
28	Cyclohexanone	98.144	20	0.9452	6.14	1.3
			25	0.943	6.11	1.3
29	Decane	142.284	25	0.7263	11.2	2.3
30	Dichloromethane	84.933	20	1.3256	10.7	2.2
31	Diethyl ether	74.122	20	0.7136	14.7	3.1
32	Dimethoxymethane	76.095	20	0.860	13.5	2.8

Table 12-3—**O$_2$ Concentration in Solvents**—Continued

No.	Solvent	Mol. mass g/mol	T °C	d g/cc[a]	$[O_2]/10^{-3}$ mol/L (1 atm O_2)[b]	$[O_2]/10^{-3}$ mol/L (0.21 atm O_2)[c]
33	**Dimethyl sulfoxide**					
		78.129	25	1.0954	2.20	0.46
34	**1,4-Dioxane**					
		88.106	20	1.0336	6.10	1.3
			25	1.0280	6.28	1.3
35	**Dodecane**					
		170.337	25	0.7452	8.14	1.7
36	**Ethane, 1,2-dichloro**					
		98.96	20	1.2521	7.40	1.6
37	**Ethanol**					
		46.069	20	0.7892	10.1	2.1
			25	0.7849	9.92	2.1
38	**Ethanol (95%)**					
		-	25	0.8074	7.84[f]	1.64[f]
39	**Ethene, tetrachloro-**					
		165.834	20	1.6228	8.27	1.7
40	**Ethylene glycol**					
		62.068	20	1.1135	0.58	0.12
41	**Glycerol**					
		92.094	20	1.2613	0.3	0.07
42	**Heptane**					
		100.203	20	0.6837	13.4	2.8
			25	0.6795	13.2	2.8
43	**Hexane**					
		86.177	20	0.6593	15.0	3.1
			25	0.6548	14.7	3.1
44	**2-Hexanone**					
		100.16	25	0.8067	9.42	2.0
45	**Methanol**					
		32.042	20	0.7910	10.3	2.2
			25	0.7864	10.2	2.1
46	**Methyl acetate**					
		74.079	20	0.9342	11.1	2.3
			25	0.9279	11.2	2.3
47	**2-Methyl-1-propanol**					
		74.122	20	0.8016	9.21	1.9
			25	0.7978	9.04	1.9

Table 12-3—**O₂ Concentration in Solvents**—Continued

No.	Solvent	Mol. mass g/mol	T °C	d g/cc[a]	$[O_2]/10^{-3}$ mol/L (1 atm O_2)[b]	$[O_2]/10^{-3}$ mol/L (0.21 atm O_2)[c]
48	Nitromethane					
		61.04	25	1.1313	8.60	1.8
49	Nonane					
		128.257	25	0.7137	11.2	2.3
50	Octane					
		114.23	20	0.7027	13.3	2.8
			25	0.6986	13.3	2.8
51	1-Octanol					
		130.23	20	0.8250	7.22	1.5
			25	0.8216	7.13	1.5
52	Pentane					
		72.15	25	0.6214	17.7	3.7
53	1-Pentanol					
		88.149	20	0.8145	8.44	1.8
54	2-Pentanone					
		86.133	25	0.8015	10.4	2.2
55	Perfluorodecalin					
		462.08	25	1.946[h]	16.1	3.4
56	Perfluoroheptane					
		388.04	25	1.7333[i]	24.8	5.2
57	Perfluoro(methylcyclohexane)					
		350.05	25	1.788[h]	23.3	4.9
58	Perfluorononane					
		488.07	25	1.84[h]	20.2	4.2
59	Perfluorooctane					
		438.06	25	1.738[h]	21.2	4.4
60	Piperidine					
		85.149	20	0.8606	7.52	1.6
			25	0.8566	7.39	1.5
61	1-Propanol					
		60.096	25	0.7996	6.69	1.4
62	2-Propanol					
		60.096	20	0.7855	10.3	2.2
			25	0.7813	10.2	2.1
63	Propylamine					
		59.111	20	0.7173	10.1	2.1

Table 12-3—**O$_2$ Concentration in Solvents**—Continued

No.	Solvent	Mol. mass g/mol	T °C	d g/cc[a]	$[O_2]$ /10^{-3} mol/L (1 atm O$_2$)[b]	$[O_2]$ /10^{-3} mol/L (0.21 atm O$_2$)[c]
64	**Pyridine**					
		79.101	20	0.9832	5.66	1.2
			25	0.9782	5.66	1.2
65	**Pyrrolidine**					
		71.122	20	0.8586	7.29	1.5
			25	0.8538	7.27	1.5
66	**Tetrahydrofuran**					
		72.107	20	0.8892	9.90	2.1
			25	0.8842[j]	10.0	2.1
67	**Toluene**					
		92.14	20	0.8668	8.63	1.8
			25	0.8622	9.88	2.1
68	**2,2,4-Trimethylpentane**					
		114.23	25	0.6878	15.5	3.3
69	**Water**					
		18.0152	20	0.9982	1.39	0.29
			25	0.9970	1.27	0.27

[a]Values selected from reference [86Z350].

[b]Values often taken from evaluated data of reference [81Z334]. These values and others chosen (from individual measurements compiled in reference [81Z334]) were almost always from experiments done at 1 atm partial pressure of O$_2$.

[c]Values for 0.21 atm partial pressure of O$_2$ calculated from b, assuming Henry's Law holds.

[d]At 25 °C. [e][717447]. [f][469001]. [g]At 20 °C. [h][77Z191]. [i][47Z002]. [j][85Z457].

Table 12-4

Low-Temperature Organic Glasses for Spectroscopy[a]

Glass[a]	Ratio	Ref.	% crack[b]	V_{77}/V_{293}[c]	η/η_{3MP}[d]
Hydrocarbons					
Pentane (tech)	-	639027	0	-	-
Petroleum ether (30-60)	-	639027	10	-	-
2-Methylpentane (2MP)	-	65M065 68M106	-	-	11
3-Methylpentane (3MP)	-	629028 64E023 68M106	-	-	1
3-Ethylpentane	-	68M107	-	-	1.1×10^6
2,3-Dimethylpentane	-	68M107	-	-	1.6×10^4
3-Methylhexane	-	68M107	-	-	1.5×10^6
4-Methylheptane	-	68M107	-	-	9.1×10^{12}
3-Methyloctane	-	68M107	-	-	5.5×10^{18}
Ethylcyclohexane	-	68M107	-	-	2.0×10^{10}
Methylcyclohexane (MCH)	-	539008 629027 67F523 68M106	-	-	$\sim\!4 \times 10^4$
Pentane / Heptane	1:1	639027	0	-	-
2MP / MCH	1:1	72M260	-	-	0.90
3MP / Isopentane (IP)	1:0	539008 629027 64E023 68M106	-	0.784	1
3MP / IP	9:1	64E023	-	-	0.25
3MP / IP	8.1:1	68M106	-	-	0.13
3MP / IP	4:1	64E023 68M106	-	-	6.6×10^{-2}
3MP / IP	7:3	64E023	-	-	1.5×10^{-2}
3MP / IP	2:1	68M106	-	-	6.0×10^{-3}
3MP / IP	3:2	539008 629027 64E023	-	0.771	5.4×10^{-3}

Table 12-4—**Low-Temperature Organic Glasses**—Continued

Glass[a]	Ratio	Ref.	% crack[b]	V_{77}/V_{293}[c]	η/η_{3MP}[d]
Hydrocarbons—Continued					
3MP / IP	1:1	64E023	-	-	6.3×10^{-4}
3MP / IP	2:3	64E023	-	-	1.9×10^{-4}
3MP / IP	3:7	64E023	-	-	5.5×10^{-5}
3MP / IP	1:3	68M106	-	-	1.6×10^{-5}
3MP / IP	1:4	64E023	-	-	1.9×10^{-5}
3MP / IP	1:6	539008 629027	-	0.760	-
3MP / IP	1:9	64E023	-	-	3.8×10^{-6}
3MP / IP	1:32	64E023	-	-	9.5×10^{-7}
MCH / Pentane	4:1 - 3:2	639027	0	-	-
MCH / IP	4:1	539008 629027	-	0.810	-
MCH / IP	1:3	539008 629027 65M065	-	0.773	1.1×10^{-2}
MCH / IP	1:4	539008 629027	-	0.769	-
MCH / IP	1:5	539008 629027	-	0.767	-
MCH / Methylcyclopentane	1:1	65M065 72M260	-	-	2.8
MCH / Methylcyclopentane	3:2	72M260	-	-	5.4
Ethers					
Diethyl ether (Et$_2$O)	-	639027	10	-	-
2-Methyltetrahydrofuran	-	629028 65M065 68M107	-	-	6.9×10^{5}
Et$_2$O / IP	1:1, 2:1	629024 629028	-	-	-
Et$_2$O / MCH	2:3	67F523	-	-	-
Propyl ether / Pentane	2:1	639027	10	-	-

Table 12-4—**Low-Temperature Organic Glasses**—Continued

Glass[a]	Ratio	Ref.	% crack[b]	V_{77}/V_{293}[c]	η/η_{3MP}[d]
Alcohols					
Ethanol (EtOH)	-	639027	10	-	-
Glycerol	-	72M260	-	-	1.2×10^{45}
1-Propanol	-	639027 72M260	20	-	1.6×10^{7}
Propylene glycol	-	72M260	-	-	4.8×10^{31}
EtOH / Methanol (MeOH)	5:1 - 9:1	639027	15	-	-
EtOH / MeOH	4:1	539008 629027	-	0.802	-
EtOH / MeOH	1:1	65M065 72M260	-	-	1.7×10^{11}
2-Propanol / IP	3:7	629028	-	-	-
1-Butanol / IP	3:7	629028	-	-	-
EtOH / Et$_2$O	1:1	629024	-	-	-
1-Propanol / 2-Propanol	1:1	72M260	-	-	1.2×10^{9}
1-Propanol / 2-Propanol	3:2	72M260	-	-	1.1×10^{8}
1-Propanol / Et$_2$O	2:5	629024	-	-	-
2-Propanol / Et$_2$O	1:3	629028	-	-	-
1-Butanol / Et$_2$O	2:5	629024	-	-	-
EtOH / MeOH / Et$_2$O	8:2:1	639027	10	-	-
EPA (EtOH / IP / Et$_2$O)	2:5:5	539008 629027 639027	0	0.778	5.5×10^{-2}
2-Propanol / IP / Et$_2$O	2:5:5	629024	-	-	-
Miscellaneous					
Perfluorodimethylcyclohexane	-	68M107	-	-	1.1×10^{35}
Et$_2$O / EtOH / Toluene	2:1:1	629028	-	-	-
Ethyl iodide / Et$_2$O / IP	1:2:1	629028	-	-	-
Ethyl iodide / EtOH / MeOH	1:16:4	629028	-	-	-
cis-2-Pentene / Et$_2$O	1:2	629024	-	-	-
Cyclohexene / IP	2:1	67F523	-	-	-

Table 12-4—**Low-Temperature Organic Glasses**—Continued

Glass[a]	Ratio	Ref.	% crack[b]	V_{77}/V_{293}[c]	η/η_{3MP}[d]
Miscellaneous—Continued					
Cyclohexene / Et_2O	3:2	67F523	-	-	-
Piperylene / Et_2O / MCH	1:2:2	67F523	-	-	-
EtOH / NH_3 (28% aq.)	<20:1	639027	30	-	-
Et_2O / EtOH / NH_3 (28% aq.)	10:9:1	639027	10	-	-
Triethylamine / Et_2O / Pentane	2:5:5	639027	0	-	-
Triethylamine / Et_2O / IP	2:5:5	629028	-	-	-
EtOH / conc. HCl	19:1	639027	30	-	-

[a]Organic solvent mixtures that form clear glasses at 77 K with a low cracking frequency. For other glasses see listed references.

[b]From reference [639027], frequency of cracking in percent.

[c]From references [539008, 629027], volume at 77 K relative to volume at 293 K.

[d]Viscosities, η, relative to 3MP at 77 K. (1) [65M065], η's relative to a value of 11 for 2MP. Value of 11 was adopted from reference [68M106] to put the η's on a common scale. Values reported have been extrapolated on the graph in reference [65M065] and are subject to considerable error. For an absolute scale, a value for 1:3 MCH / IP of 1.26×10^7 Pa·s at 77 K can be extracted from the data. (2) [68M106], using an absolute value of 2.2×10^{11} Pa·s for 3MP at 77 K. The other values for this reference have been extrapolated to 77 K. (3) [64E023], an absolute value of 9.4×10^{11} Pa·s is reported for 3MP at 77 K.

Section 13

Chemical Actinometry

This section on chemical actinometry contains instructions for the use of some commonly utilized actinometers. For other methods of using the actinometers discussed here and for discussion of other actinometers, see references [56F006, 66Z001, 68Z003, 697314, 71F593, 71Z012, 76Z005, 76Z030, 82Z269, 89Z022] and references in Table 13-5.

A. Potassium Ferrioxalate Actinometry

References

Outlined instructions for the use of potassium ferrioxalate follow. For other methods and comments on the use of potassium ferrioxalate, see references [56F006, 66Z001, 68Z003, 697314, 71F593, 71Z012, 767458, 76Z005, 76Z030, 777555, 79F262, 81F343, 82Z269, 83E027, 83F079, and 83F502].

Net Reaction

For a discussion of the reaction see [717336], but the net photochemical reaction is

$$2Fe^{3+} + C_2O_4^{2-} \xrightarrow{h\nu} 2Fe^{2+} + 2CO_2.$$

The amount of ferrous produced is measured via spectrophotometric determination of its 1,10-phenanthroline complex at 510 nm. Ferric apparently forms only a weak complex with 1,10-phenanthroline, and this complex is transparent at 510 nm.

Advantages

1) Easy and fast to use.

2) Does not depend on difference readings between large numbers to determine amount of conversion.

3) Quantum yields are accurately known.

4) Can be used in the blue region of the visible spectrum.

5) Stirring is not necessary.

6) Oxygen does not have to be excluded. [56F006, 81F343]

7) Quantum yields are relatively insensitive to wavelength, concentration, temperature, and light intensity. See Table 13-1.

Disadvantages

1) Only very short irradiation times are required to consume considerable portions of the actinometer. Therefore it should be employed primarily where light intensities do not vary significantly and measurements for short time intervals can be interpolated to cover the entire irradiation period.

2) Due to its high absorption, even in part of the visible region (see Fig. 13-1), light filtering must be properly performed or tedious corrections have to be made.

3) The refractive index of water can be considerably different than that of an organic solvent. This can cause significant errors especially when used in a slitless merry-go-round. See reference [71F595] for further discussion of this problem.

Discussion

In the procedure that follows, two modifications of the Hatchard and Parker method [56F006] were made.

First, since absorbances,

$$A = \log_{10}\{I_0/I\} , \qquad\qquad\qquad (13\text{-}1)$$

are easily measured on Cary spectrophotometers from 0 to 2, the concentration of the actinometer was raised from 0.006 to 0.02 mol/L† to allow more ferrous production without considerable depletion of the ferrioxalate. [67F523] Under these conditions, more ferrous may be produced; thus, more 1,10-phenanthroline was added. When the 1,10-phenanthroline solution recommended by Hatchard and Parker was used, the Beer's law plot was linear only to slightly above an absorbance of 1. The additional 1,10-phenanthroline extended the linearity to above 2. [67F523]

The second modification was the use of solutions for preparation of the actinometer rather than use of the solid $K_3Fe(C_2O_4)_3$. This method was suggested by Hatchard and Parker [56F006] and applied by the Hammond group. [68F301] It has been found that the use of the solutions is much more convenient than the use of the solid.

Solutions

1)   0.2% by weight 1,10-phenanthroline in water.
2)   Buffer solution.
     a)   82 g NaOAc·3H$_2$O
     b)   10 mL conc. H$_2$SO$_4$
     c)   diluted to 1 L with water
3)   Ferric sulfate solution Fe$_2$(SO$_4$)$_3$.
     a)   100 g Fe$_2$(SO$_4$)$_3$·nH$_2$O [approximately 80% Fe$_2$(SO$_4$)$_3$]
     b)   55 mL conc. H$_2$SO$_4$.
     c)   diluted to 1 L with water
4)   Standardized 0.1 mol/L EDTA.
5)   Standard solution of K$_2$C$_2$O$_4$ prepared according to instructions that follow in part B of *Preparation of Actinometer Solution*.
6)   0.08 mol/L FeSO$_4$ solution in 0.05 mol/L H$_2$SO$_4$.†

---

† Inspection of Table 13-2 indicates that the quantum yield of the 0.02 mol/L solution should be no more than 2% lower than for the 0.006 mol/L solution. This solution will absorb >99% of the light up to 400 nm. For wavelengths longer than 400 nm, corrections will have to be applied for the fraction of light absorbed or the concentration of ferrioxalate can be increased.

Table 13-1

Concentration and Temperature Dependence of $\phi_{Fe^{2+}}$ from Ferrioxalate

λ /nm	c /mol/L	T /°C[a]	$\phi_{Fe^{2+}}$	Ref.
313	0.006	5	1.27	83F079
313	0.006	22	1.24	83F079
313	0.006	40	1.25	83F079
313	0.006	60	1.28	83F079
313	0.006	80	1.24	83F079
365/6	0.0006		1.17	56F006
364	0.006		1.28	81F343
365/6	0.006		1.21	56F006
365/6	0.006		1.20	55F005
367	0.006		1.21	83E027
365/6	0.06		1.18	56F006
364	0.15		1.18	81F343
365/6	0.15		1.15	56F006
457	0.15		0.90	83E027
458	0.0154	12	1.09	81F049
458	0.0154	22	1.07	81F049
458	0.0102	23	1.12	81F049
458	0.0102	32	1.18	81F049
458	0.0154	31	1.13	81F049
458	0.0154	42	1.02	81F049
458	0.15		0.845	81F343
480	0.15		0.93	56F006
488	0.0814	13	1.09	81F049
488	0.0816	23	1.08	81F049
488	0.0816	31	1.03	81F049
488	0.0816	41	1.04	81F049
509	0.15		0.86	56F006
513	0.15		0.86	83E027
514	0.163	12	0.86	81F049
514	0.163	21	0.90	81F049
514	0.204	23	0.90	81F049
514	0.204	32	1.23	81F049
514	0.204	42	1.41	81F049

[a] Room temperature unless stated otherwise.

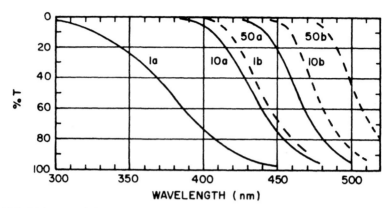

Figure 13-1. Light transmission by potassium ferrioxalate. Curves 1a, 10a and 50a are for 0.006 mol/L potassium ferrioxalate in 0.05 mol/L H_2SO_4 with optical pathlengths of 0.1, 1 and 5 cm, respectively. Curves 1b, 10b and 50b are for 0.15 mol/L potassium ferrioxalate in 0.05 mol/L H_2SO_4 with 0.1, 1 and 5 cm optical pathlengths. Reproduced from reference 56F006, courtesy of Dr. C. A. Parker and the Royal Society.

7)   Standard 0.016 mol/L potassium dichromate solution (~0.1 N).†

8)   1% Diphenylamine in conc. H_2SO_4.†

Extinction Coefficient for Ferrous 1,10-Phenanthroline Complex

      The alternative procedure is to use the Hatchard and Parker value of 1.11×10^4 L mol^{-1} cm^{-1}. [56F006]

A.   Titrate 0.08 mol/L $FeSO_4$ solution in 0.05 mol/L H_2SO_4 [60Z008] with standard 0.016 mol/L potassium dichromate solution using a few drops of diphenylamine solution as indicator.

B.   Dilute 0.08 mol/L $FeSO_4$ solution 20 times with 0.05 mol/L H_2SO_4 to give 4×10^{-4} mol/L $FeSO_4$.

C.   To a series of six 25 mL volumetric flasks add 0, 1, 3, 5, 7, 9 mL respectively of the 4×10^{-4} mol/L $FeSO_4$ solution (*do not dilute to the mark yet*).

D.   Add enough 0.05 mol/L H_2SO_4 to each of the solutions in C to bring the volume to approximately 12.5 mL.

E.   To each of the solutions in D add 2 mL of the 0.2% 1,10-phenanthroline solution.

F.   Add 6 mL of the buffer solution to each flask from E and dilute to the mark.

G.   Measure the absorbance, A, of each solution at 510 nm in a 1 cm cell vs. a cell with the solution prepared above without ferrous.

---

      † Numbers 6-8 are necessary only if an experimental determination of the extinction coefficient of the ferrous 1,10-phenanthroline complex is to be made. The alternative is to adopt the value of $\varepsilon_{510} = 1.11 \times 10^4$ L mol^{-1} cm^{-1} determined by Hatchard and Parker. [56F006]

Table 13-2

Quantum Yields of Ferrous Production

λ /nm	$\phi_{Fe^{2+}}$ 0.006 mol/L Ferrioxalate[a]	$\phi_{Fe^{2+}}$ 0.15 mol/L Ferrioxalate[b]	λ /nm	$\phi_{Fe^{2+}}$ 0.15 mol/L Ferrioxalate[b]
254	1.25	-	442	1.00
297/302	1.24	-	451	0.96
313	1.24	-	457	0.90
326	1.23[b]	1.16	458	0.845[d]
334	1.23	-	463	0.86
341	1.22[b]	1.14	468	0.91
352	1.21[b]	1.14	472	0.94
358	1.25[c]	-	480	0.93[a]
361/6	1.21	-	482	0.95
364	1.28[d]	1.18[d]	493	0.94
365/6	1.21	1.15[a]	502	0.90
367	1.21[b]	1.15	509	0.86[a]
382	1.18[b]	1.12	512	0.86
392	1.13[c]	1.10	522	0.65
402	-	1.07	530	0.53
405	1.14	-	546	0.15[a]
407	1.19[d]	-	577/9	0.013[a]
412	-	1.05		
416	1.12[c]	-		
422	-	1.04		
433	-	1.03		
436	1.11	1.01[a]		

[a]From [56F006]. Values recommended by Parker and determined by reference to uranyl oxalate and/or thermopile.
[b]From [83E087]. Determined relative to Parker's value of 1.15 for 0.15 mol/L solution at 365 nm. Data obtained by digitizing graph. [c]From [66F204]; [d]From [81F343].

H. Plot the absorbance, *A*, obtained vs. the concentrations of the complex, and determine the extinction coefficient from the slope.

Preparation of Actinometer Solution

A. Titrate approximately 0.2 mol/L $Fe_2(SO_4)_3$ solution (Solution 3) with standardized EDTA using 0.2 g salicyclic acid/100 mL solution as indicator and buffered with 0.3 g glycine/100 mL solution, to a pH of 3-4. (See p. 241 of ref. [69Z009])

B.   Prepare a standard solution (at least 100 mL) of $K_2C_2O_4$ such that its molarity is six times that of the $Fe_2(SO_4)_3$ solution (approximately 1.2 mol/L $K_2C_2O_4$).

C.   When actinometer solution is needed, pipet 5 mL of the $Fe_2(SO_4)_3$ solution and 5 mL of the $K_2C_2O_4$ solution into a 100 mL volume flask and dilute to the mark with water. Be careful not to mix the two stock solutions. The stock solutions will provide an indefinite supply of actinometer solution, and the 100 mL will supply many runs and have a shelf life of about one month.

Intensity Measurements

A.   Pipet into the reaction vessel a volume of the $K_3Fe(C_2O_4)_3$ solution equal to that of the samples to be irradiated.

B.   Irradiate for an appropriate period of time (must be determined experimentally but should give an absorbance, *A*, after work-up between 0.2 and 1.8).

C.   Mix irradiated solution thoroughly and pipet an aliquot (commonly 1 mL) of the actinometer into a 10 mL volumetric flask.

D.   Add 2 mL of the 0.2% 1,10-phenanthroline solution.

E.   Add a volume of buffer equal to one half of the aliquot of actinometer taken.

F.   Dilute to the mark with water, *mix*.

G.   Prepare a blank following C-F with a nonirradiated volume of actinometer equal to the aliquot of irradiated sample withdrawn.

H.   Measure the absorbances of the solutions F and G vs. water at 510 nm and take the difference. (Note: One can measure the absorbance of F vs. G but the absorbance of the blank should be occasionally checked to test the quality of the actinometer solution. If a value greater than $A = 0.06$ is obtained for G, a new solution should be prepared.)

Calculation of Light Intensity

A.   From Table 13-1 or 13-2 select the appropriate quantum yield for ferrous production.

B.   Using the absorbance obtained, calculate the light intensity from the following formula:

$$I \text{ (einsteins/min)} = \frac{A V_2 V_3}{\varepsilon d \phi_\lambda t V_1} \qquad (13\text{-}2)$$

where

A   Absorbance (at 510 nm) of irradiated actinometer solution corrected for absorption of blank.

d   Path length (in cm) of absorption cell used in measurement of *A*.

ε Extinction coefficient of ferrous 1,10-phenanthroline complex at 510 nm ($\sim 1.11 \times 10^4$ L mol^{-1} cm^{-1}).

ϕ_λ Quantum yield of ferrous production at wavelength of light used.

V_1 Volume (in milliliters) of irradiated actinometer solution withdrawn.

V_2 Volume (in liters) of actinometer irradiated.

V_3 Volume (in milliliters) of volumetric flask used for dilution of irradiated aliquot (10 mL, as given above).

t Irradiated time in minutes.

Corrections

A. As light filtering is generally not perfect and $K_3Fe(C_2O_4)_3$ generally absorbs at wavelengths where irradiated solutions do not, one should determine the light intensity with a cell (containing the solvent used for the sample irradiation) in front of the actinometer and then with the cell containing the sample solution. A comparison gives a percentage to be used as a correction in calculating light intensities.

B. Uranyl Oxalate Actinometry

References

The method below was adapted from Hatchard and Parker. [56F006] See references [307001, 56Z003, 66Z001, 71Z012, 76Z005, and 82Z269] for further discussion concerning the use of uranyl oxalate.

Net Reaction

The uranyl ion acts a photosensitizer to the decomposition of the oxalate ion. The overall reaction is

$$H_2C_2O_4 \xrightarrow{h\nu} H_2O + CO_2 + CO$$

The amount of oxalate before and after the measurement is determined by a titration with $KMnO_4$.

Advantages

1) Useful in range 210-435 nm, see Fig. 13-2.

2) Quantum yields are accurately known, see Table 13-3.

Disadvantages

1) Requires titrations.

2) Results are obtained from titration differences.

3) Has absorption minimum near 366 nm.

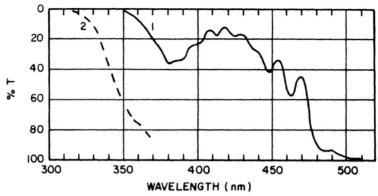

Figure 13-2. Light transmission by uranyl oxalate solutions. Curve 1 represents 0.01 uranyl sulfate in 0.05 mol/L oxalic acid (5 cm optical path). Curve 2 represents 0.001 mol/L uranyl sulfate in 0.005 mol/L oxalic acid (5 cm optical path). Reproduced from reference 56F006, courtesy of Dr. C. A. Parker and the Royal Society.

Table 13-3

Quantum Yield of Oxalate Disappearance for Uranyl Oxalate

λ /nm	$\phi_{-\text{oxalate}}$ [a]	$\phi_{-\text{oxalate}}$ [b]
254	0.602	-
265	0.582	-
302	0.570	-
313	0.561	-
366	0.492	0.492
405	0.563	-
436	0.584	0.573

[a]From 307001. [b]From 63F023.

Solutions

1)   0.01 mol/L uranyl sulfate in 0.05 mol/L oxalic acid solution.

2)   Standardized 0.008 mol/L KMnO$_4$ solution.

Procedure

A.   Titrate an appropriate volume of the actinometer with the standardized KMnO$_4$ solution. Titrate at 60 °C in approximately 0.5 mol/L H$_2$SO$_4$ solution of the oxalate.

B.   Irradiate the desired volume of actinometer for appropriate time period (~20% conversion).

C. Titrate irradiated solution with $KMnO_4$. Titration again to be done at 60 °C.

D. Calculate light intensity.

C. Benzophenone - Benzhydrol Actinometry

References

This method has been adapted from references [617010], [627011], and [697314]. Also see references [76Z005] and [82Z269].

Net Reaction

$$(C_6H_5)_2CO + (C_6H_5)_2CHOH \xrightarrow{h\nu} (C_6H_5)_2C(OH)C(OH)(C_6H_5)_2$$

Advantages

1) Does not depend on predetermined quantum yield (however limiting quantum yield is assumed to be unity).

2) Can be used for entire irradiation period and therefore takes into account light intensity variations.

3) Does not absorb above 390 nm.

Disadvantages

1) Requires degassing of several tubes.

2) Values obtained from difference in UV absorptions rather than from an absolute reading.

Discussion

The kinetic mechanism is taken to be (1) the formation of triplet benzophenone, $^3B*$, with unit quantum yield, (2) the first-order decay of $^3B*$ to benzophenone, B,

$$^3B* \rightarrow B + heat$$

with rate constant k_d, (3) the pseudo first-order reaction of $^3B*$ with benzhydrol, BH_2,

$$^3B* + BH_2 \rightarrow benzopinacol$$

with rate constant, k_r. No other deactivation of $^3B*$ is assumed to take place. For this mechanism, the quantum yield (ϕ_{-B}) for the disappearance of benzophenone is

$$\frac{1}{\phi_{-B}} = 1 + \frac{k_d}{k_r [BH_2]}. \tag{13-3}$$

If this is multiplied by the quantum yield (ϕ_{act}) of disappearance of benzophenone at a specific BH_2 concentration (hereafter considered the actinometer solution), the following results:

$$\frac{\phi_{act}}{\phi_{-B}} = \phi_{act} + \frac{\phi_{act}\, k_d}{k_r[BH_2]}.$$ (13-4)

A plot of the relative quantum yield ratio $\dfrac{\phi_{act}}{\phi_{-B}}$ vs. $\dfrac{1}{[BH_2]}$ gives an intercept of ϕ_{act}, and the light intensity can be calculated from this value.

Procedure

A. Prepare 0.1 mol/L benzophenone solutions in benzene containing varying concentrations of benzhydrol (e.g., 0.3, 0.1, 0.07 and 0.05 mol/L).

B. Degas with three freeze-pump-thaw cycles.

C. Irradiate each solution for the same time.

D. Dilute each tube 25 times and compare the benzophenone absorption of each to diluted, nonirradiated samples.

E. Plot

$$\frac{\phi_{act}}{\phi_{-B}} \quad \text{vs.} \quad \frac{1}{[BH_2]}$$

where [BH_2] is the average of the initial and final benzhydrol concentrations.

F. Determine ϕ_{act} from the intercept.

G. Calculate the light intensity from

$$I = \frac{\Delta[B]}{\phi_{act}},$$ (13-5)

where $\Delta[B]$ is the change in the benzophenone concentration of the actinometry solution.

Alternative Procedure

Use a value for ϕ_{act} of 0.68 [627011] for 0.1 mol/L benzophenone and 0.1 mol/L benzhydrol in benzene and irradiate only this tube.

D. 2-Hexanone Actinometry

References

Adapted from references 66F207 and 68F303.

Net Reaction

2-Hexanone undergoes a Norrish Type II photochemical reaction to yield acetone, see Fig. 13-3. Acetone can be easily determined

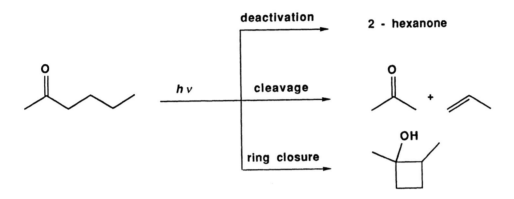

Figure 13-3. Photochemistry of 2-hexanone.

quantitatively by vapor-phase chromatography. Alternatively, the disappearance of 2-hexanone or formation of propene can also be determined, see Table 13-4.

Advantages

1) 2-Hexanone is especially useful at 313 nm as it does not absorb at longer wavelengths and filtering does not have to be perfect.

2) It can be used for the duration of the irradiation and takes into account light intensity fluctuations.

Disadvantage

As one of the products is acetone, a compound with an essentially identical absorption spectrum to 2-hexanone, only low conversions may be used or a decrease of the observed quantum yield will occur.

Procedure

A. Irradiate an appropriate volume of a 1 mol/L solution of 2-hexanone in pentane, hexane or cyclohexane.

B. Analyze on a carbowax or similar column and compare chromatographic peak areas to those obtained from a standard acetone solution.

Table 13-4

Quantum Yields for 2-Hexanone Reactions in Hydrocarbon Solvents

	Quantum Yields		
	Yang and Coulson	Wagner	Farina and Murov
$\phi_{-hexanone}$	$0.327^{a,b}$	-	-
$\phi_{acetone}$	$0.252^{a,b}$	$0.22^c (0.37)^{c,d}$	$0.22^{b,f}$
$\phi_{propene}$	$0.25^{a,b}$	-	-
$\phi_{cyclobutanol}$	$0.075^{a,b}$	-	-

[a]From [66F207]. These values should be slightly high as they were measured on a slitless merry-go-round. See reference [71F595] for precautions regarding slits. [b]Determined using ferrioxalate actimometry. [c]From [68F303]. Determined using acetone, pentadiene actinometry. Under the conditions this was performed, one calculates that about 4% of the acetone singlets were quenched by the pentadiene. [70E304, 716157, 71E377, 71E378] Also, a ϕ_{isc} of 1.00 was assumed for acetone, however, the value is probably 0.90.[e] Thus, the corrected values are 0.19 and 0.32^d. [d]In *tert*-butyl alcohol. [e]The value of ϕ_{isc} for 0.1 mol/L acetone in hexane is 1.00, but for 1 mol/L to neat acetone the value is 0.90, N.C. Yang, unpublished results. [f]Unpublished results.

E. Ketone - Pentadiene Actinometry

References

Adapted from references [65F030, 697156, 69F395, and 71F595].

Comments

The quantum yield for the benzophenone photosensitized isomerization of *cis*- and *trans*-1,3-pentadienes has been accurately measured,[65F030] and the system is useful for actinometry at 313 and 366 nm. The kinetic scheme can be sketched as is shown in Fig. 13-4. Any high-energy sensitizer (E_T > 263 kJ/mol) with a known intersystem crossing yield and whose singlet excited state cannot be quenched by 0.1 mol/L pentadiene can theoretically be used. In addition to benzophenone, acetophenone and acetone [69F395] have been used. However, if acetone is used, one must correct for singlet quenching [70E304, 716157, 71E377, 71E378] (about 4% at 0.2 mol/L *cis*-1,3-pentadiene) and the intersystem crossing yield of acetone (dependent on acetone concentration).†

† The value of ϕ_{isc} for 0.1 mol/L acetone in hexane is 1.00, but for 1 mol/L to neat acetone the value is 0.90, N.C. Yang, unpublished results.

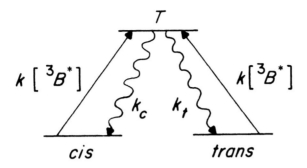

Figure 13-4. Schematic representation of a kinetic mechanism for *cis → trans* isomerization.

Advantages

1)   Can be used for long irradiation periods.

2)   At 366 nm and to some extent 313 nm, only limited light filtering is required.

Disadvantages

1)   Requires careful analysis on a long gas-liquid chromatography column.

2)   Corrections should be made for reverse reaction.

Procedure

1)   Prepare a 0.05 to 0.1 mol/L benzophenone solution with 0.05 to 0.2 mol/L *cis*-1,3-pentadiene in benzene.

2)   Irradiate.

3)   Analyze on a 20 ft × ¼ in gas-liquid chromatography column packed with 15% 1,2-bis(cyanoethoxy)ethane on Chromosorb P(45-60 mesh) at room temperature.

Quantum Yields [65F030, 71F595]

$$\phi_{c \to t} = 0.55; \quad \phi_{t \to c} = 0.44$$

In order to get the fraction of photons absorbed per *cis* molecule, F_o, from the observed fraction of *trans*-1,3-pentadiene formation, F_{obs}, it is usually necessary to correct for the back reaction. One such expression can be derived from the kinetic equations [65F030] associated with Fig. 13-4 and is given by [697156]

$$F_o = 2.303 \log_{10} \frac{\phi_{c \to t}}{\phi_{c \to t} - F_{obs}}. \tag{13-6}$$

The various parameters in Eq. (13-6) are defined in terms of the kinetic scheme of Fig. 13-4 as

$$F_0 \equiv k[^3B^*]t \, , \quad F_{obs} \equiv \frac{[trans]}{[cis]_0} \, , \quad \text{and} \quad \phi_{c \to t} = \frac{k_t}{k_t + k_c} . \tag{13-7}$$

The correction is not valid for large conversions; so it will usually not work near the photostationary state.

Table 13-5

F. Other Actinometers

Actinometer	Useful wavelengths /nm	Refs.
Azobenzene	268-365	85F276
		81F080
		767042
		84F123
Azoxybenzene	250-350	84F096
1,1'-Azoxynaphthalene	300-400	86F209
1,3-Cycloheptadiene	~254	77F762
3,4-Dimethoxynitrobenzene	254-365	86F115
1,3-Dimethyluracil	~254	77F762
Ethanol	~185	777497
		76F937
Hermatoporphyrin and 2,2,6,6-Tetramethyl-4-piperidone-*N*-oxyl[a]	366-546	79D226
Heterocoerdianthrone	400-580	82F476
Heterocoerdianthrone endoperoxide	248-334	83F240
		84F198
(*E*)-α-(2,5-Dimethyl-3-furylethylidene)(isopropylidene) succinic anhydride	313-366	81F050
cis-Cyclooctene	~185	84F149
		81F053
meso-Diphenylhelianthrene	475-610	83F206
		84F198
Orthocoumaric acid dianion	253-366	78F572
Phenylglyoxylic acid	250-400	86F159
Reinecke's Salt	316-750	76Z005
		76Z030
		82Z269
Ru(byp)$_3$(II)chloride[b]	280-560	767423
		82S163
2,2',4,4'-Tetraisopropylazobenzene	350-390	777611
Tetraphenylcyclobutane	~265	79F261

[a]Useful in an esr cavity. [b]Particularly designed for lasers.

Section 14

Transmission Characteristics of Light Filters and Glasses

Data and spectra included in this section should help facilitate the selection of appropriate filters and glasses for spectroscopic and photochemical systems. Information on solution filters, Corning and Schott colored glass filters, Corning glasses and Englehard quartz glasses are included. See also [89Z269].

Table 14-1

Solution Filters (mostly aqueous)

	Compounds	c g/L	d cm	Stability (notes)	Solubility g/L 25 °C
A.	CCl_4 (neat) †	-	0.5	-	-
B.	Cl_2 (gas, 1 atm) ‡	-	5	a	-
C.	$CoSO_4 \cdot 7H_2O$ †	45	5	-	495*
D.	$CoSO_4 \cdot 7H_2O$ ‡	84	5	b, c	495*
E.	$CuSO_4 \cdot 5H_2O$ † ‡	100	5	b	280*
F.	$CuSO_4 \cdot 5H_2O$ (2.7 mol/L NH_3) ‡	4.4	10	d	-
G.	2,7-Dimethyl-3,6-diazacyclohepta-1,6-diene perchlorate ‡	0.1	1	a	-
H.	2,7-Dimethyl-3,6-diazacyclohepta-1,6-diene iodide †	0.2	1	a	-
I.	1,4-Diphenylbutadiene (in ether) †	0.042	1	-	-
J.	I_2 and	0.108	1	e	-
	KI ‡	0.155	1	-	-
K.	I_2 (in CCl_4) ‡	7.5	1	d	-
L.	K_2CrO_4 (0.1% NaOH) ‡	0.1	5	a	-
M.	K_2CrO_4 (0.1% NaOH) †	0.2	1	a	-
N.	$NaNO_2$ ‡	75	10	d	-
O.	Naphthalene (in isooctane) † ‡	12.8	1	a	-
P.	$NiSO_4 \cdot 6H_2O$ ‡	50	5	b, f	710*
Q.	$NiSO_4 \cdot 6H_2O$ † ‡	100	5	b, f	710*
R.	$NiSO_4 \cdot 6H_2O$ †	200	5	b, f	710*
S.	$NiSO_4 \cdot 6H_2O$ ‡	275	5	b, f	710*
T.	Potassium hydrogen phthalate † ‡	5	1	g	-
U.	Quinone hydrochloride ‡	20	1	h	-

† From reference [489001].   ‡ From reference [66Z001].   * From reference [65Z004].

Notes from reference [66Z001]:

    a. Constant transmission (%T) with irradiation.

    b. Only very slight %T increase with irradiation if preirradiated to stable %T first.

    c. Do not mix with $NiSO_4$ if stability is desired.

    d. Slight %T increase with irradiation.

    e. Linear increase of %T with irradiation time.

    f. Do not mix with $CoSO_4$ if stability is desired.

    g. Inconsistent but generally significant decrease of %T with irradiation time.

    h. Large decrease of %T with irradiation time.

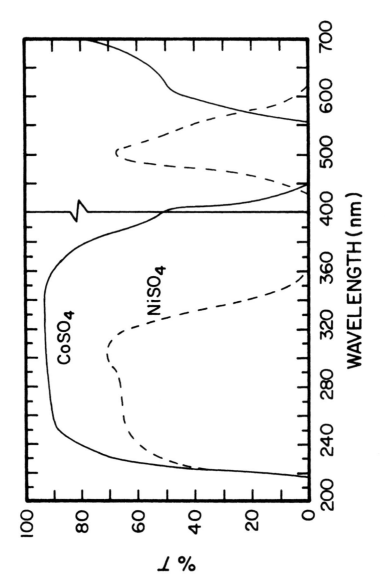

FIG 14-1. Light transmission in 1 cm pathlength cells of aqueous solutions of NiSO$_4$·6H$_2$O, 500 g/L (dashed curve) and CoSO$_4$·7H$_2$O, 240 g/L (solid curve). S.L. Murov.

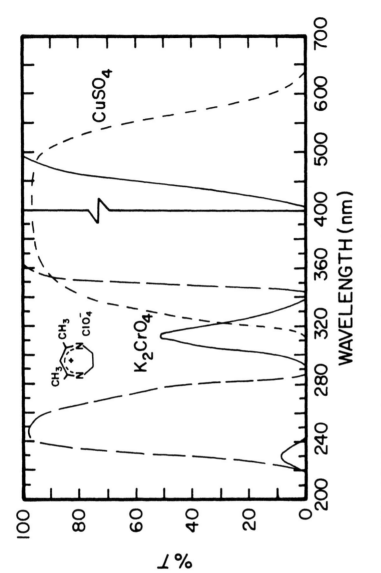

FIG 14-2. Light transmission in 1 cm pathlength cells of aqueous solutions of $CuSO_4 \cdot 5H_2O$, 250 g/L (short dashes), K_2CrO_4, 0.27 g/L + Na_2CO_3, 1 g/L (solid lines) and 2,7-dimethyl-3,6-diazacyclohepta-2,6-diene perchlorate, 0.20 g/L (long dashes). S.L. Murov.

FIGS. 14-3(a,b). Light transmission of selected Corning glass filters. Reproduced with the courtesy of the Corning Glass Works.

Curve symbol	C.S. †	Corning Glass #	Stock thickness /mm
a	9-57	7940	2.0
b	9-54	7910	2.0
c	0-53	7740	2.0
d	0-54	0160	2.0
e	0-52	7380	2.0
f	0-51	3850	4.0
g	3-75	3060	*
h	3-74	3391	*
i	3-73	3389	*
j	3-72	3387	*
k	3-71	3385	*
l	3-70	3384	*
m	3-69	3486	*
n	3-68	3484	*
o	3-67	3482	*
p	3-66	3480	*
q	2-73	2434	*
r	2-63	2424	*
s	2-62	2418	*
t	2-61	2412	*
u	2-60	2408	*
v	2-59	2404	*
w	2-58	2403	*
x	2-64	2030	*
y	7-54	9863	3.0
z	7-59	5850	4.0 (±0.5)
A	7-51	5970	5.0
B	7-60	5840	4.5
C	7-39	5874	5.0
D	7-37	5860	5.0
E	5-58	5113	4.0 (±0.5)
F	5-60	5543	5.0 (±0.6)

†Color Specification number

*Thickness generally 3 mm but may range between 1.4 mm and 4.6 mm.

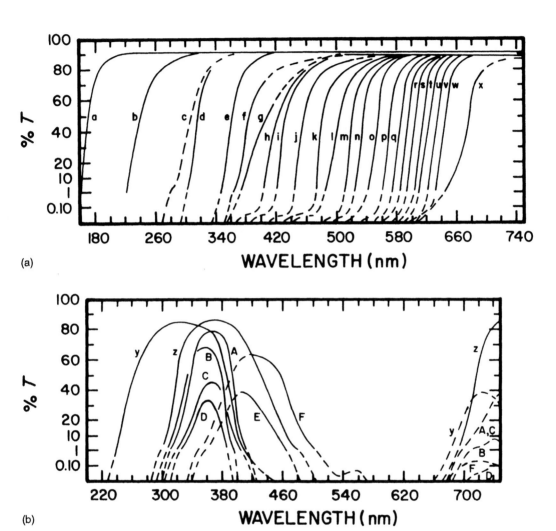

(a)

(b)

FIGS. 14-4(a,b). Light transmission of Schott glass filters. Adapted from Schott's "Color Filter Catalog".

Curve symbol	Glass type	Thickness /mm	Available /mm
a	WG 280	1	1, 2, 3
b	WG 295	1	1, 2, 3
c	WG 305	1	1, 2, 3
d	WG 320	1	1, 2, 3
e	WG 335	1	1, 2, 3
f	WG 345	1	1, 2, 3
g	WG 360	1	1, 2, 3
h	WG 395	1	1, 2, 3
i	GG 400	3	2, 3
j	GG 420	3	2, 3
k	GG 435	3	2, 3
l	GG 455	3	2, 3
m	GG 475	3	2, 3
n	GG 495	3	2, 3
o	OG 515	3	2, 3
p	OG 530	3	2, 3
q	OG 550	3	2, 3
r	OG 570	3	2, 3
s	OG 590	3	2, 3
t	RG 610	3	2, 3
u	RG 630	3	2, 3
v	RG 645	3	2, 3
w	RG 665	3	2, 3
x	RG 695	3	2, 3
y	RG 715	3	2, 3
z	UG 11	1	1, 2, 3
A	UG 1	1	1, 2, 3
B	GG 19	1	1, 2, 3

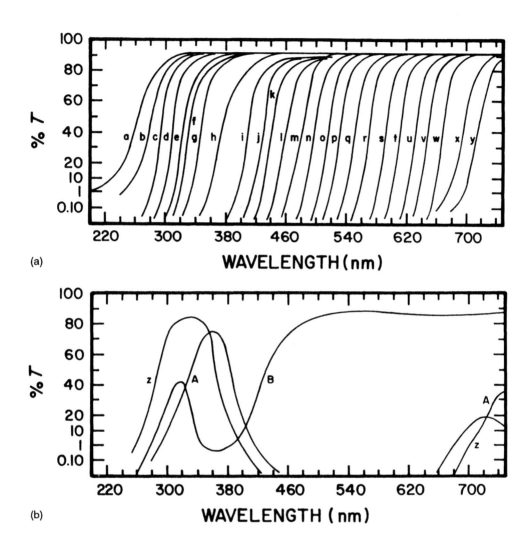

(a)

(b)

Table 14-2

Filter Combinations[a]

Region[b] /nm	Combination[c]	λ_{max}[d] /nm	%T[e]	Light[f] impurities	Ref.
245-270	$B + D + J + S$	256	15	-	66Z001
245-280	$C + I + S$	256	21	-	489001
256-290	$A + H + S + 7\text{-}54$	273	25	-	489001
305-330	$L + P + T + 7\text{-}54$	312	18	0.2% 302 0.3% 334	66Z001
290-335	$M + R + 7\text{-}54$	311	40	-	489001
305-335	$M + R + T + 7\text{-}54$	313	35	-	489001
290-335	$M + 7\text{-}54$	313	40	~4% 334 ~1% 405	*
322-364	$O + Q + 7\text{-}51$	331	17	-	66Z001
340-390	$E + G + 7\text{-}37$	367	24	-	66Z001
340-390	$0\text{-}52 + 7\text{-}37$	-	-	-	69F404
340-390	$E + H + 7\text{-}37$	363	30	-	489001
340-390	$0\text{-}52 + 7\text{-}60$	-	-	-	*
370-440	$F + K + U$	400	30-50	5% 436	66Z001
410-490	$F + N$	425	70-80	<1% 405	66Z001

*S.L. Murov

[a]The letters refer to filters in Table 14-1 and the numbers refer to Corning filters. In addition to the combinations listed, interference filters can be purchased with λ_{max} at almost any wavelength.

[b]Transmission region with extremes representing approximate onsets of transmission.

[c]Symbols refer to filters in Table 14-1 and to Corning filters.

[d]Wavelength of maximum transmission - for comments on stabilities of solution filters, see Table 14-1.

[e]Percent transmission at wavelength (nm) of maximum transmission.

[f]Percentages of indicated wavelengths transmitted that are not in desired region.

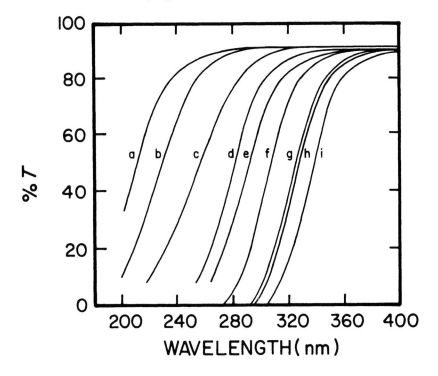

FIG. 14-5. Light transmission of various glasses (1 mm thickness). Reproduced from ref. 639028 with the courtesy of the General Electric Company.

Curve symbol	Glass type
a	Corning No. 7910 (Vycor)
b	Corning No. 9741
c	Corning No. 9720
d	Corning No. 9700*
e	Corning Corex D
f	Corning No. 7740 (Chemical Pyrex)
g	Lead Glass
h	Lime Glass
i	Corning No. 7720 (Nonex)

*However, see comments regarding this glass in ref. 72F543.

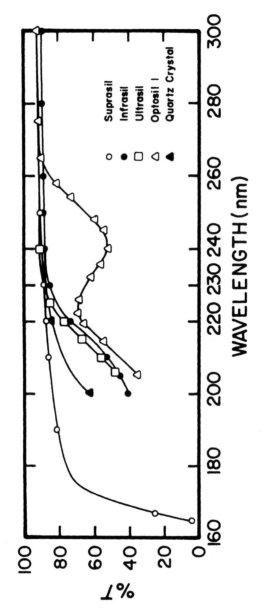

FIG. 14-6. Light transmission of various grades of quartz (10 mm thickness). Reproduced with the courtesy of the Amersil Quartz Division of Engelhard Industries, Inc.

Section 15

Spectral Distribution of Photochemical Sources

This section gives a brief survey of the spectral distributions in a variety of photon sources that are useful for photochemical studies. See also references [82Z269] and [89Z211].

Since the *First Edition* was published, lasers have come into common use in photochemical research, especially for time-resolved studies. A list of the most common wavelengths of lasers currently used in photochemistry is given in Table 15-1. The technology being made available commercially is in great flux at this time; so no attempt is made to specify any properties beyond wavelength, such as time or power specifications. In addition to these available laser wavelengths, dye lasers provide tunability over wide wavelength ranges. Although laser dyes are also undergoing rapid development, a few tuning curves are provided in Figs. 15-1 through 15-3 to illustrate current practice.

The next class sources is a group of arc lamps. The qualitative emission spectrum of a xenon arc is given in Fig. 15-4. A quantitative set of spectra is given in Fig. 15-5 for a variety of arc lamps. These plots are for *power* spectra, spectral irradiance. In order to convert these spectra to *photon* spectra, the ordinates must be multiplied by factors proportional to the wavelength (see page 253 of [68Z003]). Thus the spectra in Fig. 15-5 are greatly exaggerated in the short wavelength range if the reader is concerned with numbers of photons. Replotting Fig. 15-5 as photons per second instead of power would show much more emphasis in the red compared to the blue regions.

Although Fig. 15-5 does give one mercury arc power spectrum, it is of interest to concentrate some additional attention on the various mercury lamps available. Since mercury has a line spectrum, it is often used as a convenient calibration standard. Fig. 15-6 gives the various lines in the mercury spectrum which can be used. Figs. 15-7 through 15-9 give qualitative (photon) emission spectra for low, medium, and extra-high pressure mercury lamps. Tables 15-2 and 15-3 give corresponding energy (or power) spectra for some commonly used low and medium pressure mercury lamps, respectively.

Finally the solar power spectrum is shown in Fig. 15-10.

Table 15-1

Lasers: Common Wavelengths Used in Photochemistry †

Laser	λ /nm
F_2	157
ArF, excimer	193
KrCl, excimer	223
Ruby (tripled)	231.4
KrF, excimer	248
Nd:YAG (quadrupled)	266
XeCl, excimer	308
Nitrogen	337.1
Ruby (doubled)	347.2
Krypton ion	350.7
XeF, excimer	351
Nd:YAG (tripled)	355
Nitrogen	428
Helium-Cadmium	441.6
Argon ion	488
Argon ion	514.5
Nd:YAG (doubled)	532
Krypton ion	568.2
Helium-Neon	632.8
Krypton ion	647.1
Ruby (fundamental)	694.3
Gallium arsenide	904
Nd:glass	1060
Nd:YAG (fundamental)	1064
CO_2	10600

†For an extensive listing of gas laser wavelengths see reference [80Z182], and for a more extensive listing of output wavelengths from commercial lasers see reference [90Z543].

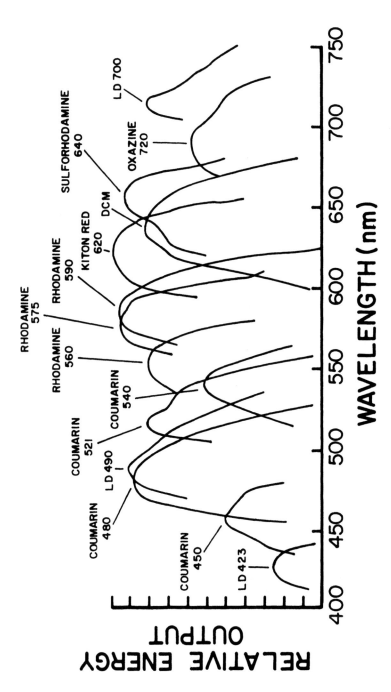

FIG. 15-1. Tuning curves for flashlamp pumped dye lasers. Reprinted with permission of Exciton. Dye names are those of Exciton.

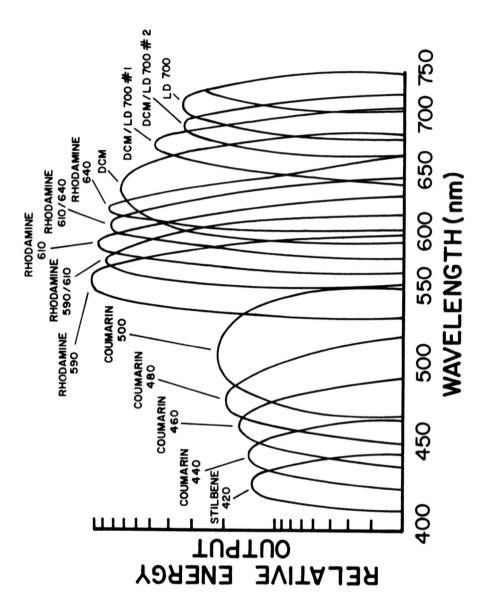

FIG. 15-2. Tuning curves for Nd:YAG pumped dye lasers. Reprinted with permission of Exciton. Dye names are those of Exciton.

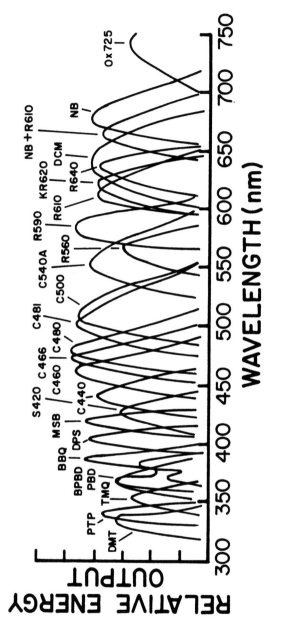

FIG. 15-3. Tuning curves for dye lasers pumped with KrF and XeCl excimer lasers. Reprinted with permission of Exciton. Dye names are those of Exciton.

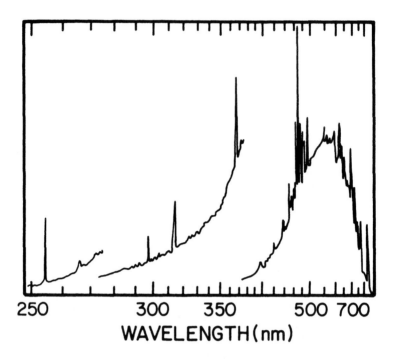

FIG. 15-4. Emission spectra of a 350 W xenon arc lamp, from Parker [68Z003]. Reprinted with permission from Elsevier.

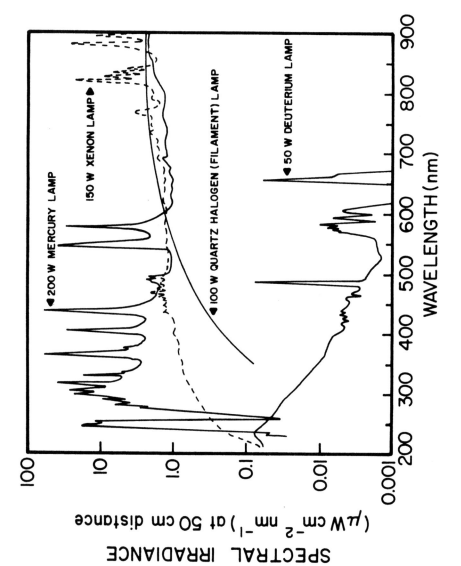

FIG. 15-5. Spectral irradiance of some arc lamp sources. Reprinted with permission of Oriel from their catalog.

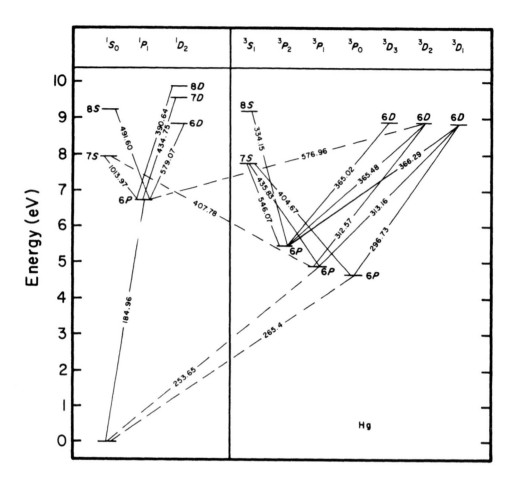

FIG. 15-6. Grotrian energy level diagram for mercury. Note the 184.96 and 253.65 nm lines are resonance lines and their intensities decrease quickly with increasing mercury pressure due to reabsorption. The 265.4 nm line is not observed in absorption due to its highly forbidden character. Adapted from ref. 66Z001, p. 52 with permission from John Wiley and Sons, Inc.

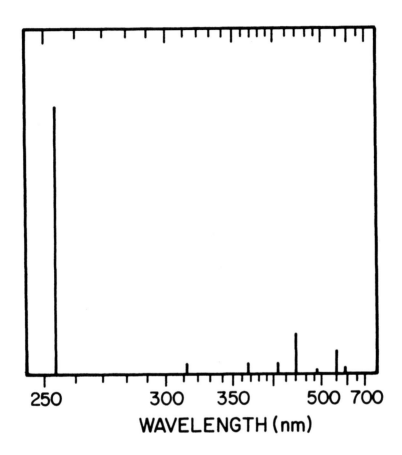

FIG. 15-7. Histogram of emission spectrum of 6 W low pressure Hg lamp, adapted from Parker [68Z003]. Reprinted with permission from Elsevier.

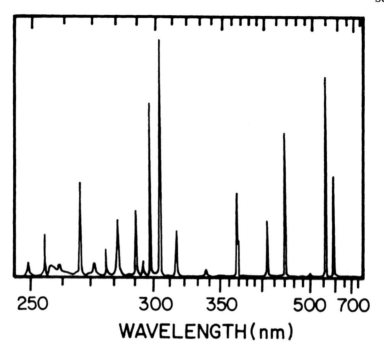

FIG. 15-8. Emission spectrum of 125 W medium pressure Hg lamp, from Parker [68Z003]. Reprinted with permission from Elsevier.

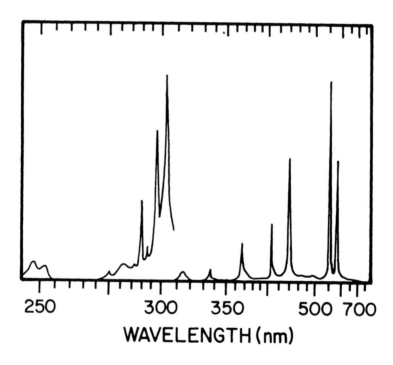

FIG. 15-9. Emission spectrum of an extra-high pressure Hg lamp, from Parker [68Z003]. Reprinted with permission from Elsevier.

Table 15-2

Relative Spectral Energy Distribution of Low Pressure Mercury Lamps

Wavelength /nm	Hanovia SC-2537[a]	Rayonet RPR-2537A[b]
248.2	0.01	-
253.7	100	100
265.2-265.5	0.05	3
275.3	0.03	-
280.4	0.02	0.1
289.4	0.04	0.1
296.7	0.20	0.5
302.2-302.8	0.06	0.3
312.6-313.2	0.60	2
334.1	0.03	-
365.0-366.3	0.54	2
404.5-407.8	0.39	2
435.8	1.00	6
546.1	0.88	3
577.0-579.0	10.14	0.7

[a]From reference [66Z001], p. 696.

[b]The Southern New England Ultraviolet Co.

Table 15-3

Spectral Power Distribution for Hanovia Medium* Pressure Mercury Lamps

Wavelength /nm	Radiated Power /W			
222.4	0.04	0.03	3.7	4.2
232.0	0.02	0.03	1.5	2.4
236.0	0.02	0.08	2.3	1.8
238.0	0.03	0.12	2.3	2.6
240.0	0.05	0.20	1.9	2.2
248.2	0.10	0.20	2.3	2.6
253.7	0.34	1.10	5.8	5.0
257.1	0.11	0.20	1.5	1.8
265.2	0.30	0.64	4.0	4.6
270.0	0.07	0.14	1.0	1.2
275.3	0.06	0.14	0.7	0.8
280.4	0.12	0.30	2.4	2.8
289.4	0.10	0.20	1.6	1.8
296.7	0.32	0.48	4.3	5.0
302.5	0.41	0.86	7.2	8.2
313.0	1.02	2.3	13.2	15.0
334.1	0.13	0.36	2.4	2.8
366.0	1.40	3.1	25.6	30.1
404.5	0.75	1.6	11.0	12.7
435.8	1.08	2.6	20.2	23.3
546.1	1.35	3.0	24.5	28.2
578.0	1.55	3.4	20.0	23.0
1014.0	0.85	1.8	10.5	12.2
1128.7	0.62	1.3	3.3	3.8
1367.3	0.65	1.0	2.6	4.6
Total Watts	11.49	25.18	175.8	202.7
Lamp Code	SOL	S	L	A
Lamp Cat. No.	608A	654A	679A	673A
Lamp Watts	100	200	450	550
Lamp Volts	100	125	135	145
Current Amps	1.2	1.9	3.6	4.4
Arc-length (inch)	2.9	4.5	4.5	4.5

*Termed "high" pressure by Hanovia. Information from Hanovia data sheet.

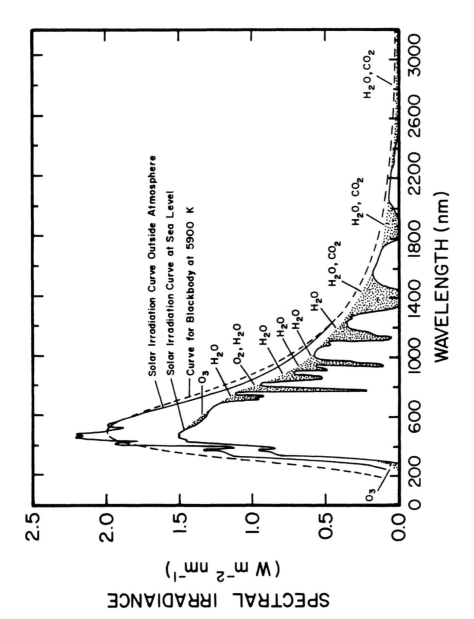

FIG. 15-10. Solar spectral irradiance. From 65Z003. Reprinted with permission of McGraw-Hill.

Section 16

Spin-Orbit Coupling

Heavy-atom effects on spectroscopic and photochemical properties have traditionally been attributed to the spin-orbit interaction. [69Z002] Some of the most widely studied areas are internal [496001] and external heavy-atom effects [526003] in photophysics. (See [81Y136] and [81Y336] for an alternative framework.)

For one-electron systems, in a central field, the relativistic effects can be studied using the spin-orbit interaction given by [62Y005]

$$H_{so} = \frac{1}{2} \alpha^2 < \frac{1}{r} \frac{\partial V}{\partial r} > \boldsymbol{l} \cdot \boldsymbol{s}. \tag{16-1}$$

In Eq. (16-1), α is the fine structure constant, $V(r)$ is the potential energy operator of the central field, $\boldsymbol{l} \cdot \boldsymbol{s}$ is the vector dot product of the electron's orbital angular momentum and spin operators, and the expectation value $<O>$ is over radial wavefunctions. In many-electron systems, problems arise concerning the nature of $V(r)$ and the extent of spin/other-orbit interactions. The interaction of an outer-electron (orbital angular momentum l) with the core is given in terms of a coupling parameter, ζ_l, which gives a measure of the strength of the spin-orbit interaction. For a one-electron, central-field system, ζ_l is just $\alpha^2 <O>$ in Eq. (16-1).

Experimentally the usual way of finding spin-orbit parameters is from the multiplet splittings in atomic spectra. The calculation of one-electron ζ_l's for various l-electrons has been reviewed in the classic treatise on atomic spectra by Condon and Shortley. [35Z001] Examples of these calculations have appeared in the photophysics literature. [496001, 61E012]

In Table 16-1, values of ζ_l for a large number of atoms have been tabulated. The theoretical estimates [76Z050] were obtained from single-configuration, spin-restricted, numerical Hartree-Fock wavefunctions. These results are supplemented with values computed from experimental spectra in cases where the ground electronic configurations contain (1) only totally-filled shells or (2) only a half-filled s-shell. In case 1, the ζ_l's were estimated by finding the ζ_p for an excited sp^n configuration. Either a p-electron (e.g. rare gases) or an s-electron (e.g. mercury) can be promoted to form such an excited configuration. Case 2 (e.g. alkalis) was treated by promoting the outer-shell s-electron to a p-shell. Experimental values are also given for the halogens since they are frequently used as heavy-atom quenchers. In the halogens, ζ_p's were calculated from the doublet splitting of the lowest configuration, $ns^2 np^5$.

In considering heavy-atom effects, it seems best to correlate the interaction of the system in question with the *core* of the heavy atom. The parameter, ζ_l, is a measure of just such an interaction; even though it is actually related to the spin-orbit interaction of an electron *in a specific open-shell orbit* of the heavy atom with its core which is in *some specific electronic configuration.*

Table 16-1

Atomic Masses and Spin-Orbit Coupling Constants[a,b]

Element	Symbol	At. Num.	At. Mass	ζ_l /cm^{-1}
Actinium	Ac	89	227*	1290
Aluminum	Al	13	26.98154	62
Americium	Am	95	243	3148
Antimony	Sb	51	121.75	2593
Argon	Ar	18	39.948	*940*
Arsenic	As	33	74.9216	1202
Astatine	At	85	206*	10608
Barium	Ba	56	137.33	*830*
Berkelium	Bk	97	247*	3852
Beryllium	Be	4	9.01218	*2.0*
Bismuth	Bi	83	208.9804	6831
Boron	B	5	10.81	10
Bromine	Br	35	79.904	*2460*
Cadmium	Cd	48	112.41	*1140*
Calcium	Ca	20	40.08	*105*
Californium	Cf	98	251*	4238
Carbon	C	6	12.011	32
Cerium	Ce	58	140.12	687
Cesium	Cs	55	132.9054	*370*
Chlorine	Cl	17	35.453	*587*
Chromium	Cr	24	51.996	248
Cobalt	Co	27	58.9332	550
Copper	Cu	29	63.546	857
Curium	Cm	96	247*	3488
Dysprosium	Dy	66	162.50	2074
Einsteinium	Es	99	252*	4645
Erbium	Er	68	167.26	2564
Europium	Eu	63	151.96	1469
Fermium	Fm	100	257*	5076
Fluorine	F	9	18.998403	*269*
Francium	Fr	87	223*	-
Gadolinium	Gd	64	157.25	1651
Gallium	Ga	31	69.72	464
Germanium	Ge	32	72.59	800
Gold	Au	79	196.9665	5104
Hafnium	Hf	72	178.49	1578
Helium	He	2	4.00260	*0.7*
Holmium	Ho	67	164.9304	2310
Hydrogen	H	1	1.00794	*0.24*
Indium	In	49	114.82	1183
Iodine	I	53	126.9045	*5069*
Iridium	Ir	77	192.22	3909

Table 16-1—**Atomic Masses and Spin-Orbit Coupling Constants**—Continued

Element	Symbol	At. Num.	At. Mass	ζ_l/cm^{-1}
Iron	Fe	26	55.847	431
Krypton	Kr	36	83.80	*3480*
Lanthanum	La	57	138.9055	556
Lawrencium	Lr	103	260*	-
Lead	Pb	82	207.2	5089
Lithium	Li	3	6.941	*0.23*
Lutetium	Lu	71	174.967	1153
Magnesium	Mg	12	24.305	*40.5*
Manganese	Mn	25	54.9380	334
Mendelevium	Md	101	258*	5533
Mercury	Hg	80	200.59	*4270*
Molybdenum	Mo	42	95.94	678
Neodymium	Nd	60	144.24	967
Neon	Ne	10	20.179	*520*
Neptunium	Np	93	237.0482	2488
Nickel	Ni	28	58.69	691
Niobium	Nb	41	92.9064	524
Nitrogen	N	7	14.0067	78
Nobelium	No	102	259*	-
Osmium	Os	76	190.2	3381
Oxygen	O	8	15.9994	154
Palladium	Pd	46	106.42	1504
Phosphorus	P	15	30.97376	230
Platinum	Pt	78	195.08	4481
Plutonium	Pu	94	244*	2810
Polonium	Po	84	209*	8509
Potassium	K	19	39.0983	*38*
Praseodymium	Pr	59	140.9077	824
Promethium	Pm	61	145*	1119
Protactinum	Pa	91	231.0359	1888
Radium	Ra	88	226.0254	-
Radon	Rn	86	222*	-
Rhenium	Re	75	186.207	2903
Rhodium	Rh	45	102.9055	1259
Rubidium	Rb	37	85.4678	*160*
Ruthenium	Ru	44	101.07	1042
Samarium	Sm	62	150.36	1286
Scandium	Sc	21	44.9559	77
Selenium	Se	34	78.96	1659
Silicon	Si	14	28.0855	130
Silver	Ag	47	107.8682	1779
Sodium	Na	11	22.98977	*11.5*
Strontium	Sr	38	87.62	*390*

Table 16-1—**Atomic Masses and Spin-Orbit Coupling Constants**—Continued

Element	Symbol	At. Num.	At. Mass	ζ_l /cm^{-1}
Sulfur	S	16	32.06	365
Tantalum	Ta	73	180.9479	1970
Technetium	Tc	43	98*	853
Tellurium	Te	52	127.60	3384
Terbium	Tb	65	158.9254	1853
Thallium	Tl	81	204.383	3410
Thorium	Th	90	232.0381	1591
Thulium	Tm	69	168.9342	2838
Tin	Sn	50	118.69	1855
Titanium	Ti	22	47.88	123
Tungsten	W	74	183.85	2433
Uranium	U	92	238.0289	2184
Vanadium	V	23	50.9415	179
Xenon	Xe	54	131.29	*6080*
Ytterbium	Yb	70	173.04	-
Yttrium	Y	39	88.9059	260
Zinc	Zn	30	65.38	*390*
Zirconium	Zr	40	91.22	387

[a]The ζ_l-values in Roman type are calculated spin-orbit coupling constants for an *l*-electron ($l = p, d, f, \ldots$) in an open shell of a neutral atom in the ground electronic configuration. Values were the results of single-configuration, numerical calculations using spin-restricted Hartree-Fock wavefunctions. [76Z050] All ζ_l-values are for an open-shell *l*-electron (other than an *s*-electron) interacting with the core.

[b]The ζ_l-values in bold italics were calculated from atomic spectra [71Z015] using the theoretical models in Chapters VII (*LS*-coupling), XI (intermediate coupling, especially the eqs. on p. 271), and XIII (intermediate coupling in the rare gases) of Condon and Shortley [35Z001]. These experimental values are for excited configurations, except in the case of the halogens. All ζ_l-values are for an open-shell *p*-electron interacting with the core, which itself may or may not be excited. All the experimental values are ζ_p's. (The values for light atoms are rough since the usual approximation for the relativistic effects breaks down, see discussion of helium in reference [35Z001], pp. 210-2.)

*Mass number of most stable isotope.

Section 17

Fundamental Constants and Conversion Factors

Included in this section are the fundamental constants and conversion factors commonly needed by photochemists and spectroscopists. The International System of Units (SI, Système International d'Unités) is used as much as possible throughout this book. Conversions are given in this section to more traditional units, especially energy conversions.

Table 17-1

Fundamental Physical Constants[a]

Quantity	Symbol	Value
Atomic mass constant	$m_u = \dfrac{m(^{12}C)}{12}$	$1.6605402 \times 10^{-27}$ kg
Avogadro constant	N_A	6.0221367×10^{23} mol^{-1}
Bohr magneton	$\mu_B = \dfrac{eh}{4\pi m_e}$	$9.2740154 \times 10^{-24}$ J T^{-1}
Bohr radius	$a_0 = \dfrac{\alpha}{4\pi R_\infty}$	$0.529177249 \times 10^{-10}$ m
Boltzmann constant	$k = \dfrac{R}{N_A}$	1.380658×10^{-23} J K^{-1}
Electron g-factor	g_e	2.002319304386
Electron rest mass	m_e	$9.1093897 \times 10^{-31}$ kg
Elementary charge	e	$1.60217733 \times 10^{-19}$ C
Faraday constant	F	9.6485309×10^4 C mol^{-1}
Fine structure constant	$\alpha = \dfrac{\mu_0 c e^2}{2h}$	7.2973506×10^{-3}
Molar gas constant	R	8.314510 J mol^{-1} K^{-1}
Neutron rest mass	m_n	$1.6749286 \times 10^{-27}$ kg
Nuclear magneton	$\mu_N = \dfrac{eh}{4\pi m_p}$	$5.0507866 \times 10^{-27}$ J T^{-1}
Permeability of vacuum	μ_0	$1.2566370614 \times 10^{-6}$ N A^{-2}
Permittivity of vacuum	$\varepsilon_0 = \dfrac{1}{\mu_0 c^2}$	$8.854187817 \times 10^{-12}$ F m^{-1}
Planck constant	h	$6.6260755 \times 10^{-34}$ J s
Proton rest mass	m_p	$1.6726231 \times 10^{-27}$ kg
Rydberg constant	$R_\infty = \dfrac{m_e c e^2}{2h}$	1.0973731534×10^7 m^{-1}
Speed of light in vacuum	c	2.99792458×10^8 m s^{-1}
Stefan-Boltzmann constant	$\sigma = \dfrac{2\pi^5 k^4}{15h^3 c^2}$	5.67051×10^{-8} W m^{-2} K^{-4}

A = ampere, C = coulomb, F = farad, J = joule, K = kelvin, N = newton, T = tesla, W = watt

[a]From reference [86Z372].

Table 17-2

Energy Conversion Factors[a,b]

		kJ/mol	cm^{-1}	Hz	eV	kcal/mol
1 kJ/mol	=	1	83.593	2.5061×10^{12}	1.0364×10^{-2}	2.3901×10^{-1}
1 cm^{-1}	=	1.1963×10^{-2}	1	2.9979×10^{10}	1.2398×10^{-4}	2.8592×10^{-3}
1 Hz	=	3.9903×10^{-13}	3.3356×10^{-11}	1	4.1357×10^{-15}	9.5371×10^{-14}
1 eV	=	96.488	8065.5	2.4180×10^{14}	1	23.061
1 kcal/mol	=	4.1840	349.75	1.0485×10^{13}	4.3363×10^{-2}	1

[a] Adapted from reference [86Z372] with factors rounded to five significant figures.

[b] To find the wavelength, λ, of a photon having an energy in the units shown in the table, it is easiest to convert to either wavenumber ($\bar{\nu}$ in cm^{-1}) or frequency (ν in Hz) and then use

$$\lambda = \frac{1}{\bar{\nu}} = \frac{c}{\nu} = \frac{hc}{E}$$

Other Important Conversion Factors

Another energy unit that is often useful is

$$1 \text{ L atm} = 9.869 \times 10^{-3} \text{ J.}$$

kT at room temperature is approximately 0.025 eV.

The SI unit of pressure is pascal (1 Pa = 1 N/m^2). A useful, but nonapproved SI unit, of pressure is

standard atmosphere = 101.325 kPa.

In these nonstandard units, the gas constant ($R = 8.31451$ J/(mol·K)) is

$$R = 0.08206 \text{ L atm/(mol·K)} = 1.9872 \text{ cal/(mol·K).}$$

Section 18

Hammett σ Constants

The Hammett equations can be expressed by

$$\rho\,\sigma \;=\; \log k_X - \log k_H$$

where k_X is a reaction parameter (either a rate constant or an equilibrium constant) for a process with the reactant, having a substituent X, and k_H is the reaction parameter for the same process involving the corresponding unsubstituted reactant. ρ depends on the type of reaction. It was originally taken to have a value of 1 for the ionizations of substituted benzoic acids. For many types of reactions, $\rho \sim 1$. The σ-values are used to make correlations between the reaction parameters, k_X and the electronic nature of the substituents, X.

For fixed ρ, it can be seen from the defining equation above that electron-releasing substituents will have negative σ's and that electron-withdrawing substituents will have positive σ's. The goal of this analysis is for typical reactions to give values for the σ's such that trends can be predicted for other types of reactions depending on the nature of the substituent.

This has not always worked that well, so a variety of σ's have been defined and used with varying success. σ's have been found to be very position dependent; so *meta-* and *para*-values are compiled separately as σ_m and σ_p, respectively. Electronic effects of *ortho*-substituents are often swamped by steric hindrance. The values of σ_m and σ_p in Table 18-1 are the preferred values chosen by the compilers of reference [79Z427] using pK_a values of substituted benzoic acids.

When the given substituent and the reaction center are conjugated, the normal σ-scales do not yield good linear correlations. For substituents that can involve direct conjugation (through resonance) with a positively charged reaction center, σ^+ has been defined. It has been based on the solvolysis of the cumyl chlorides, taking $\rho = 4.54$ from the hydrolysis of the *meta*-cumyl chlorides. For substituents that can participate in direct conjugation with a negatively charged reaction center, σ^- has been defined. Usually reactions of substituted phenols or anilines are used for defining this scale, e.g. the basicity of phenoxide ions.

The effect of polar substituents is also of interest. One σ-scale used to correlate these effects is that of σ_I. The values reported in Table 18-1 are based on the ionization of substituted acetic acids. Attempts have been made to separate these inductive (polar) effects, correlated by σ_I, with resonance effects correlated by various σ_R's, which are defined as

$$\sigma_R = \sigma_p - \sigma_I.$$

The σ_R's reported in Table 18-1 are from the σ_R^0-scale which is based on nmr and ir spectra.

Table 18-1

Hammett σ Constants

Substituent	σ_m	σ_p	σ_m^+	σ_p^+	σ_m^-	σ_p^-	σ_I	σ_R^0
H	0.00	0.00	0.00	0.00	0.00	0.00	0.00	0.00
CH_3	-0.07	-0.17	-0.07	-0.31	-0.03	-0.15	-0.04	-0.11
CH_2CH_3	-0.07	-0.15	-0.06	-0.30	-	-	-0.05	-0.14
$CH(CH_3)_2$	-0.07	-0.15	-	-0.28	-	-0.09	-0.03	-0.12
$CH_2CH(CH_3)_2$	-	-0.12	-	-	-	-	-0.03	-
$C(CH_3)_3$	-0.10	-0.20	-0.06	-0.26	-	-0.13	-0.07	-0.13
$CH=CH_2$	0.05	-0.02	-	0.10	-	-	0.09	-0.03
C_6H_5	0.06	-0.01	0.11	-0.18	-	0.10	0.10	-0.11
CH_2CN	0.16	0.01	-	0.16	-	-	0.18	-0.08
$CH_2C_6H_5$	-0.08	-0.09	-0.07	-0.20	-	-0.12	-0.08	-0.12
CH_2Cl	0.11	0.12	0.14	-0.01	-	-	0.15	-0.03
CH_2OH	0.00	0.00	-	-0.04	0.08	0.08	0.05	0.00
CH_2NH_2	-	-	-	-	-	-	0.00	-0.15
$CH_2NH_3^+$	-	-	-	-	-	-	0.36	0.00
$CH_2Si(CH_3)_3$	-0.16	-0.21	-	-0.22	-	-0.22	-0.07	-0.20
$CH=CHC_6H_5$	0.03	-0.07	-	-1.00	-	0.62	0.02	-
$CH=CHNO_2$	0.32	0.26	-	-	0.37	0.88	0.24	0.13
$COCH_3$	0.38	0.50	-	-	0.32	0.87	0.29	0.20
CHO	0.35	0.42	-	0.73	0.48	1.13	0.25	0.24
$COCF_3$	0.63	0.80	-	0.85	-	-	0.45	0.33
$COOH$	0.37	0.45	0.32	0.42	0.56	0.77	0.39	0.29
$COOCH_3$	0.37	0.45	0.37	0.49	-	0.64	0.34	0.16
$COOC_2H_5$	0.37	0.45	0.37	0.48	-	0.64	0.21	0.18
CO_2^-	-0.10	0.00	-0.03	-0.02	-	0.24	-0.17	-

$CONH_2$	0.28	-	-	-	0.63	0.27	0.01
CN	0.56	0.56	0.66	0.68	1.00	0.56	0.13
CF_3	0.43	0.52	0.61	0.41	0.65	0.42	0.10
C_6F_5	0.34	-	-	-	-	0.25	0.02
N_2^+	1.76	-	1.88	-	3.04	1.34	0.64
NH_2	-0.16	-0.16	-1.30	-0.02	-0.15	0.12	-0.48
$NHCH_3$	-0.30	-	-	-	-	-	-0.52
$N(CH_3)_2$	-0.15	-	-1.70	0.04	-0.12	0.06	-0.52
NH_3^+	0.86	-	-	-	-	0.61	-0.04
$N(CH_3)_3^+$	0.88	0.36	0.41	0.85	0.70	0.93	-0.15
$NHNH_2$	-0.02	-	-	-	-	0.14	-0.43
$NHOH$	-0.04	-	-	-	-	0.12	-0.22
$NHCOCH_3$	0.21	-	-0.60	-	-	0.24	-0.23
$N{=}N{-}C_6H_5$	0.32	-	-0.19	-	0.69	0.25	0.06
NO	0.62	-	0.79	-	1.60	0.34	0.32
NO_2	0.71	0.67	-2.30	-	1.24	0.76	0.15
O^-	-0.47	-	-0.92	-	-0.81	-0.16	-0.60
OH	0.12	-0.04	-0.78	-	-	0.29	-0.43
OCH_3	0.12	0.05	-	0.13	-0.14	0.27	-0.43
OCH_2CH_3	0.10	-	-0.50	-	-	0.27	-0.44
OC_6H_5	0.25	0.10	-	-	-0.10	0.39	-0.34
OCF_3	0.38	-	-0.06	0.47	0.27	0.39	-0.04
$OCOCH_3$	0.39	0.27	-	-	-	0.33	-0.21
SH	0.25	-	-	-	-	0.26	-0.15
SCH_3	0.15	0.16	-0.60	0.19	0.17	0.23	-0.17

Table 18-1—**Hammett σ Constants**—Continued

Substituent	σ_m	σ_p	σ_m^+	σ_p^+	σ_m^-	σ_p^-	σ_I	σ_R^0
S=O(CH$_3$)	0.52	0.49	-	-	0.45	0.62	0.49	0.00
SO$_2$CH$_3$	0.60	0.72	-	-	0.52	1.05	0.59	0.12
SO$_2$NH$_2$	0.46	0.57	-	-	0.65	0.94	0.46	0.05
SCF$_3$	0.40	0.50	-	-	0.46	0.57	0.42	0.06
S(CH$_3$)$_2$$^+$	1.00	0.90	-	-	1.00	1.16	0.89	0.17
Si(CH$_3$)$_3$	-0.04	-0.07	0.01	-0.03	0.00	0.17	-0.13	0.06
F	0.34	0.06	0.35	-0.07	-	0.05	0.52	-0.34
Cl	0.37	0.23	0.40	0.11	-	0.27	0.47	-0.23
Br	0.39	0.23	0.41	0.15	-	0.28	0.44	-0.19
I	0.35	0.18	0.36	0.14	-	-	0.39	-0.16

Bibliography and Indexes

As discussed in the Preface, we have appended several indexes to provide ready access to the data in this handbook. These have been obtained by a (semi-)automatic, computer-assisted procedure, from extensive bibliographic and numeric databases at the Radiation Chemistry Data Center (RCDC). Notice however that such automated indexing is only feasible for Sections 1-6, 8-10, and the first three parts of Section 12. The Table of Contents must serve as a locator for more general reference.

The bibliography is accessed through RCDC aquisition numbers, the first two digits of which indicate the year of publication. The third character, often alphabetic, categorizes the publication† and the last three digits reflect, serially, the date of acquisition, this order roughly corresponding to the relative time of publication within the category.

In both the compound name and molecular formula indexes, reference to tabulated data is made through the use of a Table number, which also indicates the containing Section, separated by a period from the Table Entry Number. When it is occasionally necessary to describe subtabulations, this is done by means of a hyphen, for example in Sections 10 and 12.

---

†The categories are described in the "Biweekly List of Papers on Radiation Chemistry and Photochemistry", a current-awareness bibliographic service of the Radiation Chemistry Data Center. Information on this and other services offered by the Center may be obtained by writing to Dr. Alberta B. Ross, Supervisor of the Center.

References

307001 Leighton, W.G.; Forbes, G.S., *J. Am. Chem. Soc.* **52**: 3139-52 (1930)

35Z001 Condon, E.U.; Shortley, G.H., *The Theory of Atomic Spectra*. Cambridge U. Pr., Cambridge, UK, 1935, 441p.

399001 Henri, V.; Pickett, L.W., *J. Chem. Phys.* **7**: 439-40 (1939)

44E001 Lewis, G.N.; Kasha, M., *J. Am. Chem. Soc.* **66**: 2100-16 (1944)

469001 Kretschmer, C.B.; Nowakowska, J.; Wiebe, R., *Ind. Eng. Chem.* **38**: 506-9 (1946)

479001 Johnson, W.S.; Woroch, E.; Mathews, F.J., *J. Am. Chem. Soc.* **69**: 566-71 (1947)

47Z002 Fowler, R.D.; Hamilton, J.M.,Jr.; Kasper, J.S.; Weber, C.E.; Burford, W.B.,III; Anderson, H.C., *Ind. Eng. Chem., Ind. Ed.* **39**: 375-8 (1947)

489001 Kasha, M., *J. Opt. Soc. Am.* **38**: 929-34 (1948)

496001 McClure, D.S., *J. Chem. Phys.* **17**: 905-13 (1949)

496002 Kiss, A.; Molnar, J.; Sandorfy, C., *Bull. Soc. Chim. Fr.* : 275-80 (1949)

506002 Clar, E., *Spectrochim. Acta* **4**: 116-21 (1950)

51E002 Pyatnitskii, B.A., *Izv. Akad. Nauk SSSR, Ser. Fiz.* **15**: 597-604 (1951)

51E003 Dikun, P.P.; Petrov, A.A.; Sveshnikov, B.Ya., *Zh. Eksp. Teor. Fiz.* **21**: 150-63 (1951)

51Z002 Friedel, R.A.; Orchin, M., *Ultraviolet Spectra of Aromatic Compounds*. Wiley, New York, 1951, 702p.

526003 Kasha, M., *J. Chem. Phys.* **20**: 71-4 (1952)

539008 Passerini, R.; Ross, I.G., *J. Sci. Instrum.* **30**: 274-6 (1953)

54E002 Moodie, M.M.; Reid, C., *J. Chem. Phys.* **22**: 252-4 (1954)

55E003 Livingston, R., *J. Am. Chem. Soc.* **77**: 2179-82 (1955)

55E008 Sidman, J.W.; McClure, D.S., *J. Am. Chem. Soc.* **77**: 6461-70 (1955)

55F005 Baxendale, J.H.; Bridge, N.K., *J. Phys. Chem.* **59**: 783-8 (1955)

566004 Beer, M., *J. Chem. Phys.* **25**: 745-50 (1956)

566005 Clar, E.; Zander, M., *Chem. Ber.* **89**: 749-62 (1956)

566006 Padhye, M.R.; McGlynn, S.P.; Kasha, M., *J. Chem. Phys.* **24**: 588-94 (1956)

566007 Cherkasov, A.S.; Molchanov, V.A.; Vember, T.M.; Voldaikina, K.G., *Sov. Phys. Dokl.* **1**: 427-9 (1956) Translated from: *Dokl. Akad. Nauk SSSR, Phys. Ser.* **107**: 427 (1956)

566008 Foster, R.; Hammick, D.L.; Hood, G.M.; Sanders, A.C.E., *J. Chem. Soc.* : 4865-8 (1956)

56F006 Hatchard, C.G.; Parker, C.A., *Proc. R. Soc. London, Ser. A* **235A**: 518-36 (1956)

56Z003 Masson, C.R.; Boekelheide, V.; Noyes, W.A.,Jr., In *Technique of Organic Chemistry*, A. Weissberger (ed.), Interscience Publishers, New York, 1956, Vol. II, p.257-384

573002 Dmitrievskii, O.D.; Ermolaev, V.L.; Terenin, A.N., *Dokl. Akad. Nauk SSSR* **114**: 751-3 (1957)

573003 Fujimori, E.; Livingston, R., *Nature (London)* **180**: 1036-8 (1957)

57B004 Evans, D.F., *J. Chem. Soc.* : 1351-7 (1957)

57E003 Weber, G.; Teale, F.W.J., *Trans. Faraday Soc.* **53**: 646-55 (1957)

57E004 Brody, S.S., *Rev. Sci. Instrum.* **28**: 1021-6 (1957)

58E001 Porter, G.; Windsor, M.W., *Proc. R. Soc. London, Ser. A* **245**: 238-58 (1958)

58E005 Heckman, R.C., *J. Mol. Spectrosc.* **2**: 27-41 (1958)

58E006 Goodman, L.; Kasha, M., *J. Mol. Spectrosc.* **2**: 58-65 (1958)

58R001 Linschitz, H.; Sarkanen, K., *J. Am. Chem. Soc.* **80**: 4826-32 (1958)

59B002 Claesson, S.; Lindqvist, L.; Holmstroem, B., *Nature (London)* **183**: 661-2 (1959)

59E008 Mueller, R.; Doerr, F., *Z. Elektrochem.* **63**: 1150-6 (1959)

59E009 Evans, D.F., *J. Chem. Soc.* : 2753-7 (1959)

59E011 Kanda, Y.; Shimada, R., *Spectrochim. Acta* : 211-24 (1959)

59E013 Clar, E.; Ironside, C.T.; Zander, M., *Tetrahedron* **6**: 358-63 (1959)

60A001 Lindqvist, L., *Ark. Kemi* **16**: 79-138 (1960)

60C003 Klemm, L.H.; Lind, C.D.; Spence, J.T., *J. Org. Chem.* **25**: 611-6 (1960)

60E009 Lang, K.F.; Zander, M.; Theiling, E.A., *Chem. Ber.* **93**: 321-3 (1960)

60E012 Backstrom, H.L.J.; Sandros, K., *Acta Chem. Scand.* **14**: 48-62 (1960)

60E013 Evans, D.F., *J. Chem. Soc.* : 1735-45 (1960)

60E014 Ermolaev, V.L.; Terenin, A.N., *Sov. Phys. Usp.* **3**: 423-6 (1960) Translated from: *Usp. Fiz. Nauk* **71**: 137-41 (1960)

60E016 Blackwell, L.A.; Kanda, Y.; Sponer, H., *J. Chem. Phys.* **32**: 1465-76 (1960)

60Z008 Belcher, R.; Nutten, A.J., *Quantitative Inorganic Analysis.* Butterworths, London, England, 1960, 390p.

617010 Moore, W.M.; Hammond, G.S.; Foss, R.P., *J. Am. Chem. Soc.* **83**: 2789-94 (1961)

617012 Lippert, E.; Lueder, W.; Moll, F.; Naegele, W.; Boos, H.; Prigge, H.; Seibold-Blankenstein, I., *Angew. Chem.* **73**: 695-706 (1961)

61A005 Grossweiner, L.I.; Zwicker, E.F., *J. Chem. Phys.* **34**: 1411-7 (1961)

61C002 Maki, A.H.; Geske, D.H., *J. Am. Chem. Soc.* **83**: 1852-60 (1961)

61E005 Porter, G.; Wilkinson, F., *Proc. R. Soc. London, Ser. A* **264**: 1-18 (1961)

61E009 Paris, J.P.; Hirt, R.C.; Schmitt, R.G., *J. Chem. Phys.* **34**: 1851-2 (1961)

61E010 Melhuish, W.H., *J. Phys. Chem.* **65**: 229-35 (1961)

61E011 Parker, C.A.; Hatchard, C.G., *Trans. Faraday Soc.* **57**: 1894-904 (1961)

61E012 Robinson, G.W., *J. Mol. Spectrosc.* **6**: 58-83 (1961)

61E013 Ermolaev, V.L., *Opt. Spectrosc. (USSR)* **11**: 266-9 (1961) Translated from: *Opt. Spektrosk.* **11**: 492-7 (1961)

61E014 Kanda, Y.; Shimada, R., *Spectrochim. Acta* **17**: 279-85 (1961)

61Z003 Lang, L., *Absorption Spectra in the Ultraviolet and Visible Region.* Academic Press, New York; Akademiai Kiado, Budapest, Hungary; Robert E. Krieger Publ. Co., Huntington, New York, 1961-1982, Vol. I-XXIV.

625016 Smaller, B., *J. Chem. Phys.* **37**: 1578-9 (1962)

625017 Brandon, R.W.; Gerkin, R.E.; Hutchison, C.A.,Jr., *J. Chem. Phys.* **37**: 447-8 (1962)

627011 Moore, W.M.; Ketchum, M., *J. Am. Chem. Soc.* **84**: 1368-71 (1962)

629024 Scott, D.R.; Allison, J.B., *J. Phys. Chem.* **66**: 561-2 (1962)

629025 Streitwieser, A.,Jr.; Schwager, I., *J. Phys. Chem.* **66**: 2316-20 (1962)

629027 Rosengren, K.J., *Acta Chem. Scand.* **16**: 1421-5 (1962)

629028 Smith, F.J.; Smith, J.K.; McGlynn, S.P., *Rev. Sci. Instrum.* **33**: 1367-71 (1962)

62E007 Dawson, W.; Abrahamson, E.W., *J. Phys. Chem.* **66**: 2542-7 (1962)

62E008 Rhodes, W.; El-Sayed, M.F.A., *J. Mol. Spectrosc.* **9**: 42-9 (1962)

62E009 Hoffman, M.Z.; Porter, G., *Proc. R. Soc. London, Ser. A* **268**: 46-56 (1962)

62E012 Dyck, R.H.; McClure, D.S., *J. Chem. Phys.* **36**: 2326-45 (1962)

62E014 Ware, W.R., *J. Phys. Chem.* **66**: 455-8 (1962)

62E015 Muel, B.; Hubert-Habart, M., *Adv. Mol. Spectrosc.* **2**: 647-57 (1962)

62E018 Takei, K.; Kanda, Y., *Spectrochim. Acta* **18**: 1201-16 (1962)

62E019 Parker, C.A.; Hatchard, C.G., *Analyst (London)* **87**: 664-76 (1962)

62E022 Parker, C.A., *Anal. Chem.* **34**: 502-5 (1962)

62E025 Ermolaev, V.L., *Opt. Spectrosc. (USSR)* **13**: 49-52 (1962) Translated from: *Opt. Spektrosk.* **13**: 90-5 (1962)

62Y005 Blume, M.; Watson, R.E., *Proc. Phys. Soc., London* **A270**: 127-43 (1962)

62Z002 Wetlaufer, D.B., *Adv. Protein Chem.* **17**: 303-90 (1962)

635011 Vincent, J.S.; Maki, A.H., *J. Chem. Phys.* **39**: 3088-96 (1963)

639027 Winefordner, J.D.; St. John, P.A., *Anal. Chem.* **35**: 2211-12 (1963)

639028 Mauro, J.A., *Optical Engineering Handbook.* General Electric Company, Scranton, PA, 1963

63C001 Pysh, E.S.; Yang, N.C., *J. Am. Chem. Soc.* **85**: 2124-30 (1963)

63C003 Klemm, L.H.; Kohlik, A.J., *J. Org. Chem.* **28**: 2044-9 (1963)

63E009 Dubois, J.T.; Wilkinson, F., *J. Chem. Phys.* **39**: 899-901 (1963)

63E011 Becker, R.S.; Allison, J.B., *J. Phys. Chem.* **67**: 2669-75 (1963)

63E012 Schuett, H.-U.; Zimmermann, H., *Ber. Bunsenges. Phys. Chem.* **67**: 54-62 (1963)

63E014 Kanda, Y.; Shimada, R.; Takenoshita, Y., *Spectrochim. Acta* **19**: 1249-60 (1963)

63E018 Gropper, H.; Doerr, F., *Ber. Bunsenges. Phys. Chem.* **67**: 46-54 (1963)

63E019 Doerr, F.; Gropper, H., *Ber. Bunsenges. Phys. Chem.* **67**: 193-201 (1963)

63E021 Foerster, G.v., *Z. Naturforsch., Teil A* **18**: 620-6 (1963)

63E024 Teplyakov, P.A., *Opt. Spectrosc. (USSR)* **15**: 350-2 (1963) Translated from: *Opt. Spektrosk.* **15**: 645-50 (1963)

63E027 Bergmann, K.; O'Konski, C.T., *J. Phys. Chem.* **67**: 2169-77 (1963)

63F023 Discher, C.A.; Smith, P.F.; Lippman, I.; Turse, R., *J. Phys. Chem.* **67**: 2501 (1963)

63Z003 Ermolaev, V.L., *Sov. Phys. Usp.* **6**: 333-58 (1963) Translated from: *Usp. Fiz. Nauk* **80**: 3-40 (1963)

645022 Thomson, C., *J. Chem. Phys.* **41**: 1-6 (1964)

645023 Brandon, R.W.; Gerkin, R.E.; Hutchison, C.A.,Jr., *J. Chem. Phys.* **41**: 3717-26 (1964)

647009 Kasche, V.; Lindqvist, L., *J. Phys. Chem.* **68**: 817-23 (1964)

64C001 Zweig, A.; Hodgson, W.G.; Jura, W.H., *J. Am. Chem. Soc.* **86**: 4124-9 (1964)

64C002 Zweig, A.; Lancaster, J.E.; Neglia, M.T.; Jura, W.H., *J. Am. Chem. Soc.* **86**: 4130-6 (1964)

64E011 Sandros, K., *Acta Chem. Scand.* **18**: 2355-74 (1964)

64E015 Labhart, H., *Helv. Chim. Acta* **47**: 2279-88 (1964)

64E016 Buettner, A.V., *J. Phys. Chem.* **68**: 3253-9 (1964)

64E021 Herkstroeter, W.G.; Lamola, A.A.; Hammond, G.S., *J. Am. Chem. Soc.* **86**: 4537-40 (1964)

64E022 Sponer, H.; Kanda, Y., *J. Chem. Phys.* **40**: 778-87 (1964)

64E023 Lombardi, J.R.; Raymonda, J.W.; Albrecht, A.C., *J. Chem. Phys.* **40**: 1148-56 (1964)

64E025 Zander, M., *Chem. Ber.* **97**: 2695-9 (1964)

64E026 Parker, C.A.; Hatchard, C.G.; Joyce, T.A., *J. Mol. Spectrosc.* **14**: 311-9 (1964)

64E028 Kanda, Y.; Kaseda, H.; Matumura, T., *Spectrochim. Acta* **20**: 1387-96 (1964)

64E031 Ware, W.R.; Baldwin, B.A., *J. Chem. Phys.* **40**: 1703-5 (1964)

64E034 Birks, J.B.; Dyson, D.J.; King, T.A., *Proc. R. Soc. London, Ser. A* **277**: 270-8 (1964)

64E036 Kellogg, R.E.; Schwenker, R.P., *J. Chem. Phys.* **41**: 2860-3 (1964)

64E037 Trusov, V.V.; Teplyakov, P.A., *Opt. Spectrosc. (USSR)* **16**: 27-30 (1964) Translated from: *Opt. Spektrosk.* **16**: 52-7 (1964)

64E038 McGlynn, S.P.; Azumi, T.; Kasha, M., *J. Chem. Phys.* **40**: 507-15 (1964)

64Y004 Raynes, W.T., *J. Chem. Phys.* **41**: 3020-32 (1964)

64Y005 van der Waals, J.H.; ter Maten, G., *Mol. Phys.* **8**: 301-18 (1964)

64Z006 Clar, E., *Polycyclic Hydrocarbons.* Academic Press, New York, 1964, Vol. 1, 487p.

64Z007 Clar, E., *Polycyclic Hydrocarbons.*

Academic Press, New York, 1964, Vol. 2, 487p.

655047 Vincent, J.S.; Maki, A.H., *J. Chem. Phys.* **42**: 865-8 (1965)

65A001 Shakhverdov, P.A.; Terenin, A.N., *Dokl. Phys. Chem.* **160**: 163-5 (1965) Translated from: *Dokl. Akad. Nauk SSSR* **160**: 1141-3 (1965)

65E036 Kellogg, R.E.; Simpson, W.T., *J. Am. Chem. Soc.* **87**: 4230-4 (1965)

65E038 Laposa, J.D.; Lim, E.C.; Kellogg, R.E., *J. Chem. Phys.* **42**: 3025-6 (1965)

65E042 Viktorova, E.N.; Gofman, I.A., *Russ. J. Phys. Chem.* **39**: 1416-9 (1965) Translated from: *Zh. Fiz. Khim.* **39**: 2643 (1965)

65E043 Aladekomo, J.B.; Birks, J.B., *Proc. R. Soc. London, Ser. A* **284**: 551-65 (1965)

65E046 Ware, W.R.; Baldwin, B.A., *J. Chem. Phys.* **43**: 1194-7 (1965)

65E051 Cohen, B.J.; Baba, H.; Goodman, L., *J. Chem. Phys.* **43**: 2902-3 (1965)

65F030 Lamola, A.A.; Hammond, G.S., *J. Chem. Phys.* **43**: 2129-34 (1965)

65F031 Medinger, T.; Wilkinson, F., *Trans. Faraday Soc.* **61**: 620-30 (1965)

65M065 Greenspan, H.; Fischer, E., *J. Phys. Chem.* **69**: 2466-9 (1965)

65Z001 Berlman, I.B., *Handbook of Fluorescence Spectra of Aromatic Molecules.* Academic Press, New York, 1965, 258p.

65Z003 Valley, S.L., *Handbook of Geophysics and Space Environments.* McGraw-Hill, New York, NY, 1965

65Z004 Linke, W.F., *Solubilities. Inorganic and Metal-Organic Compounds. Volumes I-II.* American Chemical Society, Washington, DC, 1958-1965

663075 Rahn, R.O.; Yamane, T.; Eisinger, J.; Longworth, J.W.; Shulman, R.G., *J. Chem. Phys.* **45**: 2947-54 (1966)

66D173 Brinen, J.S.; Koren, J.G.; Hodgson, W.G., *J. Chem. Phys.* **44**: 3095-9 (1966)

66D174 Siegel, S.; Judeikis, H.S., *J. Phys. Chem.* **70**: 2201-4 (1966)

66D175 Brinen, J.S.; Orloff, M.K., *J. Chem. Phys.* **45**: 4747-50 (1966)

66E086 Bennett, R.G.; McCartin, P.J., *J. Chem. Phys.* **44**: 1969-72 (1966)

66E089 Chessin, M.; Livingston, R.; Truscott, T.G., *Trans. Faraday Soc.* **62**: 1519-24 (1966)

66E092 Herkstroeter, W.G.; Hammond, G.S., *J. Am. Chem. Soc.* **88**: 4769-77 (1966)

66E093 Fry, A.J.; Liu, R.S.H.; Hammond, G.S., *J. Am. Chem. Soc.* **88**: 4781-2 (1966)

66E094   Kearns, D.R.; Case, W.A., *J. Am. Chem. Soc.* **88**: 5087-97 (1966)

66E095   Borkman, R.F.; Kearns, D.R., *J. Chem. Phys.* **44**: 945-9 (1966)

66E097   Yang, N.C.; Murov, S., *J. Chem. Phys.* **45**: 4358 (1966)

66E098   Parker, C.A.; Joyce, T.A., *Trans. Faraday Soc.* **62**: 2785-92 (1966)

66E099   Selinger, B.K., *Aust. J. Chem.* **19**: 825-34 (1966)

66E101   Parker, C.A.; Joyce, T.A., *Chem. Commun.* : 108-9 (1966)

66E102   Longworth, J.W.; Rahn, R.O.; Shulman, R.G., *J. Chem. Phys.* **45**: 2930-9 (1966)

66E106   Birks, J.B.; King, T.A., *Proc. R. Soc. London, Ser. A* **291**: 244-56 (1966)

66F204   Wegner, E.E.; Adamson, A.W., *J. Am. Chem. Soc.* **88**: 394-404 (1966)

66F206   Borkman, R.F.; Kearns, D.R., *J. Am. Chem. Soc.* **88**: 3467-75 (1966)

66F207   Coulson, D.R.; Yang, N.C., *J. Am. Chem. Soc.* **88**: 4511-3 (1966)

66Z001   Calvert, J.G.; Pitts, J.N.,Jr., *Photochemistry.* Wiley, New York, NY, 1966, 899p.

673066   Gueron, M.; Eisinger, J.; Shulman, R.G., *J. Chem. Phys.* **47**: 4077-91 (1967)

673067   Lamola, A.A.; Gueron, M.; Yamane, T.; Eisinger, J.; Shulman, R.G., *J. Chem. Phys.* **47**: 2210-7 (1967)

677322   Chibisov, A.K.; Skvortsov, B.V.; Karyakin, A.V.; Shvindt, N.N., *Khim. Vys. Energ.* **1**: 529-35 (1967)

677361   Usui, Y.; Koizumi, M., *Bull. Chem. Soc. Jpn.* **40**: 440-6 (1967)

677472   Lewis, H.G.; Owen, E.D., *J. Chem. Soc. B* : 422-5 (1967)

677498   Kellmann, A.; Lindqvist, L., In *The Triplet State: Proceedings of an International Symposium*, Beirut, Lebanon, 1967, p.439-45

67B017   Hochstrasser, R.M.; Marzzacco, C., *J. Chem. Phys.* **46**: 4155-6 (1967)

67C005   Zweig, A.; Maurer, A.H.; Roberts, B.G., *J. Org. Chem.* **32**: 1322-9 (1967)

67D239   Grivet, J.-P.; Ptak, M., *C. R. Hebd. Seances Acad. Sci., Ser. B* **265**: 972-5 (1967)

67E031   Bowers, P.G.; Porter, G., *Proc. R. Soc. London, Ser. A* **299**: 348-53 (1967)

67E109   Horrocks, A.R.; Medinger, T.; Wilkinson, F., *Photochem. Photobiol.* **6**: 21-8 (1967)

67E112   Marchetti, A.P.; Kearns, D.R., *J. Am. Chem. Soc.* **89**: 768-77 (1967)

67E114   Evans, T.R.; Leermakers, P.A., *J. Am. Chem. Soc.* **89**: 4380-2 (1967)

67E115   Zander, M., *Ber. Bunsenges. Phys. Chem.* **71**: 424-9 (1967)

67E116   Lim, E.C.; Chakrabarti, S.K., *Chem. Phys. Lett.* **1**: 28-31 (1967)

67E117   Cohen, B.J.; Goodman, L., *J. Chem. Phys.* **46**: 713-21 (1967)

67E118   Iwata, S.; Tanaka, J.; Nagakura, S., *J. Chem. Phys.* **47**: 2203-9 (1967)

67E119   Kuboyama, A.; Yabe, S., *Bull. Chem. Soc. Jpn.* **40**: 2475-9 (1967)

67E120   Almgren, M., *Photochem. Photobiol.* **6**: 829-40 (1967)

67E124   Zimmerman, H.E.; Binkley, R.W.; McCullough, J.J.; Zimmerman, G.A., *J. Am. Chem. Soc.* **89**: 6589-95 (1967)

67E126   Zander, M., *Fresenius' Z. Anal. Chem.* **226**: 251-9 (1967)

67E129   Parker, C.A.; Joyce, T.A., *Chem. Commun.* : 744-5 (1967)

67F508   Searle, R.; Williams, J.L.R.; DeMeyer, D.E.; Doty, J.C., *Chem. Commun.* : 1165 (1967)

67F510   Henson, R.C.; Jones, J.L.W.; Owen, E.D., *J. Chem. Soc. A* : 116-22 (1967)

67F521   Yang, N.C.; McClure, D.S.; Murov, S.L.; Houser, J.J.; Dusenbery, R., *J. Am. Chem. Soc.* **89**: 5466-8 (1967)

67F522   Lam, E.Y.Y.; Valentine, D.; Hammond, G.S., *J. Am. Chem. Soc.* **89**: 3482-7 (1967)

67F523   Murov, S.L., Ph.D., Thesis, Univ. Chicago, Chicago, IL, 1967, 227p.

67F524   Kearns, D.R.; Hollins, R.A.; Khan, A.U.; Chambers, R.W.; Radlick, P., *J. Am. Chem. Soc.* **89**: 5455-6 (1967)

680379   Dainton, F.S.; Peng, C.T.; Salmon, G.A., *J. Phys. Chem.* **72**: 3801-7 (1968)

680600   Pitts, J.N.; Burley, D.R.; Mani, J.C.; Broadbent, A.D., *J. Am. Chem. Soc.* **90**: 5902-03 (1968)

687061   Hurley, R.; Testa, A.C., *J. Am. Chem. Soc.* **90**: 1949-52 (1968)

687222   Chrysochoos, J.; Grossweiner, L.I., *Photochem. Photobiol.* **8**: 193-208 (1968)

68B014   Warwick, D.A.; Wells, C.H.J., *Spectrochim. Acta., Part A* **24A**: 589-93 (1968)

68C005   Sease, J.W.; Burton, F.G.; Nickol, S.L., *J. Am. Chem. Soc.* **90**: 2595-8 (1968)

68D217   Gondo, Y.; Maki, A.H., *J. Phys. Chem.* **72**: 3215-22 (1968)

68E098   Horrocks, A.R.; Wilkinson, F., *Proc. R. Soc. London, Ser. A* **306**: 257-73 (1968)

68E100   Knowles, A.; Roe, E.M.F., *Photochem. Photobiol.* **7**: 421-36 (1968)

68E103   Yildiz, A.; Kissinger, P.T.; Reilley, C.N., *J. Chem. Phys.* **49**: 1403-6 (1968)

68E104 Dawson, W.R., *J. Opt. Soc. Am.* **58**: 222-7 (1968)

68E105 Windsor, M.W.; Dawson, W.R., *Mol. Cryst.* **4**: 253-8 (1968)

68E113 Stevens, B.; Algar, B.E., *J. Phys. Chem.* **72**: 2582-7 (1968)

68E114 Dawson, W.R.; Windsor, M.W., *J. Phys. Chem.* **72**: 3251-60 (1968)

68E116 Lim, E.C.; Yu, J.M.H., *J. Chem. Phys.* **49**: 3878-84 (1968)

68E117 Rosenberg, H.M.; Carson, S.D., *J. Phys. Chem.* **72**: 3531-4 (1968)

68E124 Birks, J.B.; Georghiou, S., *J. Phys. B* **1**: 958-65 (1968)

68E126 Knibbe, H.; Rehm, D.; Weller, A., *Ber. Bunsenges. Phys. Chem.* **72**: 257-63 (1968)

68E128 Binet, D.J.; Goldberg, E.L.; Forster, L.S., *J. Phys. Chem.* **72**: 3017-20 (1968)

68E129 Stevens, B.; Thomaz, M.F., *Chem. Phys. Lett.* **1**: 549-50 (1968)

68E130 Vander Donckt, E.; Nasielski, J.; Greenleaf, J.R.; Birks, J.B., *Chem. Phys. Lett.* **2**: 409-10 (1968)

68E131 Brinen, J.S.; Koren, J.G., *Chem. Phys. Lett.* **2**: 671-2 (1968)

68E132 Marsh, G.; Kearns, D.R.; Schaffner, K., *Helv. Chim. Acta* **51**: 1890-9 (1968)

68E133 Wilairat, P.; Selinger, B., *Aust. J. Chem.* **21**: 733-46 (1968)

68E135 Suppan, P., *Ber. Bunsenges. Phys. Chem.* **72**: 321-6 (1968)

68F286 Stevens, B.; Algar, B.E., *J. Phys. Chem.* **72**: 3468-74 (1968)

68F291 Arnold, D.R., *Adv. Photochem.* **6**: 301-423 (1968)

68F298 Yang, N.C.; Dusenbery, R.L., *J. Am. Chem. Soc.* **90**: 5899-900 (1968)

68F301 Hyndman, H.L., Ph.D., Thesis, California Institute of Technology, Pasadena, CA, 1968

68F303 Wagner, P.J., *Tetrahedron Lett.* : 5385-8 (1968)

68M106 Ling, A.C.; Willard, J.E., *J. Phys. Chem.* **72**: 1918-23 (1968)

68M107 Ling, A.C.; Willard, J.E., *J. Phys. Chem.* **72**: 3349-51 (1968)

68Z003 Parker, C.A., *Photoluminescence of Solutions. With Applications to Photochemistry and Analytical Chemistry.* Elsevier, Amsterdam, The Netherlands, 1968, 544p.

68Z004 Gollnick, K., *Adv. Photochem.* **6**: 1-122 (1968)

68Z005 Zander, M., *Phosphorimetry.* Academic Press, New York, 1968, 206p.

68Z006 Kosower, E.M., *An Introduction to Physical Organic Chemistry.* Wiley and Sons, New York, 1968, 503p.

690087 Kemp, T.J.; Roberts, J.P., *Trans. Faraday Soc.* **65**: 725-31 (1969)

695015 Wells, C.H.J., *Chem. Commun.* : 393-4 (1969)

696019 Kropp, J.L.; Dawson, W.R.; Windsor, M.W., *J. Phys. Chem.* **73**: 1747-52 (1969)

696020 Dawson, W.R.; Kropp, J.L., *J. Phys. Chem.* **73**: 1752-58 (1969)

696034 Clark, W.D.K.; Litt, A.D.; Steel, C., *J. Am. Chem. Soc.* **91**: 5413-5 (1969)

696077 Turro, N.J.; Engel, R., *Mol. Photochem.* **1**: 143-6 (1969)

696078 Turro, N.J.; Engel, R., *Mol. Photochem.* **1**: 235-8 (1969)

696102 Walker, M.S.; Bednar, T.W.; Lumry, R., In *Molecular Luminescence*, Lim, E.C. (ed.), W.A. Benjamin, Inc., New York, 1969, p.135-52

697141 Kramer, H.E.A., *Z. Phys. Chem. (Frankfurt Am Main)* **66**: 73-85 (1969)

697155 Yang, N.C.; Dusenbery, R.L., *Mol. Photochem.* **1**: 159-71 (1969)

697156 Wagner, P.J.; Capen, G., *Mol. Photochem.* **1**: 173-88 (1969)

697266 Chibisov, A.K., *Photochem. Photobiol.* **10**: 331-47 (1969)

697272 Kemp, D.R.; Porter, G., *J. Chem. Soc. D* : 1029-30 (1969)

697314 Evans, T.R., *Tech. Org. Chem.* **14**: 297-348 (1969)

69A001 Turro, N.J.; Engel, R., *J. Am. Chem. Soc.* **91**: 7113-21 (1969)

69D288 Sharnoff, M., *J. Chem. Phys.* **51**: 451-2 (1969)

69E202 Sandros, K., *Acta Chem. Scand.* **23**: 2815-29 (1969)

69E203 Nemoto, M.; Kokubun, H.; Koizumi, M., *Bull. Chem. Soc. Jpn.* **42**: 1223-30 (1969)

69E208 Heinzelmann, W.; Labhart, H., *Chem. Phys. Lett.* **4**: 20-4 (1969)

69E215 Cadogan, K.D.; Albrecht, A.C., *J. Phys. Chem.* **73**: 1868-77 (1969)

69E216 Dawson, W.R.; Kropp, J.L., *J. Phys. Chem.* **73**: 693-9 (1969)

69E217 Sykes, A.; Truscott, T.G., *J. Chem. Soc. D* : 929-30 (1969)

69E223 Bera, S.Ch.; Mukherjee, R.; Chowdhury, M., *J. Chem. Phys.* **51**: 754-61 (1969)

69E226 Hoover, R.J.; Kasha, M., *J. Am. Chem. Soc.* **91**: 6508-10 (1969)

69E227 Lim, E.C.; Li, R.; Li, Y.H., *J. Chem. Phys.* **50**: 4925-33 (1969)

69E229 Pavlopoulos, T.G., *J. Chem. Phys.* **51**: 2936-40 (1969)

69E231 Seybold, P.G.; Gouterman, M., *J. Mol. Spectrosc.* **31**: 1-13 (1969)

69E232 Parkanyi, C.; Baum, E.J.; Wyatt, J.; Pitts, J.N.,Jr., *J. Phys. Chem.* **73**: 1132-8 (1969)

69E234 Hadley, S.G.; Keller, R.A., *J. Phys. Chem.* **73**: 4356-9 (1969)

69E235 Chambers, R.W.; Kearns, D.R., *Photochem. Photobiol.* **10**: 215-9 (1969)

69E236 Zander, M., *Z. Naturforsch., Teil A* **24A**: 1387-90 (1969)

69E238 Clarke, R.H.; Hochstrasser, R.M., *J. Mol. Spectrosc.* **32**: 309-19 (1969)

69E239 Singer, L.A., *Tetrahedron Lett.* : 923-6 (1969)

69E240 Rotkiewicz, K.; Grabowski, Z.R., *Trans. Faraday Soc.* **65**: 3263-78 (1969)

69E243 Cundall, R.B.; Voss, A.J.R., *J. Chem. Soc. D* : 116 (1969)

69E244 Eisinger, J.; Navon, G., *J. Chem. Phys.* **50**: 2069-77 (1969)

69E247 Shcheglova, N.A.; Shigorin, D.N.; Yakobson, G.G.; Tushishvili, L.Sh., *Russ. J. Phys. Chem.* **43**: 1112-7 (1969) Translated from: *Zh. Fiz. Khim.* **43**: 1984 (1969)

69E251 Thomson, A.J., *J. Chem. Phys.* **51**: 4106-16 (1969)

69F388 Stevens, B.; Algar, B.E., *J. Phys. Chem.* **73**: 1711-5 (1969)

69F395 Wagner, P.J.; Spoerke, R.W., *J. Am. Chem. Soc.* **91**: 4437-40 (1969)

69F399 Caldwell, R.A., *Tetrahedron Lett.* : 2121-4 (1969)

69F401 Scheiner, P., *Tetrahedron Lett.* : 4863-6 (1969)

69F404 Moses, F.G.; Liu, R.S.H.; Monroe, B.M., *Mol. Photochem.* **1**: 245-9 (1969)

69F410 Bylina, A.; Grabowski, Z.R., *Trans. Faraday Soc.* **65**: 458-63 (1969)

69Z002 McGlynn, S.P.; Azumi, T.; Kinoshita, M., *Molecular Spectroscopy of the Triplet State.* Prentice Hall, Englewood Cliffs, NJ, 1969, 434p.

69Z009 Schwarzenbach, G.; Flaschka, H., *Complexometric Titrations.* Methuen and Co., Ltd., London, England, 1969, 490p.

704011 Turro, N.J.; Lee, T-J., *Mol. Photochem.* **2**: 185-90 (1970)

705030 Zuclich, J., *J. Chem. Phys.* **52**: 3586-91 (1970)

706018 Hunter, T.F., *Trans. Faraday Soc.* **66**: 300-9 (1970)

706023 Maria, H.J.; McGlynn, S.P., *J. Chem. Phys.* **52**: 3399-402 (1970)

706049 Bonnier, J.-M.; Jardon, P., *J. Chim. Phys. Phys.-Chim. Biol.* **67**: 577-9 (1970)

706053 Jones, P.F.; Calloway, A.R., 20th Reunion de la Societe de Chimie Physique, Paris, May 27-30, 1969, p.110-5

706054 Kearvell, A.; Wilkinson, F., 20th Reunion de la Societe de Chimie Physique, Paris, May 27-30, 1969, p.125-32

706079 Algar, B.E.; Stevens, B., *J. Phys. Chem.* **74**: 3029-34 (1970)

706135 Kropp, J.L.; Lou, J.J., *J. Phys. Chem.* **74**: 3953-9 (1970)

706182 Patterson, L.K.; Porter, G.; Topp, M.R., *Chem. Phys. Lett.* **7**: 612-4 (1970)

706216 Rehm, D.; Weller, A., *Isr. J. Chem.* **8**: 259-71 (1970)

706229 Birks, J.B., *Photophysics of Aromatic Molecules.* Wiley-Interscience, New York, 1970, 704p.

706243 Sveshnikova, E.B.; Snegov, M.I., *Opt. Spectrosc. (USSR)* **19**: 265-8 (1970) Translated from: *Opt. Spektrosk.* **19**: 496-500 (1970)

707174 Fischer, H.; Kramer, H.E.A.; Maute, A., *Z. Phys. Chem. (Frankfurt Am Main)* **69**: 113-31 (1970)

707186 Gegiou, D.; Huber, J.R.; Weiss, K., *J. Am. Chem. Soc.* **92**: 5058-62 (1970)

707199 Heinrich, G.; Holzer, G.; Blume, H.; Schulte-Frohlinde, D., *Z. Naturforsch., Teil B* **25**: 496-9 (1970)

707232 Hadley, S.G., *Chem. Phys. Lett.* **6**: 549-50 (1970)

707240 Hadley, S.G., *J. Phys. Chem.* **74**: 3551-2 (1970)

707357 Yang, N.C.; Feit, E.D.; Hui, M.H.; Turro, N.J.; Dalton, J.C., *J. Am. Chem. Soc.* **92**: 6974-6 (1970)

707531 Gallivan, J.B., *Mol. Photochem.* **2**: 191-211 (1970)

707538 Dalton, J.C.; Turro, N.J., *Annu. Rev. Phys. Chem.* **21**: 499-560 (1970)

707561 Loeff, I.; Lutz, H.; Lindqvist, L., *Isr. J. Chem.* **8**: 141-6 (1970)

70C003 Mann, C.K.; Barnes, K.K., *Electrochemical Reactions in Nonaqueous Systems.* Marcel Dekker, Inc., New York, 1970, 560p.

70C005 Millefiori, S., *J. Heterocycl. Chem.* **7**: 145-9 (1970)

70E288 Porter, G.; Topp, M.R., *Proc. R. Soc. London, Ser. A* **315**: 163-84 (1970)

70E295 Slifkin, M.A.; Walmsley, R.H., *J. Phys. E* **3**: 160-2 (1970)

70E296 Zanker, V.; Rudolph, E.; Prell, G., *Z. Naturforsch., Teil B* **25B**: 1137-43 (1970)

70E304 Wettack, F.S.; Renkes, G.D.; Rockley, M.G.; Turro, N.J.; Dalton, J.C., *J. Am. Chem. Soc.* **92**: 1793-4 (1970)

70E309 Lim, E.C.; Li, Y.H.; Li, R., *J. Chem. Phys.* **53**: 2443-8 (1970)

70E310 Nishi, N.; Shimada, R.; Kanda, Y., *Bull. Chem. Soc. Jpn.* **43**: 41-6 (1970)

70E317 Eastwood, D.; Gouterman, M., *J. Mol. Spectrosc.* **35**: 359-75 (1970)

70E318 Hayashi, H.; Nagakura, S., *Mol. Phys.* **19**: 45-53 (1970)

70E319 Gradyushko, A.T.; Sevchenko, A.N.; Solovyov, K.N.; Tsvirko, M.P., *Photochem. Photobiol.* **11**: 387-400 (1970)

70E320 Birks, J.B.; Leite, M.S.S.C.P., *J. Phys. B.* **3**: 417-24 (1970)

70E321 El-Bayoumi, M.A.; Dalle, J.-P.; O'Dwyer, M.F., *J. Lumin.* **1-2**: 716-25 (1970)

70F737 Kelly, J.M.; Porter, G., *Proc. R. Soc. London, Ser. A* **319**: 319-29 (1970)

716017 Halpern, A.M.; Ware, W.R., *J. Chem. Phys.* **54**: 1271-6 (1971)

716113 Sykes, A.; Truscott, T.G., *Trans. Faraday Soc.* **67**: 679-86 (1971)

716157 Yang, N.C.; Hui, M.H.; Bellard, S.A., *J. Am. Chem. Soc.* **93**: 4056-8 (1971)

716169 Japar, S.M.; Abrahamson, E.W., *J. Am. Chem. Soc.* **93**: 4140-4 (1971)

716279 Geacintov, N.E.; Burgos, J.; Pope, M.; Strom, C., *Chem. Phys. Lett.* **11**: 504-8 (1971)

717003 Becker, R.S.; Inuzuka, K.; Balke, D.E., *J. Am. Chem. Soc.* **93**: 38-42 (1971)

717154 Davis, G.A.; Gresser, J.D.; Carapellucci, P.A., *J. Am. Chem. Soc.* **93**: 2179-82 (1971)

717171 Yamamuro, T.; Tanaka, I.; Hata, N., *Bull. Chem. Soc. Jpn.* **44**: 667-71 (1971)

717179 Lutz, H.; Lindqvist, L., *J. Chem. Soc. D* : 493-4 (1971)

717222 Marsh, G.; Kearns, D.R.; Schaffner, K., *J. Am. Chem. Soc.* **93**: 3129-37 (1971)

717235 Hadley, S.G., *J. Phys. Chem.* **75**: 2083-6 (1971)

717336 Cooper, G.D.; DeGraff, B.A., *J. Phys. Chem.* **75**: 2897-902 (1971)

717346 Van Thielen, J.; Van Thien, T.; De Schryver, F.C., *Tetrahedron Lett.* : 3031-4 (1971)

717384 Gallivan, J.B.; Brinen, J.S., *Chem. Phys. Lett.* **10**: 455-9 (1971)

717447 Clark, W.D.K.; Steel, C., *J. Am. Chem. Soc.* **93**: 6347-55 (1971)

717449 Pownall, H.J.; Huber, J.R., *J. Am. Chem. Soc.* **93**: 6429-36 (1971)

717459 Kearvell, A.; Wilkinson, F., *Chem. Phys. Lett.* **11**: 472-3 (1971)

717489 Porter, G.; Yip, R.W.; Dunston, J.M.; Cessna, A.J.; Sugamori, S.E., *Trans. Faraday Soc.* **67**: 3149-54 (1971)

717520 Kemp, D.R.; Porter, G., *Proc. R. Soc. London, Ser. A* **326**: 117-30 (1971)

71D299 Yamanashi, B.S.; Bowers, K.W., *J. Magn. Reson.* **5**: 109-14 (1971)

71E357 Gradyushko, A.T.; Tsvirko, M.P., *Opt. Spectrosc. (USSR)* **31**: 291-5 (1971) Translated from: *Opt. Spektrosk.* **31**: 548-56 (1971)

71E360 Bensasson, R.; Land, E.J., *Trans. Faraday Soc.* **67**: 1904-15 (1971)

71E361 Slifkin, M.A.; Walmsley, R.H., *Photochem. Photobiol.* **13**: 57-65 (1971)

71E367 Herbert, M.A.; Johns, H.E., *Photochem. Photobiol.* **14**: 693-704 (1971)

71E377 Turro, N.J.; Lee, C.; Schore, N.; Barltrop, J.; Carless, H.A.J., *J. Am. Chem. Soc.* **93**: 3079-80 (1971)

71E378 Hautala, R.R.; Turro, N.J., *J. Am. Chem. Soc.* **93**: 5595-7 (1971)

71E380 Lancelot, G.; Helene, C., *Chem. Phys. Lett.* **9**: 327-31 (1971)

71E385 Emeis, C.A.; Oosterhoff, L.J., *J. Chem. Phys.* **54**: 4809-19 (1971)

71E386 Vincett, P.S.; Voigt, E.M.; Rieckhoff, K.E., *J. Chem. Phys.* **55**: 4131-40 (1971)

71E391 Semeluk, G.P.; Stevens, R.D.S., *Can. J. Chem.* **49**: 2452-5 (1971)

71E394 Watson, F.H.,Jr.; El-Bayoumi, M.A., *J. Chem. Phys.* **55**: 5464-70 (1971)

71F587 Rusakowicz, R.; Testa, A.C., *Spectrochim. Acta, Part A* **27A**: 787-92 (1971)

71F593 Kurien, K.C., *J. Chem. Soc. B* : 2081-2 (1971)

71F595 Vesley, G.F., *Mol. Photochem.* **3**: 193-200 (1971)

71M447 Fowler, F.W.; Katritzky, A.R.; Rutherford, R.J.D., *J. Chem. Soc. B* : 460-9 (1971)

71Z001 Berlman, I.B., *Handbook of Fluorescence Spectra of Aromatic Molecules.* Academic Press, New York, 1971, Second ed., 473p.

71Z005 Gueron, M., In *Creation and Detection of the Excited State*, A.A. Lamola (ed.), Marcel Dekker, New York, 1971, Vol. 1, Pt. A, p.303-42

71Z012 Taylor, H.A., In *Analytical Photochemistry and Photochemical Analysis*, J. M. Fitzgerald (ed.), Dekker, New York, 1971, p.91-115

71Z014 Perkampus, H.-H.; Sandeman, I.; Timmons, C.J., *DMS UV Atlas of Organic Compounds.* Verlag Chemie, Weinheim; Butterworths, London, England, 1966-1971, Vol. I-V.

71Z015 Moore, C.E., *Atomic Energy Levels as Derived from the Analyses of Optical Spectra.* Natl. Bur. Stand., Washington, DC, 1971, Vol. I-III

720392 Hulme, B.E.; Land, E.J.; Phillips, G.O., *J. Chem. Soc., Faraday Trans. 1* **68**: 2003-12 (1972)

720447 Dainton, F.S.; Robinson, E.A.; Salmon, G.A., *J. Phys. Chem.* **76**: 3897-904 (1972)

720464 Tetreau, C.; Lavalette, D.; Land, E.J.; Peradejordi, F., *Chem. Phys. Lett.* **17**: 245-7 (1972)

726112 Pantke, E.R.; Labhart, H., *Chem. Phys. Lett.* **16**: 255-9 (1972)

726120 Cundall, R.B.; Pereira, L.C., *J. Chem. Soc., Faraday Trans. 2* **68**: 1152-63 (1972)

726156 Dempster, D.N.; Morrow, T.; Rankin, R.; Thompson, G.F., *J. Chem. Soc., Faraday Trans. 2* **68**: 1479-96 (1972)

726174 Blackwell, D.S.L.; Liao, C.C.; Loutfy, R.O.; de Mayo, P.; Paszyc, S., *Mol. Photochem.* **4**: 171-88 (1972)

726177 Alvarez, V.L.; Hadley, S.G., *J. Phys. Chem.* **76**: 3937-40 (1972)

726211 Almgren, M., *Mol. Photochem.* **4**: 327-34 (1972)

727047 Vander Donckt, E.; Lietaer, D., *J. Chem. Soc., Faraday Trans. 1* **68**: 112-20 (1972)

727064 Lutz, H.; Duval, M.-C.; Breheret, E.; Lindqvist, L., *J. Phys. Chem.* **76**: 821-2 (1972)

727069 Kawai, K.; Shirota, Y.; Tsubomura, H.; Mikawa, H., *Bull. Chem. Soc. Jpn.* **45**: 77-81 (1972)

727073 Soep, B.; Kellmann, A.; Martin, M.; Lindqvist, L., *Chem. Phys. Lett.* **13**: 241-4 (1972)

727134 Jackson, A.W.; Yarwood, A.J., *Can. J. Chem.* **50**: 1331-7 (1972)

727296 Yip, R.W.; Loutfy, R.O.; Chow, Y.L.; Magdzinski, L.K., *Can. J. Chem.* **50**: 3426-31 (1972)

727348 Hellner, C.; Lindqvist, L.; Roberge, P.C., *J. Chem. Soc., Faraday Trans. 2* **68**: 1928-37 (1972)

727392 Ayscough, P.B.; Sealy, R.C., *J. Photochem.* **1**: 83-5 (1972)

729040 Nishimura, A.M.; Tinti, D.S., *Chem. Phys. Lett.* **13**: 278-83 (1972)

72B002 Foerster, E.W.; Grellmann, K.H., *Chem. Phys. Lett.* **14**: 536-8 (1972)

72B007 Evans, D.F.; Tucker, J.N., *J. Chem. Soc., Faraday Trans. 2* **68**: 174-6 (1972)

72D311 Nishimura, A.M.; Vincent, J.S., *Chem. Phys. Lett.* **13**: 89-92 (1972)

72D312 Mao, S.W.; Wong, T.C.; Hirota, N., *Chem. Phys. Lett.* **13**: 199-204 (1972)

72D313 Wade, C.G.; Webber, S.E., *J. Chem. Phys.* **56**: 1619-25 (1972)

72D316 Harrigan, E.T.; Wong, T.C.; Hirota, N., *Chem. Phys. Lett.* **14**: 549-54 (1972)

72D317 Arnold, D.R.; Bolton, J.R.; Pedersen, J.A., *J. Am. Chem. Soc.* **94**: 2872-4 (1972)

72D318 Chen, C.R.; Mucha, J.A.; Pratt, D.W., *Chem. Phys. Lett.* **15**: 73-8 (1972)

72D319 Shain, A.L.; Chiang, W.-T.; Sharnoff, M., *Chem. Phys. Lett.* **16**: 206-10 (1972)

72E287 Birks, J.B., *Chem. Phys. Lett.* **17**: 370-2 (1972)

72E293 Li, R.; Lim, E.C., *J. Chem. Phys.* **57**: 605-12 (1972)

72E294 Gallivan, J.B., *Can. J. Chem.* **50**: 3601-6 (1972)

72E303 Huber, J.R.; Mantulin, W.W., *J. Am. Chem. Soc.* **94**: 3755-60 (1972)

72E313 Froehlich, P.M.; Morrison, H.A., *J. Phys. Chem.* **76**: 3566-70 (1972)

72E314 Kulberg, L.P.; Nurmukhametov, R.N.; Gorelik, M.V., *Opt. Spectrosc. (USSR)* **32**: 476-8 (1972) Translated from: *Opt. Spektrosk.* **32**: 895-9 (1972)

72E315 Vander Donckt, E.; Vogels, C., *Spectrochim. Acta, Part A* **28A**: 1969-75 (1972)

72E316 Dreeskamp, H.; Hutzinger, O.; Zander, M., *Z. Naturforsch., Teil A* **27A**: 756-9 (1972)

72E323 Takemura, T.; Yamamoto, K.; Yamazaki, I.; Baba, H., *Bull. Chem. Soc. Jpn.* **45**: 1639-42 (1972)

72E330 Hudson, B.S.; Kohler, B.E., *Chem. Phys. Lett.* **14**: 299-304 (1972)

72F543 Morrison, H.; Maleski, R., *Photochem. Photobiol.* **16**: 145-6 (1972)

72M260 Hutzler, J.S.; Colton, R.J.; Ling, A.C., *J. Chem. Eng. Data* **17**: 324-7 (1972)

733001 Truscott, T.G.; Land, E.J.; Sykes, A., *Photochem. Photobiol.* **17**: 43-51 (1973)

733187 Lakowicz, J.R.; Weber, G., *Biochemistry* **12**: 4171-9 (1973)

735067 Taylor, H.V.; Allred, A.L.; Hoffman, B.M., *J. Am. Chem. Soc.* **95**: 3215-9 (1973)

736002 Merkel, P.B.; Kearns, D.R., *J. Chem. Phys.* **58**: 398-400 (1973)

736048 Cundall, R.B.; Pereira, L.C., *Chem. Phys. Lett.* **18**: 371-4 (1973)

736051 Dempster, D.N.; Morrow, T.; Rankin, R.; Thompson, G.F., *Chem. Phys. Lett.* **18**: 488-92 (1973)

736067 Watkins, A.R., *J. Phys. Chem.* **77**: 1207-10 (1973)

736097 Gijzeman, O.L.J.; Kaufman, F.; Porter, G., *J. Chem. Soc., Faraday Trans. 2* **69**: 708-20 (1973)

736099 Gijzeman, O.L.J.; Kaufman, F.; Porter, G., *J. Chem. Soc., Faraday Trans. 2* **69**: 727-37 (1973)

736125 Olszowski, A.; Romanowski, H.; Ruziewicz, Z., *Bull. Acad. Pol. Sci., Ser. Sci., Math, Astron. Phys.* **21**: 381-7 (1973)

736162 Nakashima, N.; Mataga, N.; Yamanaka, C., *Int. J. Chem. Kinet.* **5**: 833-9 (1973)

736174 Stikeleather, J.A., *Chem. Phys. Lett.* **21**: 326-9 (1973)

737055 Bagdasaryan, Kh.S.; Kiryukhin, Yu.I.; Sinitsina, Z.A., *J. Photochem.* **1**: 225-40 (1973)

737113 Kikuchi, K.; Watarai, H.; Koizumi, M., *Bull. Chem. Soc. Jpn.* **46**: 749-54 (1973)

737138 Cundall, R.B.; Ogilvie, S.McD.; Robinson, D.A., *J. Photochem.* **1**: 417-22 (1973)

737140 Vander Donckt, E.; Barthels, M.R.; Delestinne, A., *J. Photochem.* **1**: 429-32 (1973)

737190 Yip, R.W.; Szabo, A.G.; Tolg, P.K., *J. Am. Chem. Soc.* **95**: 4471-2 (1973)

737198 Lutz, H.; Breheret, E.; Lindqvist, L., *J. Phys. Chem.* **77**: 1758-62 (1973)

737292 Porter, G.; Dogra, S.K.; Loutfy, R.O.; Sugamori, S.E.; Yip, R.W., *J. Chem. Soc., Faraday Trans. 1* **69**: 1462-74 (1973)

737439 Dekker, R.H.; Srinivasan, B.N.; Huber, J.R.; Weiss, K., *Photochem. Photobiol.* **18**: 457-66 (1973)

737541 Lamola, A.A., *Pure Appl. Chem.* **34**: 281-303 (1973)

737591 Murov, S.L., *Handbook of Photochemistry*. Dekker, New York, 1973, 272p.

73E338 Hotchandani, S.; Testa, A.C., *J. Chem. Phys.* **59**: 596-600 (1973)

73E345 Tsvirko, M.P.; Sapunov, V.V.; Solovev, K.N., *Opt. Spectrosc. (USSR)* **34**: 635-8 (1973) Translated from: *Opt. Spektrosk.* **34**: 1094-100 (1973)

73E347 Mathis, P.; Kleo, J., *Photochem. Photobiol.* **18**: 343-6 (1973)

73E352 Kothandaraman, G.; Tinti, D.S., *Chem. Phys. Lett.* **19**: 225-30 (1973)

73E353 Burke, F.P.; Small, G.J.; Braun, J.R.; Lin, T.-S., *Chem. Phys. Lett.* **19**: 574-9 (1973)

73E356 Fischer, G., *Chem. Phys. Lett.* **21**: 305-8 (1973)

73E358 Arnold, D.R.; Birtwell, R.J., *J. Am. Chem. Soc.* **95**: 4599-606 (1973)

73E359 Adams, J.E.; Mantulin, W.W.; Huber, J.R., *J. Am. Chem. Soc.* **95**: 5477-81 (1973)

73E365 Scharf, H.-D.; Leismann, H., *Z. Naturforsch., Teil B* **28B**: 662-81 (1973)

73E366 Berlman, I.B., *Energy Transfer Parameters of Aromatic Compounds*. Academic Press, New York, 1973, 379p.

73E370 Lim, E.C.; Kedzierski, M., *Chem. Phys. Lett.* **20**: 242-5 (1973)

73E372 Mantulin, W.W.; Song, P.-S., *J. Am. Chem. Soc.* **95**: 5122-9 (1973)

73E373 Sandros, K., *Acta Chem. Scand.* **27**: 3021-32 (1973)

741013 Rosenfeld, T.; Alchalel, A.; Ottolenghi, M., *J. Phys. Chem.* **78**: 336-41 (1974)

743135 Gouterman, M.; Khalil, G.-E., *J. Mol. Spectrosc.* **53**: 88-100 (1974)

743202 Vigny, P.; Duquesne, M., In *Excited States of Biological Molecules*, J.B. Birks (ed.), Wiley, London, England, 1976, p.167-77

745009 Levanon, H.; Wolberg, A., *Chem. Phys. Lett.* **24**: 96-8 (1974)

745194 Schmidt, H.; Zellhofer, R., *Z. Phys. Chem. (Frankfurt Am Main)* **91**: 204-18 (1974)

745289 Leaver, I.H., *Photochem. Photobiol.* **19**: 309-13 (1974)

745445 Van Dorp, W.G.; Soma, M.; Kooter, J.A.; Van der Waals, J.H., *Mol. Phys.* **28**: 1551-68 (1974)

745458 Scherz, A.; Orbach, N.; Levanon, H., *Isr. J. Chem.* **12**: 1037-48 (1974)

746020 Giering, L.; Berger, M.; Steel, C., *J. Am. Chem. Soc.* **96**: 953-8 (1974)

746085 Kanamaru, N.; Long, M.E.; Lim, E.C., *Chem. Phys. Lett.* **26**: 1-9 (1974)

746103 Kanamaru, N.; Bhattacharjee, H.R.; Lim, E.C., *Chem. Phys. Lett.* **26**: 174-9 (1974)

746190 Dalton, J.C.; Montgomery, F.C., *J. Am. Chem. Soc.* **96**: 6230-32 (1974)

746194 Watkins, A.R., *J. Phys. Chem.* **78**: 1885-90 (1974)

746251 Tournon, J.; Abu-Elgheit, M.; Avouris, P.; El-Bayoumi, M.A., *Chem. Phys. Lett.* **28**: 430-2 (1974)

746270 Heinrich, G.; Schoof, S.; Gusten, H., *J. Photochem.* **3**: 315-20 (1974)

746283 Watkins, A.R., *J. Phys. Chem.* **78**: 2555-8 (1974)

746288   Souto, M.A.; Wagner, P.J.; El-Sayed, M.A., *Chem. Phys.* **6**: 193-204 (1974)

746445   Waddell, W.H.; Renner, C.A.; Turro, N.J., *Mol. Photochem.* **6**: 321-4 (1974)

747022   Bent, D.V.; Schulte-Frohlinde, D., *J. Phys. Chem.* **78**: 446-50 (1974)

747025   Hochstrasser, R.M.; Lutz, H.; Scott, G.W., *Chem. Phys. Lett.* **24**: 162-7 (1974)

747039   Bonneau, R.; Fornier de Violet, Ph.; Joussot-Dubien, J., *Photochem. Photobiol.* **19**: 129-32 (1974)

747049   Dempster, D.N.; Morrow, T.; Quinn, M.F., *J. Photochem.* **2**: 329-41 (1974)

747050   Dempster, D.N.; Morrow, T.; Quinn, M.F., *J. Photochem.* **2**: 343-59 (1974)

747093   Kanamaru, N.; Nagakura, S.; Kimura, K., *Bull. Chem. Soc. Jpn.* **47**: 745-6 (1974)

747190   Kikuchi, M.; Kikuchi, K.; Kokubun, H., *Bull. Chem. Soc. Jpn.* **47**: 1331-3 (1974)

747233   Bent, D.V.; Hayon, E.; Moorthy, P.N., *Chem. Phys. Lett.* **27**: 544-7 (1974)

747304   Kraljic, I.; Lindqvist, L., *Photochem. Photobiol.* **20**: 351-5 (1974)

747307   Anderson, R.W.,Jr.; Hochstrasser, R.M.; Lutz, H.; Scott, G.W., *Chem. Phys. Lett.* **28**: 153-7 (1974)

747355   Metcalfe, J.; Rockley, M.G.; Phillips, D., *J. Chem. Soc., Faraday Trans. 2* **70**: 1660-6 (1974)

747390   Blanchi, J.-P.; Watkins, A.R., *Mol. Photochem.* **6**: 133-42 (1974)

74A006   Charney, D.R.; Dalton, J.C.; Hautala, R.R.; Snyder, J.J.; Turro, N.J., *J. Am. Chem. Soc.* **96**: 1407-10 (1974)

74E506   Harrigan, E.T.; Hirota, N., *Chem. Phys. Lett.* **27**: 405-10 (1974)

74E514   Capitanio, D.A.; Pownall, H.J.; Huber, J.R., *J. Photochem.* **3**: 225-36 (1974)

74E515   Cheng, T.H.; Hirota, N., *Mol. Phys.* **27**: 281-307 (1974)

74E517   Brocklehurst, B.; Tawn, D.N., *Spectrochim. Acta, Part A* **30A**: 1807-15 (1974)

74E524   Smagowicz, J.; Wierzchowski, K.L., *J. Lumin.* **8**: 210-32 (1974)

751124   Brede, O.; Helmstreit, W.; Mehnert, R., *Z. Phys. Chem. (Leipzig)* **256**: 505-12 (1975)

753056   Tsvirko, M.P.; Solovev, K.N.; Gradyushko, A.T.; Dvornikov, S.S., *Opt. Spectrosc. (USSR)* **38**: 400-4 (1975) Translated from: *Opt. Spektrosk.* **38**: 705-13 (1975)

755071   Chan, I.Y.; Nelson, B.N., *J. Chem. Phys.* **62**: 4080-8 (1975)

755120   Yamashita, M.; Ikeda, H.; Kashiwagi, H., *J. Chem. Phys.* **63**: 1127-31 (1975)

755396   Harrigan, E.T.; Hirota, N., *J. Am. Chem. Soc.* **97**: 6647-52 (1975)

755398   Bulska, H.; Chodkowska, A.; Grabowska, A.; Pakula, B.; Slanina, Z., *J. Lumin.* **10**: 39-57 (1975)

756028   Lui, Y.H.; McGlynn, S.P., *J. Lumin.* **9**: 449-58 (1975)

756061   Serafimov, O.; Bruehlmann, U.; Huber, J.R., *Ber. Bunsenges. Phys. Chem.* **79**: 202-5 (1975)

756077   Acuna, A.U.; Ceballos, A.; Garcia Dominguez, J.A.; Molera, M.J., *An. Quim.* **71**: 22-7 (1975)

756080   Post, M.F.M.; Langelaar, J.; Van Voorst, J.D.W., *Chem. Phys. Lett.* **32**: 59-62 (1975)

756125   DeToma, R.P.; Cowan, D.O., *J. Am. Chem. Soc.* **97**: 3283-91 (1975)

756162   Kikuchi, K.; Kokubun, H.; Kikuchi, M., *Bull. Chem. Soc. Jpn.* **48**: 1378-81 (1975)

756176   Lui, Y.H.; McGlynn, S.P., *J. Lumin.* **10**: 113-21 (1975)

756186   Marcondes, M.E.R.; Toscano, V.G.; Weiss, R.G., *J. Am. Chem. Soc.* **97**: 4485-90 (1975)

756199   Mahaney, M.; Hubert, J.R., *Chem. Phys.* **9**: 371-8 (1975)

756208   Wamser, C.C.; Medary, R.T.; Kochevar, I.E.; Turro, N.J.; Chang, P.L., *J. Am. Chem. Soc.* **97**: 4864-9 (1975)

756229   Quimby, D.J.; Longo, F.R., *J. Am. Chem. Soc.* **97**: 5111-7 (1975)

756237   Farmilo, A.; Wilkinson, F., *Chem. Phys. Lett.* **34**: 575-80 (1975)

756251   Bortolus, P.; Dellonte, S., *J. Chem. Soc., Faraday Trans. 2* **71**: 1338-42 (1975)

756270   Wierzchowski, K.L.; Berens, K.; Szabo, A.G., *J. Lumin.* **10**: 331-43 (1975)

756304   Lui, Y.H.; McGlynn, S.P., *J. Mol. Spectrosc.* **55**: 163-74 (1975)

756441   Vergragt, P.J.; van der Waals, J.H., *Chem. Phys. Lett.* **36**: 283-9 (1975)

756505   Wilkinson, F., In *Organic Molecular Photophysics*, J.B. Birks (ed.), John Wiley, New York, 1975, Vol. 2, p.95-158

757014   Arce, R.; Ramirez, L., *Photochem. Photobiol.* **21**: 13-9 (1975)

757048   Vogelmann, E.; Schmidt, H.; Steiner, U.; Kramer, H.E.A., *Z. Phys. Chem. (Frankfurt Am Main)* **94**: 101-6 (1975)

757066   Bent, D.V.; Hayon, E., *Chem. Phys. Lett.* **31**: 325-7 (1975)

757078   Schreiner, S.; Steiner, U.; Kramer, H.E.A., *Photochem. Photobiol.* **21**: 81-4 (1975)

757161   Bent, D.V.; Hayon, E., *J. Am. Chem. Soc.* **97**: 2599-606 (1975)

757162 Bent, D.V.; Hayon, E., *J. Am. Chem. Soc.* **97**: 2606-12 (1975)

757163 Bent, D.V.; Hayon, E., *J. Am. Chem. Soc.* **97**: 2612-9 (1975)

757167 Ferris, J.P.; Prabhu, K.V.; Strong, R.L., *J. Am. Chem. Soc.* **97**: 2835-9 (1975)

757247 Herkstroeter, W.G., *J. Am. Chem. Soc.* **97**: 4161-7 (1975)

757279 Alkaitis, S.A.; Graetzel, M.; Henglein, A., *Ber. Bunsenges. Phys. Chem.* **79**: 541-6 (1975)

757282 Amand, B.; Bensasson, R., *Chem. Phys. Lett.* **34**: 44-8 (1975)

757309 Bent, D.V.; Hayon, E.; Moorthy, P.N., *J. Am. Chem. Soc.* **97**: 5065-71 (1975)

757353 Alkaitis, S.A.; Beck, G.; Graetzel, M., *J. Am. Chem. Soc.* **97**: 5723-9 (1975)

757510 Salet, C.; Bensasson, R., *Photochem. Photobiol.* **22**: 231-5 (1975)

757534 Gisin, M.; Wirz, J., *Helv. Chim. Acta* **58**: 1768-71 (1975)

75C006 Siegerman, H., In *Technique of Electroorganic Synthesis.* (Techniques of Chemistry, Vol. 5, Part II), N.L. Weinberg (ed.), John Wiley, New York, 1975, p.667-1056

75E529 Bensasson, R.; Land, E.J.; Truscott, T.G., *Photochem. Photobiol.* **21**: 419-21 (1975)

761024 Barwise, A.J.G.; Gorman, A.A.; Rodgers, M.A.J., *Chem. Phys. Lett.* **38**: 313-6 (1976)

761035 Bensasson, R.; Land, E.J.; Maudinas, B., *Photochem. Photobiol.* **23**: 189-93 (1976)

761069 Zador, E.; Warman, J.M.; Hummel, A., *J. Chem. Soc., Faraday Trans. 1* **72**: 1368-76 (1976)

761088 Bensasson, R.; Land, E.J.; Lafferty, J.; Sinclair, R.S.; Truscott, T.G., *Chem. Phys. Lett.* **41**: 333-5 (1976)

761168 Land, E.J., *Photochem. Photobiol.* **24**: 475-7 (1976)

765078 Wagner, P.J.; May, M.J., *Chem. Phys. Lett.* **39**: 350-2 (1976)

765139 Clarke, R.H.; Frank, H.A., *J. Chem. Phys.* **65**: 39-47 (1976)

766054 Friedrich, J.; Vogel, J.; Windhager, W.; Doerr, F., *Z. Naturforsch., Teil A* **31**: 61-70 (1976)

766100 Morris, J.V.; Mahaney, M.A.; Huber, J.R., *J. Phys. Chem.* **80**: 969-74 (1976)

766170 Ware, W.R.; Rothman, W., *Chem. Phys. Lett.* **39**: 449-53 (1976)

766189 Harrigan, E.T.; Chakrabarti, A.; Hirota, N., *J. Am. Chem. Soc.* **98**: 3460-5 (1976)

766267 Zander, M., *Z. Naturforsch., Teil A* **31A**: 677-8 (1976)

766276 Wallace, W.L.; Van Duyne, R.P.; Lewis, F.D., *J. Am. Chem. Soc.* **98**: 5319-26 (1976)

766314 Huber, J.R.; Mahaney, M.; Morris, J.V., *Chem. Phys.* **16**: 329-35 (1976)

766343 Yang, N.C.; Shold, D.M.; Neywick, C.V., *J. Chem. Soc., Chem. Commun.* : 727-8 (1976)

766377 Fushimi, K.; Kikuchi, K.; Kokubun, H., *J. Photochem.* **5**: 457-68 (1976)

766401 Philen, D.L.; Hedges, R.M., *Chem. Phys. Lett.* **43**: 358-62 (1976)

766421 Fratev, F.; Polansky, O.E.; Zander, M., *Z. Naturforsch., Teil A* **31A**: 987-9 (1976)

766441 Pereira, L.C.; Ferreira, I.C.; Thomaz, M.P.F., *Chem. Phys. Lett.* **43**: 157-61 (1976)

766464 Kobayashi, T.; Nagakura, S., *Chem. Phys. Lett.* **43**: 429-34 (1976)

766469 Hirata, Y.; Tanaka, I., *Chem. Phys. Lett.* **43**: 568-70 (1976)

766474 Haink, H.J.; Huber, J.R., *Chem. Phys. Lett.* **44**: 117-20 (1976)

767042 Gauglitz, G., *J. Photochem.* **5**: 41-7 (1976)

767094 Wilkinson, F.; Farmilo, A., *J. Chem. Soc., Faraday Trans. 2* **72**: 604-18 (1976)

767144 Amouyal, E.; Bensasson, R., *J. Chem. Soc., Faraday Trans. 1* **72**: 1274-87 (1976)

767159 Ohno, T.; Kato, S., *Chem. Lett.* : 263-6 (1976)

767171 Garner, A.; Wilkinson, F., *J. Chem. Soc., Faraday Trans. 2* **72**: 1010-20 (1976)

767177 Alkaitis, S.A.; Graetzel, M., *J. Am. Chem. Soc.* **98**: 3549-54 (1976)

767180 Bensasson, R.; Salet, C.; Balzani, V., *J. Am. Chem. Soc.* **98**: 3722-4 (1976)

767189 Treinin, A.; Hayon, E., *J. Am. Chem. Soc.* **98**: 3884-91 (1976)

767246 Vogelmann, E.; Kramer, H.E.A., *Photochem. Photobiol.* **23**: 383-90 (1976)

767269 Capellos, C.; Suryanarayanan, K., *Int. J. Chem. Kinet.* **8**: 529-39 (1976)

767270 Capellos, C.; Suryanarayanan, K., *Int. J. Chem. Kinet.* **8**: 541-8 (1976)

767343 Metcalfe, J.; Chervinsky, S.; Oref, I., *Chem. Phys. Lett.* **42**: 190-2 (1976)

767370 Arnold, D.R.; Maroulis, A.J., *J. Am. Chem. Soc.* **98**: 5931-7 (1976)

767423 Demas, J.N.; McBride, R.P.; Harris, E.W., *J. Phys. Chem.* **80**: 2248-53 (1976)

767458 Bowman, W.D.; Demas, J.N., *J. Phys. Chem.* **80**: 2434-5 (1976)

767471 Schaffner, K., *Tetrahedron* **32**: 641-53 (1976)

767546 Wagner, P.J.; Thomas, M.J.; Harris, E., *J. Am. Chem. Soc.* **98**: 7675-9 (1976)

767574 Kayser, R.H.; Young, R.H., *Photochem. Photobiol.* **24**: 395-401 (1976)

767584 Vogelmann, E.; Schreiner, S.; Rauscher, W.; Kramer, H.E.A., *Z. Phys. Chem. (Frankfurt Am Main)* **101**: 321-36 (1976)

767661 Vogelmann, E.; Kramer, H.E.A., *Photochem. Photobiol.* **24**: 595-7 (1976)

76E686 Lancelot, G., *Mol. Phys.* **31**: 241-54 (1976)

76E691 Hotchandani, S.; Testa, A.C., *Spectrochim. Acta, Part A* **32A**: 1659-63 (1976)

76E692 Clar, E.; Schmidt, W., *Tetrahedron* **32**: 2563-6 (1976)

76E696 Thulstrup, E.W.; Nepras, M.; Dvorak, V.; Michl, J., *J. Mol. Spectrosc.* **59**: 265-85 (1976)

76F937 Davies, A.K.; Khan, K.A.; McKellar, J.F.; Phillips, G.O., *Mol. Photochem.* **7**: 389-98 (1976)

76M467 Aue, D.H.; Webb, H.M.; Bowers, M.T., *J. Am. Chem. Soc.* **98**: 311-7 (1976)

76Z005 de Mayo, P.; Shizuka, H., In *Creation and Detection of the Excited State*, W.R. Ware (ed.), Dekker, New York, 1976, Vol. 4, p.139-215

76Z030 Demas, J.N., In *Creation and Detection of the Excited State*, W.R. Ware (ed.), Dekker, New York, 1976, Vol. 4, p.1-62

76Z049 Benson, S.W., *Thermochemical Kinetics. Methods for the Estimation of Thermochemical Data and Rate Parameters. Second Edition*. Wiley and Sons, New York, 1976, 320p.

76Z050 Fraga, S.; Saxena, K.M.S.; Karwowski, J., *Handbook of Atomic Data. Physical Sciences Data, Vol. 5*. Elsevier, Amsterdam, The Netherlands, 1976, 551p.

771014 Garner, A.; Wilkinson, F., *Chem. Phys. Lett.* **45**: 432-5 (1977)

771021 Wilkinson, F.; Garner, A., *J. Chem. Soc., Faraday Trans. 2* **73**: 222-33 (1977)

771048 Irie, M.; Yorozu, T.; Yoshida, K.; Hayashi, K., *J. Phys. Chem.* **81**: 973-6 (1977)

771078 Chantrell, S.J.; McAuliffe, C.A.; Munn, R.W.; Pratt, A.C.; Land, E.J., *J. Chem. Soc., Faraday Trans. 1* : 858-65 (1977)

775025 Sheng, S.J.; El-Sayed, M.A., *Chem. Phys.* **20**: 61-9 (1977)

775062 Vyas, H.M.; Wan, J.K.S., *Can. J. Chem.* **55**: 1175-80 (1977)

775124 Tria, J.J.; Johnsen, R.H., *J. Phys. Chem.* **81**: 1274-8 (1977)

775232 Chodkowska, A.; Grabowska, A.; Herbich, J., *Chem. Phys. Lett.* **51**: 365-9 (1977)

776013 Motten, A.G.; Kwiram, A.L., *Chem. Phys. Lett.* **45**: 217-20 (1977)

776016 Wolf, M.W.; Brown, R.E.; Singer, L.A., *J. Am. Chem. Soc.* **99**: 526-31 (1977)

776060 Crosby, P.M.; Dyke, J.M.; Metcalfe, J.; Rest, A.J.; Salisbury, K.; Sodeau, J.R., *J. Chem. Soc., Perkin Trans. 2* : 182-5 (1977)

776085 Anton, M.F.; Moomaw, W.R., *J. Chem. Phys.* **66**: 1808-18 (1977)

776118 Archer, M.D.; Ferreira, M.I.C.; Porter, G.; Tredwell, C.J., *Nouv. J. Chim.* **1**: 9-12 (1977)

776194 Capellos, C.; Suryanarayanan, K., *Int. J. Chem. Kinet.* **9**: 399-407 (1977)

776222 Acuna, A.U.; Ceballos, A.; Molera, M.J., *J. Phys. Chem.* **81**: 1090-3 (1977)

776226 Zander, M., *Z. Naturforsch., Teil A* **32A**: 339-40 (1977)

776251 Fleming, G.R.; Knight, A.W.E.; Morris, J.M.; Morrison, R.J.S.; Robinson, G.W., *J. Am. Chem. Soc.* **99**: 4306-11 (1977)

776258 Kellmann, A., *J. Phys. Chem.* **81**: 1195-8 (1977)

776328 Moore, W.M.; McDaniels, J.C.; Hen, J.A., *Photochem. Photobiol.* **25**: 505-12 (1977)

776369 Loutfy, R.O.; Yip, R.W.; Dogra, S.K., *Tetrahedron Lett.* : 2843-6 (1977)

776378 Aloisi, G.G.; Mazzucato, U.; Birks, J.B.; Minuti, L., *J. Am. Chem. Soc.* **99**: 6340-7 (1977)

776387 Mardelli, M.; Olmsted, J.,III., *J. Photochem.* **7**: 277-85 (1977)

776412 Bensasson, R.; Dawe, E.A.; Long, D.A.; Land, E.J., *J. Chem. Soc., Faraday Trans. 1* **73**: 1319-25 (1977)

777004 Schuster, D.I.; Goldstein, M.D.; Bane, P., *J. Am. Chem. Soc.* **99**: 187-93 (1977)

777037 Pernot, C.; Lindqvist, L., *J. Photochem.* **6**: 215-20 (1976/77)

777041 Korobov, V.E.; Shubin, V.V.; Chibisov, A.K., *Chem. Phys. Lett.* **45**: 498-501 (1977)

777063 Bonneau, R., *Photochem. Photobiol.* **25**: 129-32 (1977)

777201 Bandyopadhyay, B.N.; Harriman, A., *J. Chem. Soc., Faraday Trans. 1* **73**: 663-74 (1977)

777242 Steiner, U.; Winter, G.; Kramer, H.E.A., *J. Phys. Chem.* **81**: 1104-10 (1977)

777265 Fouassier, J.-P.; Lougnot, D.-J.; Wieder, F.; Faure, J., *J. Photochem.* **7**: 17-28 (1977)

777315 Ferreira, M.I.C.; Harriman, A., *J. Chem. Soc., Faraday Trans. 1* **73**: 1085-92 (1977)

777316 Dunne, A.; Quinn, M.F., *J. Chem. Soc., Faraday Trans. 1* **73**: 1104-10 (1977)

777432 Volkert, W.A.; Kuntz, R.R.; Ghiron, C.A.; Evans, R.F.; Santus, R.; Bazin, M., *Photochem. Photobiol.* **26**: 3-9 (1977)

777434 Harriman, A.; Liu, R.S.H., *Photochem. Photobiol.* **26**: 29-32 (1977)

777439 Chapple, A.P.; Vikesland, J.P.; Wilkinson, F., *Chem. Phys. Lett.* **50**: 81-4 (1977)

777497 von Sonntag, C.; Schuchmann, H.-P., *Adv. Photochem.* **10**: 59-145 (1977)

777555 Nicodem, D.E.; Cabral, M.L.P.F.; Ferreira, J.C.N., *Mol. Photochem.* **8**: 213-38 (1977)

777602 Amouyal, E.; Bensasson, R., *J. Chem. Soc., Faraday Trans. 1* **73**: 1561-8 (1977)

777603 Brown, R.G.; Porter, G., *J. Chem. Soc., Faraday Trans. 1* **73**: 1569-73 (1977)

777611 Frank, R.; Gauglitz, G., *J. Photochem.* **7**: 355-7 (1977)

777617 Grodowski, M.S.; Veyret, B.; Weiss, K., *Photochem. Photobiol.* **26**: 341-52 (1977)

779025 Morgan, D.D.; Warshawsky, D.; Atkinson, T., *Photochem. Photobiol.* **25**: 31-8 (1977)

77A178 Yamamoto, S.; Kikuchi, K.; Kokubun, H., *Chem. Lett.* : 1173-6 (1977)

77A203 Kikuchi, K.; Tamura, S.-I.; Iwanaga, C.; Kokubun, H.; Usui, Y., *Z. Phys. Chem. (Frankfurt am Main)* **106**: 17-24 (1977)

77C008 Meites, L.; Zuman, P., *CRC Handbook Series in Organic Electrochemistry, Volume I-V.* CRC Press, Inc., Boca Raton, FL, 1977-1982

77E581 Hotchandani, S.; Testa, A.C., *J. Chem. Phys.* **67**: 5201-6 (1977)

77E638 Jardon, P., *J. Chim. Phys. Phys.-Chim. Biol.* **74**: 1177-84 (1977)

77E662 Abdul-Rasoul, F.; Catherall, C.L.R.; Hargreaves, J.S.; Mellor, J.M.; Phillips, D., *Eur. Polym. J.* **13**: 1019-23 (1977)

77E663 Ivanov, V.L.; Martynov, I.Yu.; Uzhinov, B.M.; Kuz'min, M.G., *High Energy Chem.* **11**: 361-4 (1977) Translated from: *Khim. Vys. Energ.* **11**: 327-31 (1977)

77E801 Kunavin, N.I.; Nurmukhametov, R.N.; Khachaturova, G.T., *J. Appl. Spectrosc.* **26**: 735-9 (1977) Translated from: *Zh. Prikl. Spektrosk.* **26**: 1023-7 (1977)

77F762 Numao, N.; Hamada, T.; Yonemitsu, O., *Tetrahedron Lett.* : 1661-4 (1977)

77F954 Gonzenbach, H.-U.; Tegmo-Larsson, I.-M.; Grosclaude, J.-P.; Schaffner, K., *Helv. Chim. Acta* **60**: 1091-123 (1977)

77Z191 Wesseler, E.P.; Iltis, R.; Clark, L.C.,Jr., *J. Fluorine Chem.* **9**: 137-46 (1977)

78A163 Levin, G., *J. Phys. Chem.* **82**: 1584-8 (1978)

78A170 Wilkinson, F.; Garner, A., *Photochem. Photobiol.* **27**: 659-70 (1978)

78A183 Berger, M.; McAlpine, E.; Steel, C., *J. Am. Chem. Soc.* **100**: 5147-51 (1978)

78A304 Korobov, V.E.; Chibisov, A.K., *J. Photochem.* **9**: 411-24 (1978)

78A307 Schoof, S.; Guesten, H.; von Sonntag, C., *Ber. Bunsenges. Phys. Chem.* **82**: 1068-73 (1978)

78A333 Marshall, E.J.; Pilling, M.J., *J. Chem. Soc., Faraday Trans. 2* **74**: 579-90 (1978)

78A343 Gorman, A.A.; Parekh, C.T.; Rodgers, M.A.J.; Smith, P.G., *J. Photochem.* **9**: 11-7 (1978)

78A345 Bauer, H.; Reske, G., *J. Photochem.* **9**: 43-54 (1978)

78A368 Yamamoto, S.-A.; Kikuchi, K.; Kokubun, H., *Z. Phys. Chem. (Wiesbaden)* **109**: 47-58 (1978)

78A378 McVie, J.; Sinclair, R.S.; Truscott, T.G., *J. Chem. Soc., Faraday Trans. 2* **74**: 1870-9 (1978)

78A447 Tamura, S.-I.; Kikuchi, K.; Kokubun, H.; Usui, Y., *Z. Phys. Chem. (Wiesbaden)* **111**: 7-18 (1978)

78B145 Fang, H.L.-B.; Thrash, R.J.; Leroi, G.E., *Chem. Phys. Lett.* **57**: 59-63 (1978)

78C018 Park, S.-M., *J. Electrochem. Soc.* **125**: 216-22 (1978)

78D004 Kleibeuker, J.F.; Platenkamp, R.J.; Schaafsma, T.J., *Chem. Phys.* **27**: 51-64 (1978)

78D026 Haegele, W.; Schmid, D.; Wolf, H.C., *Z. Naturforsch., Teil A* **33A**: 83-93 (1978)

78D082 Niizuma, S.; Kwan, L.; Hirota, N., *Mol. Phys.* **35**: 1029-46 (1978)

78E021 Kuz'min, V.A.; Tatikolov, A.S.; Borisevich, Yu.E., *Chem. Phys. Lett.* **53**: 52-5 (1978)

78E060 Huggenberger, C.; Labhart, H., *Helv. Chim. Acta* **61**: 250-7 (1978)

78E065 Niizuma, S.; Hirota, N., *J. Phys. Chem.* **82**: 453-9 (1978)

78E067 Latas, K.J.; Nishimura, A.M., *J. Phys. Chem.* **82**: 491-5 (1978)

78E088 Coyle, J.D.; Newport, G.L.; Harriman, A., *J. Chem. Soc., Perkin Trans. 2* : 133-7 (1978)

78E131 Barwise, A.J.G.; Gorman, A.A.; Leyland, R.L.; Smith, P.G.; Rodgers, M.A.J., *J. Am. Chem. Soc.* **100**: 1814-20 (1978)

78E157 Bensasson, R.V.; Land, E.J.; Salet, C., *Photochem. Photobiol.* **27**: 273-80 (1978)

78E222 Noe, L.J.; Degenkolb, E.O.; Rentzepis, P.M., *J. Chem. Phys.* **68**: 4435-8 (1978)

78E263 Gorman, A.A.; Lovering, G.; Rodgers, M.A.J., *J. Am. Chem. Soc.* **100**: 4527-32 (1978)

78E273 Pownall, H.J.; Schaffer, A.M.; Becker, R.S.; Mantulin, W.W., *Photochem. Photobiol.* **27**: 625-8 (1978)

78E306 Yagi, M.; Nishi, N.; Kinoshita, M.; Nagakura, S., *Mol. Phys.* **35**: 1369-79 (1978)

78E312 Madej, S.L.; Gillispie, G.D.; Lim, E.C., *Chem. Phys.* **32**: 1-10 (1978)

78E314 Braeuchle, C.; Kabza, H.; Voitlaender, J.; Clar, E., *Chem. Phys.* **32**: 63-73 (1978)

78E359 Croteau, R.; Leblanc, R.M., *Photochem. Photobiol.* **28**: 33-8 (1978)

78E394 Romashov, L.V.; Borovkova, V.A.; Kiryukhin, Yu.I.; Bagdasar'yan, Kh.S., *High Energy Chem.* **12**: 132-4 (1978) Translated from: *Khim. Vys. Energ.* **12**: 156-9 (1978)

78E414 Soboleva, I.V.; Sadovskii, N.A.; Kuz'min, M.G., *Dokl. Phys. Chem.* **238**: 70-3 (1978) Translated from: *Dokl. Akad. Nauk SSSR* **238**: 400-3 (1978)

78E431 Das, P.K.; Becker, R.S., *J. Phys. Chem.* **82**: 2081-93 (1978)

78E432 Das, P.K.; Becker, R.S., *J. Phys. Chem.* **82**: 2093-105 (1978)

78E467 Bensasson, R.; Land, E.J., *Nouv. J. Chim.* **2**: 503-7 (1978)

78E489 Damschen, D.E.; Merritt, C.D.; Perry, D.L.; Scott, G.W.; Talley, L.D., *J. Phys. Chem.* **82**: 2268-72 (1978)

78E495 Andrews, L.J.; Deroulede, A.; Linschitz, H., *J. Phys. Chem.* **82**: 2304-9 (1978)

78E504 Gradyushko, A.T.; Knyukshto, V.N.; Solovev, K.N.; Shulga, A.M., *Opt. Spectrosc. (USSR)* **44**: 268-72 (1978) Translated from: *Opt. Spektrosk.* **44**: 458-65 (1978)

78E534 Raemme, G., *J. Photochem.* **9**: 439-47 (1978)

78E538 Morris, J.M.; Yoshihara, K., *Mol. Phys.* **36**: 993-1003 (1978)

78E542 Siegmund, M.; Bendig, J., *Ber. Bunsenges. Phys. Chem.* **82**: 1061-8 (1978)

78E589 Brown, R.G.; Harriman, A.; Harris, L., *J. Chem. Soc., Faraday Trans. 2* **74**: 1193-9 (1978)

78E649 Eriksen, J.; Foote, C.S., *J. Phys. Chem.* **82**: 2659-62 (1978)

78E721 Becker, R.S.; Bensasson, R.V.; Lafferty, J.; Truscott, T.G.; Land, E.J., *J. Chem. Soc., Faraday Trans. 2* **74**: 2246-55 (1978)

78E761 Zander, M., *Z. Naturforsch., Teil A* **33A**: 998-1000 (1978)

78E878 Val'kova, G.A.; Shcherbo, S.N.; Shigorin, D.N., *Dokl. Phys. Chem.* **240**: 491-3 (1978) Translated from: *Dokl. Akad. Nauk SSSR* **240**: 884-7 (1978)

78E889 Basara, H.; Ruziewicz, Z.; Zawadzka, H., *J. Lumin.* **17**: 283-90 (1978)

78E891 Lui, Y.H.; McGlynn, S.P., *Spectrosc. Lett.* **11**: 47-58 (1978)

78E894 Basara, H.; Ruziewicz, Z., *Acta Phys. Pol., A* **A54**: 689-94 (1978)

78F030 Grellmann, K.H.; Hentzschel, P., *Chem. Phys. Lett.* **53**: 545-51 (1978)

78F572 Perbet, G.; Coulangeon, L.M.; Boule, P.; Lemaire, J., *J. Chim. Phys. Phys.-Chim. Biol.* **75**: 1096-104 (1978)

78M177 Flicker, W.M.; Mosher, O.A.; Kuppermann, A., *J. Chem. Phys.* **69**: 3311-20 (1978)

79A093 Wilkinson, F.; Schroeder, J., *J. Chem. Soc., Faraday Trans. 2* **75**: 441-50 (1979)

79A114 Sloper, R.W.; Truscott, T.G.; Land, E.J., *Photochem. Photobiol.* **29**: 1025-9 (1979)

79A237 Grellmann, K.-H.; Hentzschel, P.; Wismontski-Knittel, T.; Fischer, E., *J. Photochem.* **11**: 197-213 (1979)

79A241 George, M.V.; Kumar, Ch.V.; Scaiano, J.C., *J. Phys. Chem.* **83**: 2452-5 (1979)

79A260 Heelis, P.F.; Parsons, B.J.; Phillips, G.O., *Biochim. Biophys. Acta* **587**: 455-62 (1979)

79A284 Encinas, M.V.; Scaiano, J.C., *J. Photochem.* **11**: 241-7 (1979)

79B007 Greene, B.I.; Hochstrasser, R.M.; Weisman, R.B., *J. Chem. Phys.* **70**: 1247-59 (1979)

79B037 Baugher, J.; Hindman, J.C.; Katz, J.J., *Chem. Phys. Lett.* **63**: 159-62 (1979)

79B042 Beaumont, P.C.; Parsons, B.J.; Phillips, G.O.; Allen, J.C., *Biochim. Biophys. Acta* **562**: 214-21 (1979)

79B044 Kuz'min, V.A.; Darmanyan, A.P.; Levin, P.P., *Chem. Phys. Lett.* **63**: 509-14 (1979)

79B061 Gschwind, R.; Haselbach, E., *Helv. Chim. Acta* **62**: 941-55 (1979)

79B086 Pileni, M.P.; Giraud, M.; Santus, R., *Photochem. Photobiol.* **30**: 251-6 (1979)

79B087 Salet, C.; Bensasson, R.; Becker, R.S., *Photochem. Photobiol.* **30**: 325-9 (1979)

79B163 Simons, W.W., *Sadtler Handbook of Ultraviolet Spectra.* Sadtler Res. Lab., Philadelphia, PA, 1979, 1016p.

79C010 Bock, C.R.; Connor, J.A.; Gutierrez, A.R.; Meyer, T.J.; Whitten, D.G.; Sullivan, B.P.; Nagle, J.K., *J. Am. Chem. Soc.* **101**: 4815-24 (1979)

79D031 Connors, R.E.; Comer, J.C.; Durand, R.R.,Jr., *Chem. Phys. Lett.* **61**: 270-4 (1979)

79D033 Kim, S.S., *Chem. Phys. Lett.* **61**: 327-30 (1979)

79D171 Latas, K.J.; Power, R.K.; Nishimura, A.M., *Chem. Phys. Lett.* **65**: 272-7 (1979)

79D226 Moan, J.; Hovik, B.; Wold, E., *Photochem. Photobiol.* **30**: 623-4 (1979)

79D290 Hirota, N.; Baba, M.; Hirata, Y.; Nagaoka, S., *J. Phys. Chem.* **83**: 3350-4 (1979)

79E086 Heisel, F.; Miehe, J.A.; Sipp, B., *Chem. Phys. Lett.* **61**: 115-8 (1979)

79E099 Werner, T., *J. Phys. Chem.* **83**: 320-5 (1979)

79E108 Dreeskamp, H.; Pabst, J., *Chem. Phys. Lett.* **61**: 262-5 (1979)

79E109 Delouis, J.F.; Delaire, J.A.; Ivanoff, N., *Chem. Phys. Lett.* **61**: 343-6 (1979)

79E140 Braeuchle, Chr.; Kabza, H.; Voitlaender, J., *Z. Naturforsch., Teil A* **34A**: 6-12 (1979)

79E156 Niizuma, S.; Hirota, N., *J. Phys. Chem.* **83**: 706-13 (1979)

79E210 Groenen, E.J.J.; Koelman, W.N., *J. Chem. Soc., Faraday Trans. 2* **75**: 69-78 (1979)

79E219 Vogelmann, E.; Rauscher, W.; Kramer, H.E.A., *Photochem. Photobiol.* **29**: 771-6 (1979)

79E243 Arvis, M.; Mialocq, J.-C., *J. Chem. Soc., Faraday Trans. 2* **75**: 415-21 (1979)

79E265 Condirston, D.A.; Laposa, J.D., *Chem. Phys. Lett.* **63**: 313-7 (1979)

79E282 Land, E.J.; Truscott, T.G., *Photochem. Photobiol.* **29**: 861-6 (1979)

79E378 Goerner, H.; Schulte-Frohlinde, D., *J. Am. Chem. Soc.* **101**: 4388-90 (1979)

79E412 Alder, L.; Gloyna, D.; Wegener, W.; Pragst, F.; Henning, H.-G., *Chem. Phys. Lett.* **64**: 503-6 (1979)

79E415 Carsey, T.P.; Findley, G.L.; McGlynn, S.P., *J. Am. Chem. Soc.* **101**: 4502-10 (1979)

79E505 Bluemer, G.-P.; Zander, M., *Z. Naturforsch., Teil A* **34A**: 909-10 (1979)

79E543 Goerner, H.; Schulte-Frohlinde, D., *Chem. Phys. Lett.* **66**: 363-9 (1979)

79E546 Das, P.K.; Becker, R.S., *J. Am. Chem. Soc.* **101**: 6348-53 (1979)

79E560 Olmsted, J.,III, *J. Phys. Chem.* **83**: 2581-4 (1979)

79E564 Eweg, J.K.; Mueller, F.; Visser, A.J.W.G.; Veeger, C.; Bebelaar, D.; van

Voorst, J.D.W., *Photochem. Photobiol.* **30**: 463-71 (1979)

79E640 Goerner, H.; Schulte-Frohlinde, D., *J. Phys. Chem.* **83**: 3107-18 (1979)

79E677 Bendig, J.; Siegmund, M., *J. Prakt. Chem.* **321**: 587-600 (1979)

79E678 Sa E Melo, M.T.; Averbeck, D.; Bensasson, R.V.; Land, E.J.; Salet, C., *Photochem. Photobiol.* **30**: 645-51 (1979)

79E690 Encinas, M.V.; Scaiano, J.C., *J. Am. Chem. Soc.* **101**: 7740-1 (1979)

79E793 Heinrich, G.; Guesten, H., *Z. Phys. Chem. (Wiesbaden)* **118**: 31-41 (1979)

79E838 Dvornikov, S.S.; Knyukshto, V.N.; Solovev, K.N.; Tsvirko, M.P., *Opt. Spectrosc. (USSR)* **46**: 385-8 (1979) Translated from: *Opt. Spektrosk.* **46**: 689-95 (1979)

79E967 Dinse, K.P.; Winscom, C.J., *J. Lumin.* **18-19**: 500-4 (1979)

79F100 Bruhlmann, U.; Huber, J.R., *J. Photochem.* **10**: 205-13 (1979)

79F261 Takamuku, S.; Beck, G.; Schnabel, W., *J. Photochem.* **11**: 49-52 (1979)

79F262 Fernandez, E.; Figuera, J.M.; Tobar, A., *J. Photochem.* **11**: 69-71 (1979)

79M363 Griffiths, T.R.; Pugh, D.C., *Coord. Chem. Rev.* **29**: 129-211 (1979)

79N005 Moroi, Y.; Braun, A.M.; Graetzel, M., *J. Am. Chem. Soc.* **101**: 567-72 (1979)

79P084 Encina, M.V.; Lissi, E.A., *J. Polym. Sci., Polym. Chem. Ed.* **17**: 1645-53 (1979)

79Z027 Schulte-Frohlinde, D.; Goerner, H., *Pure Appl. Chem.* **51**: 279-97 (1979)

79Z427 Hansch, C.; Leo, A., *Substituent Constants for Correlation Analysis in Chemistry and Biology.* Wiley, New York, 1979, 339p.

80A038 Osif, T.L.; Lichtin, N.N.; Hoffman, M.Z.; Ray, S., *J. Phys. Chem.* **84**: 410-4 (1980)

80A171 Ohno, T.; Lichtin, N.N., *J. Am. Chem. Soc.* **102**: 4636-43 (1980)

80A235 Grieser, F.; Thomas, J.K., *J. Chem. Phys.* **73**: 2115-9 (1980)

80A274 Saltiel, J.; Shannon, P.T.; Zafiriou, O.C.; Uriarte, A.K., *J. Am. Chem. Soc.* **102**: 6799-808 (1980)

80A338 Scaiano, J.C., *J. Am. Chem. Soc.* **102**: 7747-53 (1980)

80A369 Winter, G.; Steiner, U., *Ber. Bunsenges. Phys. Chem.* **84**: 1203-14 (1980)

80B017 Sinclair, R.S.; Tait, D.; Truscott, T.G., *J. Chem. Soc., Faraday Trans. 1* **76**: 417-25 (1980)

80B021 Toth, M., *Chem. Phys.* **46**: 437-43 (1980)

80B040 Scott, G.W.; Talley, L.D.; Anderson, R.W.,Jr., *J. Chem. Phys.* **72**: 5002-13 (1980)

80B055 Bonneau, R., *J. Am. Chem. Soc.* **102**: 3816-22 (1980)

80B057 Nishida, Y.; Kikuchi, K.; Kokubun, H., *J. Photochem.* **13**: 75-81 (1980)

80B077 Arce, R.; Jimenez, L.A.; Rivera, V.; Torres, C., *Photochem. Photobiol.* **32**: 91-5 (1980)

80B112 Ronfard-Haret, J.-C.; Bensasson, R.V.; Amouyal, E., *J. Chem. Soc., Faraday Trans. 1* **76**: 2432-6 (1980)

80B135 Bennett, J.A.; Birge, R.R., *J. Chem. Phys.* **73**: 4234-46 (1980)

80C005 Scaiano, J.C.; Neta, P., *J. Am. Chem. Soc.* **102**: 1608-11 (1980)

80D063 Van Strien, A.J.; Schmidt, J., *Chem. Phys. Lett.* **70**: 513-7 (1980)

80D074 Shinohara, H.; Hirota, N., *J. Chem. Phys.* **72**: 4445-57 (1980)

80D098 Higuchi, J.; Yagi, M.; Iwaki, T.; Bunden, M.; Tanigaki, K.; Ito, T., *Bull. Chem. Soc. Jpn.* **53**: 890-5 (1980)

80D102 Schaaf, R.; Perkampus, H.-H., *Chem. Phys. Lett.* **71**: 467-70 (1980)

80D111 Yagi, M.; Higuchi, J., *Chem. Phys. Lett.* **72**: 135-8 (1980)

80D122 Braeuchle, C.; Kabza, H.; Voitlaender, J., *Chem. Phys.* **48**: 369-85 (1980)

80D124 Grebel, V.; Levanon, H., *Chem. Phys. Lett.* **72**: 218-24 (1980)

80D131 Frank, H.A.; Bolt, J.D.; Costa, S.M. de B.; Sauer, K., *J. Am. Chem. Soc.* **102**: 4893-8 (1980)

80D196 van der Velden, G.P.M.; de Boer, E.; Veeman, W.S., *J. Phys. Chem.* **84**: 2634-41 (1980)

80E014 Szabo, A.G.; Rayner, D.M., *J. Am. Chem. Soc.* **102**: 554-63 (1980)

80E025 Becker, R.S; Kogan, G., *Photochem. Photobiol.* **31**: 5-13 (1980)

80E109 Saucin, M.; Van de Vorst, A., *Radiat. Environ. Biophys.* **17**: 159-68 (1980)

80E113 Saltiel, J.; Khalil, G.-E.; Schanze, K., *Chem. Phys. Lett.* **70**: 233-5 (1980)

80E116 Watkins, A.R., *Chem. Phys. Lett.* **70**: 262-5 (1980)

80E137 Takemura, T.; Chihara, K.; Becker, R.S.; Das, P.K.; Hug, G.L., *J. Am. Chem. Soc.* **102**: 2604-9 (1980)

80E163 Bulska, H.; Chodkowska, A., *J. Am. Chem. Soc.* **102**: 3259-61 (1980)

80E169 Bartocci, G.; Mazzucato, U.; Masetti, F.; Galiazzo, G., *J. Phys. Chem.* **84**: 847-51 (1980)

80E200 Bonnett, R.; Charalambides, A.A.; Land, E.J.; Sinclair, R.S.; Tait, D.; Truscott, T.G., *J. Chem. Soc., Faraday Trans. 1* **76**: 852-9 (1980)

80E223 Leigh, W.J.; Arnold, D.R., *J. Chem. Soc., Chem. Commun.* : 406-8 (1980)

80E233 Gupta, A.K.; Basu, S.; Rohatgi-Mukherjee, K.K., *Can. J. Chem.* **58**: 1046-50 (1980)

80E253 Perichet, G.; Chapelon, R.; Pouyet, B., *J. Photochem.* **13**: 67-74 (1980)

80E262 Visser, R.J.; Varma, C.A.G.O., *J. Chem. Soc., Faraday Trans. 2* **76**: 453-71 (1980)

80E296 Jones, G.,II; Jackson, W.R.; Halpern, A.M., *Chem. Phys. Lett.* **72**: 391-5 (1980)

80E429 Davidson, R.S.; Bonneau, R.; Joussot-Dubien, J.; Trethewey, K.R., *Chem. Phys. Lett.* **74**: 318-20 (1980)

80E439 Takahashi, T.; Kikuchi, K.; Kokubun, H., *J. Photochem.* **14**: 67-76 (1980)

80E540 Harriman, A., *J. Chem. Soc., Faraday Trans. 1* **76**: 1978-85 (1980)

80E563 Wilson, T.; Halpern, A.M., *J. Am. Chem. Soc.* **102**: 7279-83 (1980)

80E593 Matthews, J.I.; Braslavsky, S.E.; Camilleri, P., *Photochem. Photobiol.* **32**: 733-8 (1980)

80E596 Siegmund, M.; Bendig, J., *Z. Naturforsch., Teil A* **35**: 1076-86 (1980)

80E627 Najbar, J.; Trzcinska, B.M.; Urbanek, Z.H.; Proniewicz, L.M., *Acta Phys. Pol., A* **A58**: 331-44 (1980)

80E641 Braeuchle, C.; Deeg, F.W.; Voitlaender, J., *Chem. Phys.* **53**: 373-81 (1980)

80E792 Val'kova, G.A.; Sakhno, T.V.; Shcherbo, S.N.; Shigorin, D.N.; Andrievskii, A.M.; Poplavskii, A.N.; Dyumaev, K.M., *Russ. J. Phys. Chem.* **54**: 1382-3 (1980) Translated from: *Zh. Fiz. Khim.* **54**: 2416-8 (1980)

80F190 Wagner, P.J.; Lam, H.M.H., *J. Am. Chem. Soc.* **102**: 4167-72 (1980)

80F269 Turro, N.J.; Shima, K.; Chung, C.-J.; Tanielian, C.; Kanfer, S., *Tetrahedron Lett.* **21**: 2775-8 (1980)

80F373 Kalyanasundaram, K.; Dung, D., *J. Phys. Chem.* **84**: 2251-6 (1980)

80F439 Turro, N.J.; Tanimoto, Y., *J. Photochem.* **14**: 199-203 (1980)

80F701 Losev, A.P.; Zen'kevich, E.I.; Sagun, E.I., *Bull. Acad. Sci. USSR, Phys. Ser.* **44**: 84-8 (1980) Translated from: *Izv. Akad. Nauk SSSR, Ser. Fiz.* **44**: 783-8 (1980)

80N087 Pileni, M.-P.; Graetzel, M., *J. Phys. Chem.* **84**: 1822-5 (1980)

80Z097 Lee, E.K.C.; Loper, G.L., In *Radiationless Transitions*, S.H. Lin (ed.), Academic Press, New York, 1980, p.1-80

80Z182 Beck, R.; Englisch, W.; Guers, K., *Table of Laser Lines in Gases and Vapors.* Springer-Verlag, New York, 3rd ed., 1980, 247p. (Springer Series in Optical Sciences, Vol. 2)

81A016 Griller, D.; Howard, J.A.; Marriott, P.R.; Scaiano, J.C., *J. Am. Chem. Soc.* **103**: 619-23 (1981)

81A070 Davidson, R.S.; Bonneau, R.; Fornier de Violet, P.; Joussot-Dubien, J., *Chem. Phys. Lett.* **78**: 475-8 (1981)

81A078 Das, P.K., *Tetrahedron Lett.* **22**: 1307-10 (1981)

81A087 Kamat, P.V.; Lichtin, N.N., *J. Phys. Chem.* **85**: 814-8 (1981)

81A114 Das, P.K.; Bhattacharyya, S.N., *J. Phys. Chem.* **85**: 1391-5 (1981)

81A140 Kemp, T.J.; Martins, L.J.A., *J. Chem. Soc., Faraday Trans. 1* **77**: 1425-35 (1981)

81A170 Zacharova, G.V.; Lifanov, Yu.I.; Chibisov, A.K., *High Energy Chem.* **15**: 56-60 (1981) Translated from: *Khim. Vys. Energ.* **15**: 68-72 (1981)

81A174 Das, P.K.; Encinas, M.V.; Scaiano, J.C., *J. Am. Chem. Soc.* **103**: 4154-62 (1981)

81A275 Encinas, M.V.; Scaiano, J.C., *J. Am. Chem. Soc.* **103**: 6393-7 (1981)

81A294 Amirzadeh, G.; Schnabel, W., *Makromol. Chem.* **182**: 2821-35 (1981)

81B008 Wilbrandt, R.; Jensen, N.-H., *J. Am. Chem. Soc.* **103**: 1036-41 (1981)

81B064 Capellos, C., *J. Photochem.* **17**: 213-25 (1981)

81B115 Poletti, A.; Murgia, S.M.; Cannistraro, S., *Photobiochem. Photobiophys.* **2**: 167-72 (1981)

81C032 Leigh, W.J.; Arnold, D.R.; Baines, K.M., *Tetrahedron Lett.* **22**: 909-12 (1981)

81D022 Taherian, M.-R.; Maki, A.H., *Chem. Phys.* **55**: 85-96 (1981)

81D168 Schmidt, H.; Roedder, H.D.; Dietzel, U., *Photogr. Sci. Eng.* **25**: 21-8 (1981)

81E012 Monti, S.; Gardini, E.; Bortolus, P.; Amouyal, E., *Chem. Phys. Lett.* **77**: 115-9 (1981)

81E041 Traber, R.; Vogelmann, E.; Schreiner, S.; Werner, T.; Kramer, H.E.A., *Photochem. Photobiol.* **33**: 41-8 (1981)

81E042 Becker, R.S.; Bensasson, R.V.; Salet, C., *Photochem. Photobiol.* **33**: 115-6 (1981)

81E082 Klein, R.; Tatischeff, I.; Bazin, M.; Santus, R., *J. Phys. Chem.* **85**: 670-7 (1981)

81E084 Nahor, G.S.; Rabani, J.; Grieser, F., *J. Phys. Chem.* **85**: 697-702 (1981)

81E099 van der Velden, G.P.M.; de Boer, E.; Veeman, W.S., *Chem. Phys.* **56**: 181-8 (1981)

81E147 Kasama, K.; Kikuchi, K.; Yamamoto, S.; Uji-ie, K.; Nishida, Y.; Kokubun, H., *J. Phys. Chem.* **85**: 1291-6 (1981)

81E151 Chahidi, C.; Aubailly, M.; Monzikoff, A.; Bazin, M.; Santus, R., *Photochem. Photobiol.* **33**: 641-9 (1981)

81E183 Hirayama, S., *J. Am. Chem. Soc.* **103**: 2934-8 (1981)

81E192 Palewska, K., *Chem. Phys.* **58**: 21-8 (1981)

81E214 Goerner, H.; Schulte-Frohlinde, D., *J. Phys. Chem.* **85**: 1835-41 (1981)

81E261 Jain, K.M.; Misra, T.N., *Spectrosc. Lett.* **14**: 157-62 (1981)

81E270 Gorman, A.A.; Gould, I.R.; Hamblett, I., *J. Am. Chem. Soc.* **103**: 4553-8 (1981)

81E271 Harriman, A., *J. Chem. Soc., Faraday Trans. 2* **77**: 1281-91 (1981)

81E297 Motten, A.G.; Kwiram, A.L., *J. Chem. Phys.* **75**: 2608-15 (1981)

81E346 Herkstroeter, W.G.; Merkel, P.B., *J. Photochem.* **16**: 331-41 (1981)

81E374 Wilkinson, F.; Tsiamis, C., *J. Chem. Soc., Faraday Trans. 2* **77**: 1681-93 (1981)

81E375 Harriman, A.; Hosie, R.J., *J. Chem. Soc., Faraday Trans. 2* **77**:1695-702 (1981)

81E433 Mordzinski, A.; Grabowska, A., *J. Lumin.* **23**: 393-404 (1981)

81E442 Kossanyi, J.; Sabbah, S.; Chaquin, P.; Ronfart-Haret, J.C., *Tetrahedron* **37**: 3307-15 (1981)

81E444 Bendig, J.; Henkel, B.; Kreysig, D., *Ber. Bunsenges. Phys. Chem.* **85**: 38-44 (1981)

81E457 Jacques, P.; Braun, A.M., *Helv. Chim. Acta* **64**: 1800-6 (1981)

81E545 Lewanowicz, A.; Lipinski, J.; Ruziewicz, Z., *J. Lumin.* **26**: 159-75 (1981)

81E552 Kasama, K.; Kikuchi, K.; Nishida, Y.; Kokubun, H., *J. Phys. Chem.* **85**: 4148-53 (1981)

81E561 Ghoshal, S.K.; Sarkar, S.K.; Kastha, G.S., *Bull. Chem. Soc. Jpn.* **54**: 3556-61 (1981)

81E594 Kuboyama, A.; Matsuzaki, S.Y., *Bull. Chem. Soc. Jpn.* **54**: 3635-8 (1981)

81E644 Sveshnikova, E.B.; Kondakova, V.P., *Opt. Spectrosc. (USSR)* **50**: 477-9 (1981) Translated from: *Opt. Spektrosk.* **50**: 870-4 (1981)

81E648 Davydov, S.N.; Rodionov, A.N.; Shigorin, D.N.; Syutkina, O.P.; Krasnova, T.L., *Russ. J. Phys. Chem.* **55**: 444-5 (1981) Translated from: *Zh. Fiz. Khim.* **55**: 784-7 (1981)

81E716 Darmanyan, A.P.; Kuz'min, V.A., *Dokl. Phys. Chem.* **260**: 938-41 (1981) Translated from: *Dokl. Akad. Nauk SSSR* **260**: 1167-70 (1981)

81E790 Jordan, A.D.; Fischer, G.; Ross, I.G., *J. Mol. Spectrosc.* **87**: 345-56 (1981)

81F049 Langford, C.H.; Holubov, C.A., *Inorg. Chim. Acta* **53**: L59-L60 (1981)

81F050 Heller, H.G.; Langan, J.R., *J. Chem. Soc., Perkin Trans. 2* : 341-3 (1981)

81F053 Schuchmann, H.-P.; von Sonntag, C., *J. Photochem.* **15**: 159-62 (1981)

81F070 Davidson, R.S.; Goodwin, D.; Fornier de Violet, Ph., *Chem. Phys. Lett.* **78**: 471-4 (1981)

81F080 Gauglitz, G.; Hubig, S., *J. Photochem.* **15**: 255-7 (1981)

81F121 Asano, M.; Koningstein, J.A., *Chem. Phys.* **57**: 1-10 (1981)

81F130 Harriman, A.; Mills, A., *Photochem. Photobiol.* **33**: 619-25 (1981)

81F215 Das, P.K.; Bobrowski, K., *J. Chem. Soc., Faraday Trans. 2* **77**: 1009-27 (1981)

81F218 Mahaney, M.; Huber, J.R., *J. Mol. Spectrosc.* **87**: 438-48 (1981)

81F275 Goerner, H.; Schulte-Frohlinde, D., *J. Photochem.* **16**: 169-77 (1981)

81F343 Demas, J.N.; Bowman, W.D.; Zalewski, E.F.; Velapoldi, R.A., *J. Phys. Chem.* **85**: 2766-71 (1981)

81F364 Stevens, B.; Marsh, K.L.; Barltrop, J.A., *J. Phys. Chem.* **85**: 3079-82 (1981)

81F390 Demuth, M.; Amrein, W.; Bender, C.O.; Braslavsky, S.E.; Burger, U.; George, M.V.; Lemmer, D.; Schaffner, K., *Tetrahedron* **37**: 3245-61 (1981)

81F452 Wagner, P.J.; Siebert, E.J., *J. Am. Chem. Soc.* **103**: 7329-35 (1981)

81F509 Anderson, R.W.; Knox, W., *J. Lumin.* **24-25**: 647-50 (1981)

81Y136 Strek, W.; Wierzchaczewski, M., *Chem. Phys.* **58**: 185-93 (1981)

81Y336 Strek, W.; Wierzchaczewski, M., *Acta Phys. Pol. A* **60**: 857-65 (1981)

81Z334 Battino, R., *Solubility Data Series: Volume 7, Oxygen and Ozone.* Pergamon, Oxford, England, 1981, 519p.

81Z335 Kamlet, M.J.; Abboud, J.L.M.; Taft, R.W., In *Progress in Physical Organic Chemistry*, R.W. Taft (ed.), Wiley and Sons, New York, 1981, Vol. 13, p.485-630

82A082 Shizuka, H.; Obuchi, H., *J. Phys. Chem.* **86**: 1297-302 (1982)

82A153 Martins, L.J.A.; Kemp, T.J., *J. Chem. Soc., Faraday Trans. 1* **78**: 519-31 (1982)

82A154 Martins, L.J.A., *J. Chem. Soc., Faraday Trans. 1* **78**: 533-43 (1982)

82A205 Lo, K.K.N.; Land, E.J.; Truscott, T.G., *Photochem. Photobiol.* **36**: 139-45 (1982)

82A259 Teply, J.; Mehnert, R.; Brede, O.; Fojtik, A., *Radiochem. Radioanal. Lett.* **53**: 141-51 (1982)

82A288 Das, P.K.; Hug, G.L., *Photochem. Photobiol.* **36**: 455-61 (1982)

82A292 Kobashi, H.; Ikawa, H.; Kondo, R.; Morita, T., *Bull. Chem. Soc. Jpn.* **55**: 3013-8 (1982)

82A370 Becker, H.G.O.; Jirkovsky, J.; Fojtik, A.; Kleinschmidt, J., *J. Prakt. Chem.* **324**: 505-11 (1982)

82B102 Scaiano, J.C.; Lee, C.W.B.; Chow, Y.L.; Buono-Core, G.E., *J. Photochem.* **20**: 327-34 (1982)

82B121 O'Dowd, R.F.; O'Hare, A.; Cooke, J.; Taaffe, J.K., *J. Phys. E* **15**: 736-40 (1982)

82B140 American Petroleum Institute Project-44, *Selected Ultraviolet Spectral Data.* Thermodyanamics Research Center Hydrocarbon Project, College Station, TX, 1945-1982, Vol. I-IV.

82D180 Murai, H.; Imamura, T.; Obi, K., *J. Phys. Chem.* **86**: 3279-81 (1982)

82D324 Stoesser, R.; Thurner, J.-U.; Hanke, T.; Sarodnick, G., *J. Prakt. Chem.* **324**: 761-8 (1982)

82E042 Hirano, H.; Azumi, T., *Chem. Phys. Lett.* **86**: 109-12 (1982)

82E051 Deeg, F.W.; Braeuchle, Chr.; Voitlaender, J., *Chem. Phys.* **64**: 427-36 (1982)

82E055 Yagi, M.; Matsunaga, M.; Higuchi, J., *Chem. Phys. Lett.* **86**: 219-22 (1982)

82E059 Guesten, H.; Heinrich, G., *J. Photochem.* **18**: 9-17 (1982)

82E060 Koehler, G.; Kittel, G.; Getoff, N., *J. Photochem.* **18**: 19-27 (1982)

82E067 Wong, P.C., *Can. J. Chem.* **60**: 339-41 (1982)

82E072 Darmanyan, A.P., *Chem. Phys. Lett.* **86**: 405-10 (1982)

82E086 Velsko, S.P.; Fleming, G.R., *Chem. Phys.* **65**: 59-70 (1982)

82E129 Gurinovich, G.P.; Zenkevich, E.I.; Sagun, E.I., *J. Lumin.* **26**: 297-317 (1982)

82E133 Ronfard-Haret, J.C.; Averbeck, D.; Bensasson, R.V.; Bisagni, E.; Land, E.J., *Photochem. Photobiol.* **35**: 479-89 (1982)

82E181 Bonneau, R., *J. Am. Chem. Soc.* **104**: 2921-3 (1982)

82E203 Boldridge, D.W.; Scott, G.W.; Spiglanin, T.A., *J. Phys. Chem.* **86**: 1976-9 (1982)

82E204 Goerner, H., *J. Phys. Chem.* **86**: 2028-35 (1982)

82E207 Arce, R.; Rivera, M., *Photochem. Photobiol.* **35**: 737-40 (1982)

82E214 Taherian, M.-R.; Maki, A.H., *Chem. Phys.* **68**: 179-89 (1982)

82E232 Kamat, P.V.; Lichtin, N.N., *J. Photochem.* **18**: 197-209 (1982)

82E257 Cundall, R.B.; Grant, D.J.W.; Shulman, N.H., *J. Chem. Soc., Faraday Trans. 2* **78**: 737-50 (1982)

82E258 Smith, G.J., *J. Chem. Soc., Faraday Trans. 2* **78**: 769-73 (1982)

82E271 Specht, D.P.; Martic, P.A.; Farid, S., *Tetrahedron* **38**: 1203-11 (1982)

82E303 Huppert, D.; Rand, S.D.; Reynolds, A.H.; Rentzepis, P.M., *J. Chem. Phys.* **77**: 1214-24 (1982)

82E338 Hirayama, S., *J. Chem. Soc., Faraday Trans. 1* **78**: 2411-21 (1982)

82E341 Wolleben, J.; Testa, A.C., *J. Photochem.* **19**: 267-9 (1982)

82E343 Inoue, H.; Hida, M.; Nakashima, N.; Yoshihara, K., *J. Phys. Chem.* **86**: 3184-8 (1982)

82E344 Shizuka, H.; Ueki, Y.; Iizuka, T.; Kanamaru, N., *J. Phys. Chem.* **86**: 3327-33 (1982)

82E355 Suga, K.; Kinoshita, M., *Bull. Chem. Soc. Jpn.* **55**: 1695-704 (1982)

82E365 Chattopadhyay, S.K.; Das, P.K.; Hug, G.L., *J. Am. Chem. Soc.* **104**: 4507-14 (1982)

82E367 Kokai, F.; Azumi, T., *J. Chem. Phys.* **77**: 2757-62 (1982)

82E429 Goerner, H., *J. Photochem.* **19**: 343-56 (1982)

82E456 Kamat, P.V.; Lichtin, N.N., *Isr. J. Chem.* **22**: 113-6 (1982)

82E474 Yokoyama, K., *Chem. Phys. Lett.* **92**: 93-6 (1982)

82E484 Caldwell, R.A.; Singh, M., *J. Am. Chem. Soc.* **104**: 6121-2 (1982)

82E516 Clarke, R.H.; Mitra, P.; Vinodgopal, K., *J. Chem. Phys.* **77**: 5288-97 (1982)

82E526 Caldwell, R.A.; Cao, C.V., *J. Am. Chem. Soc.* **104**: 6174-80 (1982)

82E585 Yamauchi, S.; Ueno, T.; Hirota, N., *Mol. Phys.* **47**: 1333-48 (1982)

82E586 Tetreau, C.; Lavalette, D.; Cabaret, D.; Geraghty, N.; Welvart, Z., *Nouv. J. Chim.* **6**: 461-5 (1982)

82E622 Kalyanasundaram, K.; Neumann-Spallart, M., *J. Phys. Chem.* **86**: 5163-9 (1982)

82E624 Tway, P.C.; Love, L.J.C., *J. Phys. Chem.* **86**: 5223-6 (1982)

82E632 Bolotko, L.M.; Gruzinskii, V.V.; Danilova, V.I.; Kopylova, T.N., *Opt. Spectrosc. (USSR)* **52**: 379-81 (1982) Translated from: *Opt. Spektrosk.* **52**: 635-8 (1982)

82E660 Lessing, H.E.; Richardt, D.; von Jena, A., *J. Mol. Struct.* **84**: 281-92 (1982)

82E663 Zander, M., *Z. Naturforsch., Teil A* **37A**: 1348-52 (1982)

82E680 Ortmann, W.; Kassem, A.; Hinzmann, S.; Fanghaenel, E., *J. Prakt. Chem.* **324**: 1017-25 (1982)

82E832 Verheijdt, P.L.; Cerfontain, H., *J. Chem. Soc., Perkin Trans 2* : 1541-7 (1982)

82F056 Velsko, S.P.; Fleming, G.R., *J. Chem. Phys.* **76**: 3553-62 (1982)

82F161 Venediktov, E.A.; Krasnovsky, A.A.,Jr., *Zh. Prikl. Spektrosk.* **36**: 152-4 (1982)

82F225 Maharaj, U.; Winnik, M. A., *Tetrahedron Lett.* **23**: 3035-8 (1982)

82F252 Bartocci, G.; Mazzucato, U., *J. Lumin.* **27**: 163-75 (1982)

82F433 Simpson, J.T.; Krantz, A.; Lewis, F.D.; Kokel, B., *J. Am. Chem. Soc.* **104**: 7155-61 (1982)

82F476 Brauer, H.-D.; Drews, W.; Schmidt, R.; Gauglitz, G.; Hubig, S., *J. Photochem.* **20**: 335-40 (1982)

82N068 Graetzel, C.K.; Graetzel, M., *J. Phys. Chem.* **86**: 2710-4 (1982)

82S163 Dressick, W.J.; Meyer, T.J.; Durham, B., *Isr. J. Chem.* **22**: 153-7 (1982)

82Z015 Mirbach, M.F.; Mirbach, M.J.; Saus, A., *Chem. Rev.* **82**: 59-76 (1982)

82Z025 Maki, A.H., In *Triplet State ODMR Spectroscopy: Techniques and Applications to Biophysical Systems*, R.H. Clarke (ed.), Wiley, New York, 1982, p.479-557

82Z053 Darwent, J.R.; Douglas, P.; Harriman, A.; Porter, G.; Richoux, M.-C., *Coord. Chem. Rev.* **44**: 83-126 (1982)

82Z102 Mazzucato, U., *Pure Appl. Chem.* **54**: 1705-21 (1982)

82Z269 Rabek, J.F., *Experimental Methods in Photochemistry and Photophysics, Pt.2.* Wiley, Chichester, UK, 1982, 506p.

82Z365 Levin, R.D.; Lias, S.G., NSRDS-NBS 71, 1982, 634p. (National Bureau of Standards, Washington, DC)

83A007 Johansen, O.; Mau, A.W.-H.; Sasse, W.H.F., *Chem. Phys. Lett.* **94**: 107-12 (1983)

83A110 Stewart, L.C.; Carlsson, D.J.; Wiles, D.M.; Scaiano, J.C., *J. Am. Chem. Soc.* **105**: 3605-9 (1983)

83A113 Das, P.K., *J. Chem. Soc., Faraday Trans. 1* **79**: 1135-45 (1983)

83A213 Pepmiller, C.; Bedwell, E.; Kuntz, R.R.; Ghiron, C.A., *Photochem. Photobiol.* **38**: 273-80 (1983)

83B054 Loeff, I.; Treinin, A.; Linschitz, H., *J. Phys. Chem.* **87**: 2536-44 (1983)

83B067 Hurley, J.K.; Sinai, N.; Linschitz, H., *Photochem. Photobiol.* **38**: 9-14 (1983)

83B068 Craw, M.; Lambert, C., *Photochem. Photobiol.* **38**: 241-3 (1983)

83B121 Cogdell, R.J.; Land, E.J.; Truscott, T.G., *Photochem. Photobiol.* **38**: 723-5 (1983)

83C028 Swaminathan, M.; Dogra, S.K., *Spectrochim. Acta, Part A* **39A**: 973-7 (1984)

83D200 Hilburn, S.G.; Power, R.K.; Martin, K.A.; Nishimura, A.M., *Chem. Phys. Lett.* **100**: 429-35 (1983)

83D207 Baiardo, J.; Vala, M.; Trabjerg, I., *Chem. Phys.* **80**: 305-15 (1983)

83D218 Noda, M.; Hirota, N., *J. Am. Chem. Soc.* **105**: 6790-4 (1983)

83E018 Maciejewski, A., *Chem. Phys. Lett.* **94**: 344-9 (1983)

83E026 Das, P.K.; Hug, G.L., *J. Phys. Chem.* **87**: 49-54 (1983)

83E027 Hamai, S.; Hirayama, F., *J. Phys. Chem.* **87**: 83-9 (1983)

83E031 Lee, J.; Li, F.; Bernstein, E.R., *J. Phys. Chem.* **87**: 260-5 (1983)

83E054 Wilkinson, F.; Tsiamis, C., *J. Am. Chem. Soc.* **105**: 767-74 (1983)

83E064 Murray, D.; Becker, R.S., *J. Phys. Chem.* **87**: 625-8 (1983)

83E087 Williams, J.O.; Jones, A.C.; Davies, M.J., *J. Chem. Soc., Faraday Trans. 2* **79**: 263-9 (1983)

83E156 Leismann, H.; Scharf, H.-D.; Strassburger, W.; Wollmer, A., *J. Photochem.* **21**: 275-80 (1983)

83E169 Taherian, M.R.; Maki, A.H., *Chem. Phys. Lett.* **96**: 541-6 (1983)

83E176 Brenner, K.; Lipinski, J.; Ruziewicz, Z., *J. Lumin.* **28**: 13-26 (1983)

83E180 Basara, H., *J. Lumin.* **28**: 73-86 (1983)

83E230 Safarzadeh-Amiri, A.; Verrall, R.E.; Steer, R.P., *Can. J. Chem.* **61**: 894-900 (1983)

83E258 Gorman, A.A.; Hamblett, I., *Chem. Phys. Lett* **97**: 422-6 (1983)

83E281 Chattopadhyay, S.K.; Kumar, Ch.V.; Das, P.K., *Chem. Phys. Lett.* **98**: 250-4 (1983)

83E311 Fisher, G.J.; Land, E.J., *Photochem. Photobiol.* **37**: 27-32 (1983)

83E324 Craw, M.; Bensasson, R.V.; Ronfard-Haret, J.C.; Sa E Melo, M.T.; Truscott, T.G., *Photochem. Photobiol.* **37**: 611-5 (1983)

83E387 Kumar, C.V.; Chattopadhyay, S.K.; Das, P.K., *Photochem. Photobiol.* **38**: 141-52 (1983)

83E417 Najbar, J.; Jarzeba, W.; Urbanek, Z.H., *Chem. Phys.* **79**: 245-53 (1983)

83E440 Wermuth, G., *Z. Naturforsch., Teil A* **38A**: 368-77 (1983)

83E462 Kalyanasundaram, K., *J. Chem. Soc., Faraday Trans. 2* **79**: 1365-74 (1983)

83E489 Savory, B.; Turnbull, J.H., *J. Photochem.* **23**: 171-81 (1983)

83E527 Meech, S.R.; Phillips, D., *J. Photochem.* **23**: 193-217 (1983)

83E543 Meech, S.R.; O'Connor, D.V.; Phillips, D.; Lee, A.G., *J. Chem. Soc., Faraday Trans. 2* **79**: 1563-84 (1983)

83E597 von Borczyskowski, C.; Fallmer, E., *Chem. Phys. Lett.* **102**: 433-7 (1983)

83E625 Salet, C.; Bensasson, R.V.; Favre, A., *Photochem. Photobiol.* **38**: 521-5 (1983)

83E630 Sarkar, S.K.; Ghoshal, S.K.; Kastha, G.S., *Proc. - Indian Acad. Sci., Chem. Sci.* **92**: 47-58 (1983)

83E638 Slama, H.; Braeuchle, C.; Voitlaender, J., *Chem. Phys. Lett.* **102**: 307-11 (1983)

83E835 Petrushenko, K.B.; Vokin, A.I.; Turchaninov, V.K.; Frolov, Yu.L., *Bull. Acad. Sci. USSR, Div. Chem. Sci.* **32**: 2151-2 (1983) Translated from: *Izv. Akad. Nauk SSSR, Ser. Khim.* **32**: 2387-8 (1983)

83E890 Visser, R.J.; Varma, C.A.G.O.; Konijnenberg, J.; Weisenborn, P.C.M., *J. Mol. Struct.* **114**: 105-12 (1984)

83E901 Sarphatie, L.A.; Verheijdt, P.L.; Cerfontain, H., *Recl. Trav. Chim. Pays-Bas* **102**: 9-13 (1983)

83F055 Ito, Y.; Nishimura, H.; Umehara, Y.; Yamada, Y.; Tone, M.; Matsuura, T., *J. Am. Chem. Soc.* **105**: 1590-7 (1983)

83F075 Darmanyan, A.P., *Chem. Phys. Lett.* **96**: 383-9 (1983)

83F079 Nicodem, D.E.; Aquilera, O.M.V., *J. Photochem.* **21**: 189-93 (1983)

83F123 Chow, Y.L.; Buono-Core, G.E.; Marciniak, B.; Beddard, C., *Can. J. Chem.* **61**: 801-8 (1983)

83F178 Bunce, N.J.; Hayes, P.J.; Lemke, M.E., *Can. J. Chem.* **61**: 1103-4 (1983)

83F182 Harriman, A.; Porter, G.; Wilowska, A., *J. Chem. Soc., Faraday Trans. 2* **79**: 807-16 (1983)

83F206 Brauer, H.-D.; Schmidt, R.; Gauglitz,

G.; Hubig, S., *Photochem. Photobiol.* **37**: 595-8 (1983)

83F240 Brauer, H.-D.; Schmidt, R., *Photochem. Photobiol.* **37**: 587-91 (1983)

83F297 Nishimoto, S.; Izukawa, T.; Kagiya, T., *J. Chem. Soc., Perkin Trans. 2* : 1147-52 (1983)

83F476 Jardon, P.; Azarnouche, B.; Corval, A.; Gautron, R., *J. Chim. Phys. Phys.-Chim. Biol.* **80**: 603-8 (1983)

83F502 Kirk, A.D.; Namasivayam, C., *Anal. Chem.* **55**: 2428-9 (1983)

83Z077 Korobov, V.E.; Chibisov, A.K., *Russ. Chem. Rev.* **52**: 27-42 (1983) Translated from: *Usp. Khim.* **52**: 43-71 (1983)

84A201 Scaiano, J.C.; Stewart, L.C.; Livant, P.; Majors, A.W., *Can. J. Chem.* **62**: 1339-43 (1984)

84A221 Kumar, C.V.; Qin, L.; Das, P.K., *J. Chem. Soc., Faraday Trans. 2* **80**: 783-93 (1984)

84A286 Wilkinson, F.; Farmilo, A., *J. Chem. Soc., Faraday Trans. 2* **80**: 1117-24 (1984)

84A344 Das, P.K.; Griffin, G.W., *J. Org. Chem.* **49**: 3452-7 (1984)

84A363 Beecroft, R.A.; Davidson, R.S.; Goodwin, D.; Pratt, J.E., *Tetrahedron* **40**: 4487-96 (1984)

84A458 Mehnert, R.; Brede, O.; Cserep, G., *Radiat. Phys. Chem.* **24**: 455-7 (1984)

84B007 Lazare, S.; Bonneau, R.; Lapouyade, R., *J. Phys. Chem.* **88**: 18-23 (1984)

84B033 Baral-Tosh, S.; Chattopadhyay, S.K.; Das, P.K., *J. Phys. Chem.* **88**: 1404-8 (1984)

84B051 Naito, I.; Schnabel, W., *Bull. Chem. Soc. Jpn.* **57**: 771-5 (1984)

84B061 Nakamura, S.; Kanamaru, N.; Nohara, S.; Nakamura, H.; Saito, Y.; Tanaka, J.; Sumitani, M.; Nakashima, N.; Yoshihara, K., *Bull. Chem. Soc. Jpn.* **57**: 145-50 (1984)

84B066 Das, P.K.; Muller, A.J.; Griffin, G.W., *J. Org. Chem.* **49**: 1977-85 (1984)

84B110 Hamanoue, K.; Tai, S.; Hidaka, T.; Nakayama, T.; Kimoto, M.; Teranishi, H., *J. Phys. Chem.* **88**: 4380-4 (1984)

84D056 Murai, H.; Hayashi, T.; I'Haya, Y.J., *Chem. Phys. Lett.* **106**: 139-42 (1984)

84D190 Chandrashekar, T.K.; van Willigen, H.; Ebersole, M.H., *J. Phys. Chem.* **88**: 4326-32 (1984)

84D226 Noda, M.; Nagaoka, S.; Hirota, N., *Bull. Chem. Soc. Jpn.* **57**: 2376-82 (1984)

84E018 Wismontski-Knittel, T.; Kilp, T., *J. Phys. Chem.* **88**: 110-5 (1984)

84E036 Bensasson, R.V.; Land, E.J.; Liu, R.S.H.; Lo, K.K.N.; Truscott, T.G., *Photochem. Photobiol.* **39**: 263-5 (1984)

84E054 Iwasaki, N.; Misra, T.N.; Kinoshita, M., *J. Lumin.* **29**: 83-92 (1984)

84E082 Shizuka, H.; Sato, Y.; Ueki, Y.; Ishikawa, M.; Kumada, M., *J. Chem. Soc., Faraday Trans. 1* **80**: 341-57 (1984)

84E090 Sarkar, S.K.; Maiti, A.; Kastha, G.S., *Chem. Phys. Lett.* **105**: 355-8 (1984)

84E092 Urruti, E.H.; Kilp, T., *Macromolecules* **17**: 50-4 (1984)

84E102 Savory, B.; Turnbull, J.H., *J. Photochem.* **24**: 355-71 (1984)

84E107 Maeda, Y.; Okada, T.; Mataga, N.; Irie, M., *J. Phys. Chem.* **88**: 1117-9 (1984)

84E144 Kumar, C.V.; Chattopadhyay, S.K.; Das, P.K., *Chem. Phys. Lett.* **106**: 431-6 (1984)

84E180 Chattopadhyay, S.K.; Kumar, C.V.; Das, P.K., *J. Chem. Soc., Faraday Trans. 1* **80**: 1151-61 (1984)

84E203 Lee, W.A.; Graetzel, M.; Kalyanasundaram, K., *Chem. Phys. Lett.* **107**: 308-13 (1984)

84E208 Wilson, T.; Frye, S.L.; Halpern, A.M., *J. Am. Chem. Soc.* **106**: 3600-6 (1984)

84E216 Grajcar, L.; Ivanoff, N.; Delouis, J.F.; Faure, J., *J. Chim. Phys. Phys.-Chim. Biol.* **81**: 33-8 (1984)

84E236 Davis, H.F.; Chattopadhyay, S.K.; Das, P.K., *J. Phys. Chem.* **88**: 2798-803 (1984)

84E237 Wismontski-Knittel, T.; Das, P.K., *J. Phys. Chem.* **88**: 2803-8 (1984)

84E319 Chattopadhyay, S.K.; Kumar, C.V.; Das, P.K., *J. Photochem.* **26**: 39-47 (1984)

84E322 Diverdi, L.A.; Topp, M.R., *J. Phys. Chem.* **88**: 3447-51 (1984)

84E335 Eftink, M.R.; Ghiron, C.A., *Biochemistry* **23**: 3891-9 (1984)

84E342 Kumar, C.V.; Davis, H.F.; Das, P.K., *Chem. Phys. Lett.* **109**: 184-9 (1984)

84E346 Kalyanasundaram, K., *Inorg. Chem.* **23**: 2453-9 (1984)

84E390 Menzel, R.; Rapp, W., *Chem. Phys.* **89**: 445-55 (1984)

84E393 Darmanyan, A.P., *Chem. Phys. Lett.* **110**: 89-94 (1984)

84E405 Hashimoto, S.; Thomas, J.K., *J. Phys. Chem.* **88**: 4044-9 (1984)

84E477 Yamauchi, S.; Miyake, K.; Hirota, N., *Mol. Phys.* **53**: 479-91 (1984)

84E491 Gorman, A.A.; Hamblett, I.; Harrison, R.J., *J. Am. Chem. Soc.* **106**: 6952-5 (1984)

84E529 Previtali, C.M.; Ebbesen, T.W., *J. Photochem.* **27**: 9-15 (1984)

84E530 Yip, R.W.; Sharma, D.K.; Giasson, R.; Gravel, D., *J. Phys. Chem.* **88**: 5770-2 (1984)

84E533 Masuhara, H.; Shioyama, H.; Saito, T.; Hamada, K.; Yasoshima, S.; Mataga, N., *J. Phys. Chem.* **88**: 5868-73 (1984)

84E581 Biczok, L.; Berces, T.; Forgeteg, S.; Marta, F., *J. Photochem.* **27**: 41-8 (1984)

84E582 Laszlo, B.; Forgeteg, S.; Berces, T.; Marta, F., *J. Photochem.* **27**: 49-59 (1984)

84E583 Malkin, Y.N.; Dvornikov, A.S.; Kuz'min, V.A., *J. Photochem.* **27**: 343-54 (1984)

84E612 Zander, M., *Z. Naturforsch., Teil A* **39A**: 1145-6 (1984)

84E679 Ghoshal, S.K.; Maiti, A.K.; Kastha, G.S., *J. Lumin.* **31-32**: 541-5 (1984)

84E785 Bubekov, Yu.I.; Kabelka, V.; Lysak, N.A.; Milyauskas, A.; Tolstorozhev, G.B., *Bull. Acad. Sci. USSR, Phys. Ser.* **48**: 137-41 (1984) Translated from: *Izv. Akad. Nauk SSSR, Ser. Fiz.* **48**: 554-8 (1984)

84E794 Craw, M.; Truscott, T.G.; Dall'Acqua, F.; Guiotto, A.; Vedaldi, D.; Land, E.J., *Photobiochem. Photobiophys.* **7**: 359-65 (1984)

84E803 Usacheva, M.N.; Osipov, V.V.; Drozdenko, I.V.; Dilung, I.I., *Russ. J. Phys. Chem.* **58**: 1550-3 (1984) Translated from: *Zh. Fiz. Khim.* **58**: 2559-63 (1984)

84E843 Tine, A.; Aaron, J.-J., *Can. J. Spectrosc.* **29**: 121-30 (1984)

84E863 Paone, S.; Moule, D.C.; Bruno, A.E.; Steer, R.P., *J. Mol. Spectrosc.* **107**: 1-11 (1984)

84E864 Maiti, A.; Sarkar, S.K.; Kastha, G.S., *Proc. - Indian Acad. Sci., Chem. Sci.* **93**: 1-11 (1984)

84F005 Chattopadhyay, S.K.; Kumar, C.V.; Das, P.K., *J. Photochem.* **24**: 1-9 (1984)

84F039 Kikuchi, K.; Yamamoto, S.; Kokubun, H., *J. Photochem.* **24**: 271-83 (1984)

84F051 Scaiano, J.C.; Lissi, E.A.; Stewart, L.C., *J. Am. Chem. Soc.* **106**: 1539-42 (1984)

84F060 Shizuka, H.; Obuchi, H.; Ishikawa, M.; Kumada, M., *J. Chem. Soc., Faraday Trans. 1* **80**: 383-401 (1984)

84F074 Das, P.K.; Muller, A.J.; Griffin, G.W.; Gould, I.R.; Tung, C.-H.; Turro, N.J., *Photochem. Photobiol.* **39**: 281-5 (1984)

84F096 Bunce, N.J.; LaMarre, J.; Vaish, S.P., *Photochem. Photobiol.* **39**: 531-3 (1984)

84F123 Drabek, J.; Cepciansky, I.; Poskocil, J., *Chem. Listy* **78**: 94-8 (1984)

84F149 Adam, W.; Oppenlaender, T., *Photochem. Photobiol.* **39**: 719-23 (1984)

84F198 Schmidt, R.; Brauer, H.-D., *J. Photochem.* **25**: 489-99 (1984)

84F248 Malkin, Ya.N.; Pirogov, N.O.; Kuz'min, V.A., *J. Photochem.* **26**: 193-202 (1984)

84F268 Takuma, K.; Kirmura, T.; Sonoda, T.; Kobayashi, H., *Chem. Lett.* : 881-4 (1984)

84F284 Orr, U.; Traber, R.; Hemmerich, P.; Kramer, H.E.A., *Photochem. Photobiol.* **40**: 309-18 (1984)

84F375 Jones, G.,II; Bergmark, W.R.; Jackson, W.R., *Opt. Commun.* **50**: 320-4 (1984)

84F449 Pac, C.; Fukunaga, T.; Ohtsuki, T.; Sakurai, H., *Chem. Lett.* : 1847-50 (1984)

84F488 Goerner, H., *Ber. Bunsenges. Phys. Chem.* **88**: 1199-208 (1984)

84F625 Sundararajan, K.; Ramakrishnan, V.; Kuriacose, J.C., *Indian J. Chem., Sect. B* **23B**: 1068-70 (1984)

84Z133 Christodoulides, A.A.; McCorkle, D.L.; Christophorou, L.G., In *Electron-Molecule Interactions and Their Applications*, L.G. Christophorou (ed.), Academic Press, New York, 1984, Vol. 2, p.423-641

84Z150 Newton, M.D.; Sutin, N., *Ann. Rev. Phys. Chem.* **35**: 437-80 (1984)

84Z353 Bensasson, R.V., *NATO ASI Ser., Ser. A* **85**: 241-54 (1985)

85A009 Abdullah, K.A.; Kemp, T.J., *J. Photochem.* **28**: 61-9 (1985)

85A166 Mau, A.W.-H.; Johansen, O.; Sasse, W.H.F., *Photochem. Photobiol.* **41**: 503-9 (1985)

85A186 Yoshimura, A.; Kato, S., *Bull. Chem. Soc. Jpn.* **58**: 1556-9 (1985)

85A268 Chattopadhyay, S.K.; Kumar, C.V.; Das, P.K., *J. Photochem.* **30**: 81-91 (1985)

85A300 Bhattacharyya, K.; Kumar, C.V.; Das, P.K.; Jayasree, B.; Ramamurthy, V., *J. Chem. Soc., Faraday Trans. 2* **81**: 1383-93 (1985)

85A336 Petrushenko, K.B.; Vokin, A.I.; Turchaninov, V.K.; Gorshkov, A.G.; Frolov, Yu.L., *Bull. Acad. Sci. USSR, Div. Chem. Sci.* **34**: 242-6 (1985) Translated from: *Izv. Akad. Nauk SSSR, Ser. Khim.* : 267-71 (1985)

85A361 Lougnot, D.J.; Jacques, P.; Fouassier, J.P.; Casal, H.L.; Nguyen, K.-T.; Scaiano, J.C., *Can. J. Chem.* **63**: 3001-6 (1985)

85A374 Heelis, P.F.; De la Rosa, M.A.; Phillips, G.O., *Photobiochem. Photobiophys.* **9**: 57-63 (1985)

85A406 Shizuka, H.; Hagiwara, H.; Fukushima, M., *J. Am. Chem. Soc.* **107**: 7816-23 (1985)

85B074 Kanemoto, A.; Kikuchi, K.; Kokubun, H., *J. Phys. Chem.* **89**: 3567-70 (1985)

85B078 Kumar, C.V.; Ramaiah, D.; Das, P.K.; George, M.V., *J. Org. Chem.* **50**: 2818-25 (1985)

85C011 Grimsrud, E.P.; Caldwell, G.; Chowdhury, S.; Kebarle, P., *J. Am. Chem. Soc.* **107**: 4627-34 (1985)

85D003 Slama, H.; Braeuchle, C.; Voitlaender, J., *Chem. Phys.* **92**: 91-6 (1985)

85D016 Murai, H.; Minami, M.; Hayashi, T.; I'Haya, Y.J., *Chem. Phys.* **93**: 333-8 (1985)

85D181 Terazima, M.; Yamauchi, S.; Hirota, N., *J. Chem. Phys.* **83**: 3234-43 (1985)

85D207 Terazima, M.; Yamauchi, S.; Hirota, N., *Chem. Phys. Lett.* **120**: 321-6 (1985)

85E006 Visser, R.J.; Weisenborn, P.C.M.; Varma, C.A.G.O., *Chem. Phys. Lett.* **113**: 330-6 (1985)

85E025 Jones, G.,II; Jackson, W.R.; Choi, C.; Bergmark, W.R., *J. Phys. Chem.* **89**: 294-300 (1985)

85E054 Mordzinski, A.; Komorowski, S.J., *Chem. Phys. Lett.* **114**: 172-7 (1985)

85E096 Hirayama, S.; Lampert, R.A.; Phillips, D., *J. Chem. Soc., Faraday Trans. 2* **81**: 371-82 (1985)

85E190 Bhattacharyya, K.; Das, P.K., *Chem. Phys. Lett.* **116**: 326-32 (1985)

85E246 Visser, R.-J.; Weisenborn, P.C.M.; van Kan, P.J.M.; Huizer, B.H.; Varma, C.A.G.O.; Warman, J.M.; de Haas, M.P., *J. Chem. Soc., Faraday Trans. 2* **81**: 689-704 (1985)

85E272 Zander, M., *Z. Naturforsch., A, Phys., Phys. Chem., Kosmophys.* **40A**: 497-502 (1985)

85E281 Gorman, A.A.; Hamblett, I.; Irvine, M.; Raby, P.; Standen, M.C.; Yeates, S., *J. Am. Chem. Soc.* **107**: 4404-11 (1985)

85E293 Chattopadhyay, S.K.; Kumar, C.V.; Das, P.K., *Photochem. Photobiol.* **42**: 17-24 (1985)

85E351 Abdullah, K.A.; Kemp, T.J., *J. Chem. Soc., Perkin Trans 2* : 1279-83 (1985)

85E370 Suter, G.W.; Wild, U.P.; Brenner, K.; Ruziewicz, Z., *Chem. Phys.* **98**: 455-63 (1985)

85E384 Baba, M., *J. Chem. Phys.* **83**: 3318-26 (1985)

85E406 Alfassi, Z.B.; Previtali, C.M., *J. Photochem.* **30**: 127-32 (1985)

85E408 Komorowski, S.J.; Grabowski, Z.R.; Zielenkiewicz, W., *J. Photochem.* **30**: 141-51 (1985)

85E449 Scaiano, J.C.; Leigh, W.J.; Meador, M.A.; Wagner, P.J., *J. Am. Chem. Soc.* **107**: 5806-7 (1985)

85E452 Rohatgi-Mukherjee, K.K.; Bhattacharyya, K.; Das, P.K., *J. Chem. Soc., Faraday Trans. 2* **81**: 1331-44 (1985)

85E488 Previtali, C.M.; Ebbesen, T.W., *J. Photochem.* **30**: 259-67 (1985)

85E519 Jinguji, M.; Ashizawa, M.; Nakazawa, T.; Tobita, S.; Hikida, T.; Mori, Y., *Chem. Phys. Lett.* **121**: 400-4 (1985)

85E555 Jardon, P.; Gautron, R., *J. Chim. Phys. Phys.-Chim. Biol.* **82**: 353-60 (1985)

85E575 Ikeyama, T.; Azumi, T., *J. Phys. Chem.* **89**: 5332-3 (1985)

85E653 Previtali, C.M., *J. Photochem.* **31**: 233-8 (1985)

85E712 Zalesskaya, G.A.; Blinov, S.I., *Sov. Phys. Dokl.* **30**: 297-9 (1985) Translated from: *Dokl. Akad. Nauk SSSR* **281**: 1102-5 (1985)

85E766 Val'kova, G.A.; Sakhno, T.V.; Shigorin, D.N.; Davydov, S.N.; Andrievskii, A.N.; Eremenko, L.V., *Russ. J. Phys. Chem.* **59**: 1050-3 (1985) Translated from: *Zh. Fiz. Khim.* **59**: 1782-6 (1985)

85E800 Bryukhanov, V.V.; Levshin, L.V.; Smagulov, Zh.K.; Muldakhmetov, Z.M., *Opt. Spectrosc. (USSR)* **59**: 540-2 (1985) Translated from: *Opt. Spektrosk.* **59**: 896-9 (1985)

85E829 Maiti, A.K.; Sarkar, S.K.; Kastha, G.S., *Proc. - Indian Acad. Sci., Chem. Sci.* **95**: 409-19 (1985)

85F138 Baumann, H.; Becker, H.G.O.; Kronfeld, K.P.; Pfeifer, D.; Timpe, H.-J., *J. Photochem.* **28**: 393-403 (1985)

85F172 Tinnemans, A.H.A.; den Ouden, B.; Bos, H.J.T.; Mackor, A., *Recl., J.R. Neth. Chem. Soc.* **104**: 109-16 (1985)

85F276 Gauglitz, G.; Hubig, S., *J. Photochem.* **30**: 121-5 (1985)

85F371 Arnold, B.; Donald, L.; Jurgens, A.; Pincock, J.A., *Can. J. Chem.* **63**: 3140-6 (1985)

85F416 Zimmerman, H.E.; Caufield, C.E.; King, R.K., *J. Am. Chem. Soc.* **107**: 7732-44 (1985)

85F417 Zimmerman, H.E.; Lynch, D.C., *J. Am. Chem. Soc.* **107**: 7745-56 (1985)

85F488 Palmer, T.F.; Parmar, S.S., *J. Photochem.* **31**: 273-88 (1985)

85R139 van Willigen, H.; Das, U.; Ojadi, E.; Linschitz, H., *J. Am. Chem. Soc.* **107**: 7784-5 (1985)

85Z114 Ramamurthy, V., *Org. Photochem.* **7**: 231-338 (1985)

85Z456 Weast, R.C., *Handbook of Physics and Chemistry. Sixty Sixth Edition.* CRC Press, Inc., Boca Raton, FL, 1985

85Z457 Marcus, Y., *Ion Solvation*. Wiley and Sons, New York, 1985, 306p.

86A043 Goerner, H.; Schulte-Frohlinde, D., *Chem. Phys. Lett.* **124**: 321-5 (1986)

86A164 Scaiano, J.C.; Encinas, M.V.; Lissi, E.A.; Zanocco, A.; Das, P.K., *J. Photochem.* **33**: 229-36 (1986)

86A166 Wang, G.-C.; Winnik, M.A.; Schaefer, H.J.; Schmidt, W., *J. Photochem.* **33**: 291-6 (1986)

86A205 Hoshi, M.; Kikuchi, K.; Kokubun, H.; Yamamoto, S.-A., *J. Photochem.* **34**: 63-71 (1986)

86A240 Bhattacharyya, K.; Das, P.K.; Ramamurthy, V.; Rao, V.P., *J. Chem. Soc., Faraday Trans. 2* **82**: 135-47 (1986)

86A248 Bhattacharyya, K.; Das, P.K., *J. Phys. Chem.* **90**: 3987-93 (1986)

86A322 Herkstroeter, W.G.; Farid, S., *J. Photochem.* **35**: 71-85 (1986)

86A357 Evans, C.; Weir, D.; Scaiano, J.C.; Mac Eachern, A.; Arnason, J.T.; Morand, P.; Hollebone, B.; Leitch, L.C.; Philogene, B.J.R., *Photochem. Photobiol.* **44**: 441-51 (1986)

86A400 Wagner, P.J.; Truman, R.J.; Puchalski, A.E.; Wake, R., *J. Am. Chem. Soc.* **108**: 7727-38 (1986)

86A401 Wagner, P.J.; Thomas, M.J.; Puchalski, A.E., *J. Am. Chem. Soc.* **108**: 7739-44 (1986)

86A507 Becker, H.G.O.; Schuetz, R.; Tillack, B.; Rehak, V., *J. Prakt. Chem.* **328**: 661-72 (1986)

86B042 Boldridge, D.W.; Scott, G.W., *J. Chem. Phys.* **84**: 6790-8 (1986)

86D018 Yagi, M., *Chem. Phys. Lett.* **124**: 459-62 (1986)

86D051 Nagaoka, S.; Harrigan, E.T.; Noda, M.; Hirota, N.; Higuchi, J., *Bull. Chem. Soc. Jpn.* **59**: 355-61 (1986)

86D067 Arce, R.; Rodriguez, G., *J. Photochem.* **33**: 89-97 (1986)

86D238 van Willigen, H.; Chandrashekar, T.K.; Das, U.; Ebersole, M.H., *ACS Symp. Ser.* **321**: 140-53 (1986)

86E020 Feitelson, J.; Barboy, N., *J. Phys. Chem.* **90**: 271-4 (1986)

86E058 Suter, G.W.; Wild, U.P.; Holzwarth, A.R., *Chem. Phys.* **102**: 205-14 (1986)

86E128 Kikuchi, K.; Takahashi, T.; Koike, T.; Kokubun, H., *J. Photochem.* **32**: 341-9 (1986)

86E207 Waluk, J.; Komorowski, S.J., *J. Mol. Struct.* **142**: 159-62 (1986)

86E320 Wasielewski, M.R.; Kispert, L.D., *Chem. Phys. Lett.* **128**: 238-43 (1986)

86E458 Lopez-Arbeloa, I.; Rohatgi-Mukherjee, K.K., *Chem. Phys. Lett.* **129**: 607-14 (1986)

86E481 Khasawneh, I.M.; Winefordner, J.D., *Anal. Chim. Acta* **184**: 307-10 (1986)

86E546 Hofstra, U.; Koehorst, R.B.M.; Schaafsma, T.J., *Chem. Phys. Lett.* **130**: 555-9 (1986)

86E567 Bhattacharyya, K.; Ramaiah, D.; Das, P.K.; George, M.V., *J. Phys. Chem.* **90**: 5984-9 (1986)

86E628 Palit, D.K.; Mukherjee, T.; Mittal, J.P., *J. Indian Chem. Soc.* **63**: 35-42 (1986)

86E633 Aramendia, P.F.; Redmond, R.W.; Nonell, S.; Schuster, W.; Braslavsky, S.E.; Schaffner, K.; Vogel, E., *Photochem. Photobiol.* **44**: 555-9 (1986)

86E675 Bhattacharyya, K.; Ramamurthy, V.; Das, P.K.; Sharat, S., *J. Photochem.* **35**: 299-309 (1986)

86E676 Meier, K.; Zweifel, H., *J. Photochem.* **35**: 353-66 (1986)

86E782 Lewitzka, F.; Loehmannsroeben, H.-G., *Z. Phys. Chem. (Munich)* **150**: 69-86 (1986)

86E784 Kim, D., *Bull. Korean Chem. Soc.* **7**: 416-21 (1986)

86E917 Maiti, A.; Kastha, G.S., *Indian J. Phys., B* **60B**: 336-46 (1986)

86F115 Pavlickova, L.; Kuzmic, P.; Soucek, M., *Collect. Czech. Chem. Commun.* **51**: 368-74 (1986)

86F159 Defoin, A.; Defoin-Straatmann, R.; Hildenbrand, K.; Bittersmann, E.; Kreft, D.; Kuhn, H.J., *J. Photochem.* **33**: 237-55 (1986)

86F209 Bunce, N.J.; Debrabandere, G.G.; Jacobs, K.B.; Lemke, M.E.; Montgomery, C.R.; Nakai, J.S.; Stewart, E.J., *J. Photochem.* **34**: 105-15 (1986)

86F287 Matsushima, R.; Sakai, K., *J. Chem. Soc., Perkin Trans. 2* : 1217-22 (1986)

86F465 Ndiaye, S.A.; Aaron, J.J.; Garnier, F., *J. Photochem.* **35**: 389-94 (1986)

86R013 Jabben, M.; Garcia, N.A.; Braslavsky, S.E.; Schaffner, K., *Photochem. Photobiol.* **43**: 127-31 (1986)

86Z026 Carmichael, I.; Hug, G.L., *J. Phys. Chem. Ref. Data* **15**: 1-250 (1986)

86Z077 Kramer, H.E.A., *Chimia* **40**: 160-9 (1986)

86Z350 Riddick, J.A.; Bunger, W.B.; Sakano, T.K., *Organic Solvents. Physical Properties and Methods of Purification. Fourth Edition.* Techniques of Chemistry, Wiley and Sons, New York, 1986, Vol. II, 1325p.

86Z372 Cohen, E.R.; Taylor, B.N., *CODATA Bull.* : 1-36 (1986)

87A031 Gersdorf, J.; Mattay, J.; Goerner, H., *J. Am. Chem. Soc.* **109**: 1203-9 (1987)

87A090 Kemp, T.J.; Parker, A.W.; Wardman, P., *J. Chem. Soc., Perkin Trans. 2* : 397-403 (1987)

87A340 Bhattacharyya, K.; Ramamurthy, V.; Das, P.K., *J. Phys. Chem.* **91**: 5626-31 (1987)

87B054 Hamanoue, K.; Nakayama, T.; Kajiwara, Y.; Yamaguchi, T.; Teranishi, H., *J. Chem. Phys.* **86**: 6654-9 (1987)

87B098 Malkin, Ya.N.; Ruziev, Sh.; Pirogov, N.O.; Kuz'min, V.A., *Bull. Acad. Sci. USSR, Div. Chem. Sci.* **36**: 51-6 (1987) Translated from: *Izv. Akad. Nauk SSSR, Ser. Khim.* : 62-7 (1987)

87D045 Tro, N.J.; Tro, J.J.; Marten, D.F.; Nishimura, A.M., *J. Photochem.* **36**: 141-8 (1987)

87D046 Yamauchi, S.; Hirota, N., *J. Phys. Chem.* **91**: 1754-60 (1987)

87D053 Akiyama, K.; Ikegami, Y.; Tero-Kubota, S., *J. Am. Chem. Soc.* **109**: 2538-9 (1987)

87D086 Yamauchi, S.; Hirota, N., *J. Chem. Phys.* **86**: 5963-70 (1987)

87D090 Ofir, H.; Regev, A.; Levanon, H.; Vogel, E.; Koecher, M.; Balci, M., *J. Phys. Chem.* **91**: 2686-8 (1987)

87E138 Shim, S.C.; Shin, E.J.; Kang, H.K.; Park, S.K., *J. Photochem.* **36**: 163-75 (1987)

87E187 Shim, S.C.; Kang, H.K.; Park, S.K.; Shin, E.J., *J. Photochem.* **37**: 125-37 (1987)

87E199 Kemp, T.J.; Parker, A.W.; Wardman, P., *Photochem. Photobiol.* **45**: 663-6 (1987)

87E433 Olba, A.; Tomas, F.; Zabala, I.; Medina, P., *J. Photochem.* **39**: 263-72 (1987)

87E509 Duguid, R.; Maxwell, B.D.; Munoz-sola, Y.; Muthuramu, K.; Rasbury, V.; Singh, T.-V.; Morrison, H.; Das, P.K.; Hug, G.L., *Chem. Phys. Lett.* **139**: 475-8 (1987)

87E518 Rotkiewicz, K.; Koehler, G., *J. Lumin.* **37**: 219-25 (1987)

87E535 Terazima, M., *J. Chem. Phys.* **87**: 3789-95 (1987)

87E642 Terazima, M.; Azumi, T., *Chem. Phys. Lett.* **141**: 237-40 (1987)

87E785 Palewska, K.; Ruziewicz, Z.; Chojnacki, H., *J. Lumin.* **39**: 75-85 (1987)

87E893 Rudenko, N.A.; Val'kova, G.A.; Shigorin, D.N.; Poplavskii, A.N.; Andrievskii, A.M., *Russ. J. Phys. Chem.* **61**: 871-3 (1987) Translated from: *Zh. Fiz. Khim.* **61**: 1668-71 (1987)

87F258 Usui, Y.; Misawa, H.; Sakuragi, H.; Tokumaru, K., *Bull. Chem. Soc. Jpn.* **60**: 1573-8 (1987)

87F366 Ono, I.; Hata, N., *Bull. Chem. Soc. Jpn.* **60**: 2891-7 (1987)

87F368 Arai, T.; Oguchi, T.; Wakabayashi, T.; Tsuchiya, M.; Nishimura, Y.; Oishi, S.; Sakuragi, H.; Tokumaru, K., *Bull. Chem. Soc. Jpn.* **60**: 2937-43 (1987)

87Z100 Carmichael, I.; Helman, W.P.; Hug, G.L., *J. Phys. Chem. Ref. Data* **16**: 239-60 (1987)

88A385 Yoshimura, A.; Ohno, T., *Photochem. Photobiol.* **48**: 561-5 (1988)

88D018 Petrin, M.J.; Ghosh, S.; Maki, A.H., *Chem. Phys.* **120**: 299-309 (1988)

88D039 Yamauchi, S.; Hirota, N., *J. Am. Chem. Soc.* **110**: 1346-51 (1988)

88D063 Murai, H.; Minami, M.; I'Haya, Y.J., *J. Phys. Chem.* **92**: 2120-4 (1988)

88D064 Yamauchi, S.; Hirota, N.; Higuchi, J., *J. Phys. Chem.* **92**: 2129-33 (1988)

88D092 Slama, H.; Basche, T.; Braeuchle, C.; Voitlaender, J., *Photochem. Photobiol.* **47**: 661-7 (1988)

88D117 Yagi, M.; Komura, A.; Higuchi, J., *Chem. Phys. Lett.* **148**: 37-40 (1988)

88D270 Suisalu, A.P.; Aslanov, L.A.; Kamyshnyi, A.L.; Zakharov, V.N.; Avarmaa, R.A., *Bull. Acad. Sci. USSR, Phys. Ser.* **52**: 26-9 (1988) Translated from: *Izv. Akad. Nauk SSSR, Ser. Fiz.* **52**: 445-8 (1988)

88E219 Thompson, R.B.; Gratton, E., *Anal. Chem.* **60**: 670-4 (1988)

88E230 Terazima, M.; Azumi, T., *Chem. Phys. Lett.* **145**: 286-8 (1988)

88E253 Saini, R.D.; Dhanya, S.; Bhattacharyya, P.K., *J. Photochem. Photobiol., A* **43**: 91-103 (1988)

88E481 Kunjappu, J.T.; Rao, K.N., *Indian J. Chem., Sect. A* **27A**: 1-3 (1988)

88E643 Sa e Melo, T.; Macanita, A.; Prieto, M.; Bazin, M.; Ronfard-Haret, J.C.; Santus, R., *Photochem. Photobiol.* **48**: 429-37 (1988)

88E648 Kuz'min, V.A.; Levin, P.P., *Bull. Acad. Sci. USSR, Div. Chem. Sci.* **37**: 429-32 (1988) Translated from: *Izv. Akad. Nauk SSSR, Ser. Khim.* : 515-9 (1988)

88E675 Weir, D.; Scaiano, J.C.; Schuster, D.I., *Can. J. Chem.* **66**: 2595-600 (1988)

88E754 Druzhinin, S.I.; Rodchenkov, G.M.; Uzhinov, B.M., *Chem. Phys.* **128**: 383-94 (1988)

88E763 Ryzhikov, M.B.; Rodionov, A.N.; Nesterova, O.V.; Shigorin, D.N., *Russ. J. Phys. Chem.* **62**: 552-4 (1988) Translated from: *Zh. Fiz. Khim.* **62**: 1097-100 (1988)

88E804 Bright, F.V., *Appl. Spectrosc.* **42**: 1531-7 (1988)

88E862 Rtishchev, N.I.; Lebedeva, G.K.; Kvitko, I.Ya.; El'tsov, A.V., *J. Gen. Chem., USSR* **58**: 1914-27 (1988) Translated from: *Zh. Obshch. Khim.* **58**: 2148-63 (1988)

88R129 Hedstrom, J.; Sedarous, S.; Prendergast, F.G., *Biochemistry* **27**: 6203-8 (1988)

88R179 Ghiron, C.A.; Bazin, M.; Santus, R., *Biochim. Biophys. Acta* **957**: 207-16 (1988)

88Z003 Saltiel, J.; Atwater, B.W., *Adv. Photochem.* **14**: 1-90 (1988)

88Z502 Lias, S.G.; Bartmess, J.E.; Liebman, J.F.; Holmes, J.L.; Levin, R.D.; Mallard, W.G., *J. Phys. Chem. Ref. Data* **17**: 861p. (1988)

89A110 Guerin, B.; Johnston, L.J., *Can. J. Chem.* **67**: 473-80 (1989)

89A120 Timpe, H.-J.; Kronfeld, K.-P., *J. Photochem. Photobiol., A* **46**: 253-67 (1989)

89A179 Boate, D.R.; Johnston, L.J.; Scaiano, J.C., *Can. J. Chem.* **67**: 927-32 (1989)

89A343 Gopidas, K.R.; Kamat, P.V., *J. Photochem. Photobiol., A* **48**: 291-301 (1989)

89A345 Netto-Ferreira, J.C.; Weir, D.; Scaiano, J.C., *J. Photochem. Photobiol., A* **48**: 345-52 (1989)

89B155 Kikuchi, K., *Triplet-Triplet Absorption Spectra.* Bunshin Publishing Co., Tokyo, Japan, 1989, 189p.

89D004 Shioya, Y.; Yagi, M.; Higuchi, J., *Chem. Phys. Lett.* **154**: 25-8 (1989)

89D071 Tanigaki, K.; Taguchi, N.; Yagi, M.; Higuchi, J., *Bull. Chem. Soc. Jpn.* **62**: 668-73 (1989)

89D094 Shimoishi, H.; Tero-Kubota, S.; Akiyama, K.; Ikegami, Y., *J. Phys. Chem.* **93**: 5410-4 (1989)

89D101 Murai, H.; I'Haya, Y.J., *Chem. Phys.* **135**: 131-7 (1989)

89D116 Gundel, D.; Frick, J.; Krzystek, J.; Sixl, H.; von Schuetz, J.U.; Wolf, H.C., *Chem. Phys.* **132**: 363-72 (1989)

89D149 Tanigaki, K.; Yagi, M.; Higuchi, J., *J. Magn. Reson.* **84**: 282-95 (1989)

89E090 Goerner, H., *J. Phys. Chem.* **93**: 1826-32 (1989)

89E158 Bruce, J.M.; Gorman, A.A.; Hamblett, I.; Kerr, C.W.; Lambert, C.; McNeeney, S.P., *Photochem. Photobiol.* **49**: 439-45 (1989)

89E178 Zander, M.; Kirsch, G., *Z. Naturforsch., A, Phys. Sci.* **44A**: 205-9 (1989)

89E424 Schoof, S.; Guesten, H., *Ber. Bunsenges. Phys. Chem.* **93**: 864-70 (1989)

89E447 Maiti, A.K.; Kastha, G.S., *J. Lumin.* **43**: 383-5 (1989)

89F011 Ito, Y.; Uozu, Y.; Arai, H.; Matsuura, T., *J. Org. Chem.* **54**: 506-9 (1989)

89M184 Schafer, O.; Allan, M.; Haselbach, E.; Davidson, R.S., *Photochem. Photobiol.* **50**: 717-9 (1989)

89N023 Kirstein, S.; Moehwald, H.; Shimomura, M., *Chem. Phys. Lett.* **154**: 303-8 (1989)

89Z021 Hoffman, M.Z.; Bolletta, F.; Moggi, L.; Hug, G.L., *J. Phys. Chem. Ref. Data* **18**: 219-543 (1989)

89Z022 Kuhn, H.J.; Braslavsky, S.E.; Schmidt, R., *Pure Appl. Chem.* **61**: 187-210 (1989)

89Z211 Scaiano, J.C., *CRC Handbook of Organic Photochemistry.* CRC Press, Inc., Boca Raton, FL, 1989, Vol. I, 451p., Vol. II, 481p.

89Z269 Mizuno, K.; Otsuji, Y., *Yuki Gosei Kagaku Kyokaishi* **47**: 916-30 (1989)

90D074 Tominaga, K.; Yamauchi, S.; Hirota, N., *J. Phys. Chem.* **94**: 4425-31 (1990)

90E280 Goerner, H., *J. Photochem. Photobiol., B* **5**: 359-77 (1990)

90Z543 Andrews, D.L., *Lasers in Chemistry. Second Edition.* Springer-Verlag, Berlin, Federal Republic Germany, 1990, 188p.

90Z548 Lias, S.G.; Bartmess, J.E.; Liebman, J.F.; Holmes, J.L.; Levin, R.D.; Mallard, W.G., *NIST Positive Ion Energetics Database: Ver. 1.1.* National Institute of Standards and Technology, Gaithersburg, MD, 1990.

91A349 Tanigaki, K.; Ebbesen, T.W.; Kuroshima, S., *Chem. Phys. Lett.* **185**: 189-92 (1991)

91B003 Bonneau, R.; Carmichael, I.; Hug, G.L., *Pure Appl. Chem.* **63**: 289-99 (1991)

91D034 Wasielewski, M.R.; O'Neil, M.P.; Lykke, K.R.; Pellin, M.J.; Gruen, D.M., *J. Am. Chem. Soc.* **113**: 2774-6 (1991)

91D221 Wagner, P.J.; May, M.L., *J. Phys. Chem.* **95**: 10317-21 (1991)

91D249 Yagi, M.; Shioya, Y.; Higuchi, J., *J. Photochem. Photobiol., A* **62**: 65-73 (1991)

91E003 Arbogast, J.W.; Darmanyan, A.P.; Foote, C.S.; Rubin, Y.; Diederich, F.N.; Alvarez, M.M.; Anz, S.J.; Whetten, R.L., *J. Phys. Chem.* **95**: 11-2 (1991)

91E302 Ebbesen, T.W.; Tanigaki, K.; Kuroshima, S., *Chem. Phys. Lett.* **181**: 501-4 (1991)

91E368 Hung, R.R.; Grabowski, J.J., *J. Phys. Chem.* **95**: 6073-5 (1991)

91E594 Arbogast, J.W.; Foote, C.S., *J. Am. Chem. Soc.* **113**: 8886-9 (1991)

92D032 Ros, M.; Hogenboom, M.A.; Kok, P.; Groenen, E.J.J., *J. Phys. Chem.* **96**: 2975-82 (1992)

92E049 Terazima, M.; Hirota, N., *Chem. Phys. Lett.* **189**: 560-4 (1992)

92E142 Hung, R.R.; Grabowski, J.J., *Chem. Phys. Lett.* **192**: 249-53 (1992)

92E205 Dimitrijevic, N.M.; Kamat, P.V., *J. Phys. Chem.* **96**: 4811-4 (1992)

92E260 Palit, D.K.; Sapre, A.V.; Mittal, J.P.; Rao, C.N.R., *Chem. Phys. Lett.* **195**: 1-6 (1992)

92N199 Haley, J.L.; Fitch, A.N.; Goyal, R.; Lambert, C.; Truscott, T.G.; Chacon, J.N.; Stirling, D.; Schalch, W., *J. Chem. Soc., Chem. Commun.* : 1175-6 (1992)

Compound Name Index

Molecular Formula Index

$C_{44}H_{24}Cl_4N_4Zn$
    Zinc(II) tetrakis(2-chlorophenyl)porphyrin *2.46*
    Zinc(II) tetrakis(4-chlorophenyl)porphyrin *2.44*

$C_{44}H_{24}N_4O_{12}S_4Zn^{4-}$
    Tetrakis(4-sulfonatophenyl)porphinatozincate(II) ion *1.280, 3.145, 6.161*

$C_{44}H_{26}N_4O_{12}S_4^{4-}$
    Tetrakis(4-sulfonatophenyl)porphine *1.279, 3.144, 6.162, 8.95, 9.193*

$C_{44}H_{28}MgN_4$ Tetraphenylporphinatomagnesium(II) *1.284, 2.25, 3.149, 4.130, 6.166*

$C_{44}H_{28}N_4Zn$ Tetraphenylporphinatozinc(II) *1.285, 2.39, 3.150, 4.131, 6.167*

$C_{44}H_{30}N_4$ Tetraphenylporphine *1.283, 2.20, 3.148, 4.129, 6.165, 10b-8.22*

$C_{44}H_{36}N_8Zn^{4+}$
    Tetrakis(1-methylpyridinium-4-yl)porphinatozinc(II) ion *1.278*

$C_{44}H_{38}N_8^{4+}$ Tetrakis(1-methylpyridinium-4-yl)porphine *1.277, 6.160*

$C_{48}H_{24}$ Hexabenzo[*a,d,g,j,m,p*]coronene *2.152*

$C_{48}H_{36}N_4Zn$ Zinc(II) tetrakis(2-methylphenyl)porphyrin *2.45*

$C_{48}H_{38}N_4$ Tetrakis(4-methylphenyl)porphine *6.159*

$C_{55}H_{70}MgN_4O_6$
    Chlorophyll *b* *1.148, 2.14, 3.80, 4.63, 6.79, 9.85*

$C_{55}H_{70}N_4O_6Zn$
    Zinc(II) chlorophyll *b* *1.369, 2.22, 4.207*

$C_{55}H_{72}MgN_4O_5$
    Chlorophyll *a* *1.147, 2.11, 3.79, 4.62, 6.78, 9.84*

$C_{55}H_{72}N_4O_5Zn$
    Zinc(II) chlorophyll *a* *1.368, 2.15, 4.206*

$C_{55}H_{72}N_4O_6$ Pheophytin *b* *1.264, 2.18, 3.136, 4.118, 6.153*

$C_{55}H_{74}MgN_4O_6$
    Bacteriochlorophyll *a* *6.28*

$C_{55}H_{74}N_4O_5$ Pheophytin *a* *1.263, 2.13, 3.135, 4.117, 6.152*

$C_{55}H_{76}N_4O_6$ Bacteriopheophytin *a* *6.29*

$C_{56}H_{60}N_8Zn^{4+}$
    Tetrakis-4-(*N,N,N*-trimethylammonio)phenylporphinezinc(II) ion *1.282, 3.147, 6.163*

$C_{56}H_{62}N_8^{4+}$ Tetrakis(4-trimethylammoniophenyl)porphine *1.281, 3.146, 6.164*

C_{60} C_{60} *1.139, 2.35, 3.71, 6.73*

$C_{60}H_{80}$ Dodecapreno-β-carotene *9.113*

C_{70} C_{70} *1.140, 2.29, 3.72, 4.59, 6.74*

H_2O Water *12-1.67, 12-3.69*

NH_3 Ammonia *10a-1.1*

O_2 Oxygen *12-3*